Advanced Processes and Technologies for Wastewater: Collection, Treatment, and Resource

Advanced Processes and Technologies for Wastewater: Collection, Treatment, and Resource

Topic Editors

Zhiqiang Zhang
Heliang Pang

MDPI

Basel • Beijing • Wuhan • Barcelona • Belgrade • Novi Sad • Cluj • Manchester

Topic Editors

Zhiqiang Zhang
School of Environmental and Municipal Engineering
Xi'an University of Architecture and Technology
Xi'an
China

Heliang Pang
School of Environmental and Municipal Engineering
Xi'an University of Architecture and Technology
Xi'an
China

Editorial Office
MDPI AG
Grosspeteranlage 5
4052 Basel, Switzerland

This is a reprint of the Topic, published open access by the journals *Environments* (ISSN 2076-3298), *Processes* (ISSN 2227-9717), *Separations* (ISSN 2297-8739) and *Water* (ISSN 2073-4441), freely accessible at: https://www.mdpi.com/topics/2WJ8N43828.

For citation purposes, cite each article independently as indicated on the article page online and as indicated below:

Lastname, A.A.; Lastname, B.B. Article Title. *Journal Name* **Year**, *Volume Number*, Page Range.

ISBN 978-3-7258-7144-5 (Hbk)
ISBN 978-3-7258-7145-2 (PDF)
https://doi.org/10.3390/books978-3-7258-7145-2

Contents

About the Editors

Zhiqiang Zhang

Zhiqiang Zhang is a professor of Environmental Science and Engineering at the School of Environmental and Municipal Engineering, Xi'an University of Architecture and Technology, the People's Republic of China. He is also the executive director of the Shaanxi Key Laboratory of Environmental Engineering. His research primarily focuses on risk prevention technologies for urban drainage network systems. He pays special attention to the control of problems such as odors, corrosion, and blockages in drainage pipes. His main research interests also include the theory and technology of green environmental governance using micro-nanobubbles. By leveraging the unique physical, chemical, and micro-interface properties of micro-nanobubbles, it is possible to effectively regulate the micro-environment of water bodies and sediments (such as environmental factors and microbial community structures), thereby enhancing the removal efficiency of environmental pollutants. In 2024, he was selected for the "Excellent Young Talent Support Program" of Shaanxi, and he was the winner of the "Young Excellent Paper Award", awarded by the China Urban Water Association.

Heliang Pang

Heliang Pang is a professor of Environmental Science and Engineering at the School of Environmental and Municipal Engineering, Xi'an University of Architecture and Technology, the People's Republic of China. His main research interests lie in the resource utilization of urban solid waste (including activated sludge, organic solid waste, etc.), and his expertise covers anaerobic digestion theory and technology, organic solid waste carbon source/energy recovery, microbial community research, etc. In addition, he also conducts research on the formation mechanism and control of sedimentation in sewage pipelines, thereby providing solutions for the control of sewage pipeline blockage. In 2022, he was selected for the "Shaanxi High-level Talent Introduction Program".

water

MDPI

Article

Multi-Objective Optimization Based on Simulation Integrated Pareto Analysis to Achieve Low-Carbon and Economical Operation of a Wastewater Treatment Plant

Jianbo Liao [1], Shuang Li [1], Yihong Liu [1], Siyuan Mao [1], Tuo Tian [2], Xueyan Ma [2], Bing Li [1,*] and Yong Qiu [2,*]

1 School of Energy and Environmental Engineering, University of Science and Technology Beijing, Beijing 100083, China

2 State Key Joint Laboratory of Environment Simulation and Pollution Control, School of Environment, Tsinghua University, Beijing 100084, China; tiantuo@mail.tsinghua.edu.cn (T.T.); maxueyan@mail.tsinghua.edu.cn (X.M.)

* Correspondence: libing@ustb.edu.cn (B.L.); qiuyong@tsinghua.edu.cn (Y.Q.); Tel.: +86-13810103096 (Y.Q.)

Abstract: It is essential to reduce carbon emissions in wastewater treatment plants (WWTPs) to achieve carbon neutrality in society. However, current optimization of WWTPs prioritizes the operation cost index (OCI) and effluent quality index (EQI) over greenhouse gas (GHG) emissions. This study aims to conduct a multi-objective optimization of a WWTP, considering GHG emissions, EQI, and OCI. The anaerobic-anoxic-oxic integrated membrane bioreactor (AAO-MBR) process in an actual WWTP was selected as a typical case, tens of thousands of scenarios with combinations of six operational parameters (dissolved oxygen (DO), external carbon resource (ECR), poly aluminum chloride (PAC), internal reflux ratio (IRR), external reflux ratio (ERR), and sludge discharge (SD)) were simulated by GPS-X software (Hydromantics 8.0.1). It was shown that ECR has the greatest impact on optimization objectives. In the optimal scenario, the main parameters of ATDO, MTDO, IRR, and ERR were 0.1 mg/L, 4 mg/L, 50%, and 100%, respectively. The EQI, OCI, and GHG of the best scenario were 0.046 kg/m^3, 0.27 ¥/m^3, and 0.51 $kgCO_2/m^3$, which were 2.1%, 72.2%, and 34.6% better than the current situation of the case WWTP, respectively. This study provides an effective method for realizing low-carbon and economical operation of WWTPs.

Keywords: multi-objective optimization; Pareto analysis; greenhouse gas emissions; AAO-MBR; operational parameters

1. Introduction

Reducing greenhouse gas emissions is of utmost importance in the global fight against climate change. The Paris Agreement, established in 2015, has motivated numerous countries to establish objectives for carbon peaking and carbon neutrality [1]. As a result, industries around the world have adapted their development strategies and implemented emission-reducing measures. According to statistics from major developed countries, the wastewater treatment industry is among the top ten contributors to total carbon emissions, accounting for 1% to 2% [2]. It is expected that the total greenhouse gas emissions (in terms of CO_2 equivalent) from China's wastewater treatment industry will be as high as 365 million t by 2030. It accounts for 2.95% of the country's total emissions [3,4]. On 29 December 2023, the Chinese government released the policy entitled "Opinions on the Implementation of Promoting Synergistic Efficiency in Wastewater Treatment for Reducing Pollution and Reducing Carbon". It announced Chinese plans to build 100 green and low-carbon benchmark plants for wastewater treatment in 2025 with high-efficiency recycling in energy and resources. Therefore, measures such as adopting new processes and improving operation levels to achieve carbon reduction in the wastewater treatment industry are of great significance in achieving carbon neutrality.

Due to the diversity of municipal wastewater treatment processes, there is a large level of complexity and uncertainty when it comes to balancing multiple objectives such as operational efficiency and low-carbon emissions. In practice, improving effluent quality and minimizing operating costs are still the primary concerns of operation managers. To attain these objectives, it is crucial to explore avenues for reducing the operation cost index (OCI) and greenhouse gas (GHG) while upholding the standards of the effluent quality index (EQI). This is a problem involving the optimization of multiple objectives in nature.

In terms of single-objective and bi-objective optimization, there have been numerous reports. Wu et al. [5] used WEST 2012® software to simulate and optimize an industrial wastewater effluent plant with complex influent compositions, which led to the reduction in OCI from 6.2 to 5.5 €/m^3. Cao et al. [6] screened and analyzed the sensitivity of 61 parameters. They developed a quadratic polynomial response surface model of six key process parameters. Finally, the water quality improvement was achieved by optimizing two process parameters, dissolved oxygen (DO) and solids retention time (SRT). All of the above studies focus on the single-objective optimization problem of wastewater treatment process. However, they have not formed an integrated and comprehensive study. Vega et al. [7] combined real-time optimization and hierarchical control with nonlinear model predictive control, and evaluated the wastewater treatment process EQI and OCI through the control structure. Zhang et al. [8] proposed a multi-objective optimization and control method of BP neural network combined with a genetic algorithm, which effectively solves the problems of EQI and OCI, a pair of mutually constrained optimization objectives. Guerrero et al. [9] used both OCI and EQI as the control optimization objectives, which produced a set of optimal operating setpoints that could be approximated by a Pareto surface. These optimization studies can guide wastewater treatment plants (WWTPs) to effectively reduce OCI without consideration about GHG emissions.

Regarding the tri-objective optimization of EQI, OCI, and GHG, although there are some research reports in recent years, there are still some deficiencies that need to be improved. Lu et al. [10] assessed the effectiveness of gauging a dynamic simulation model in regulating GHG in WWTP. The evaluation puts forth a fresh approach for achieving optimal control of such plants, taking into account EQI, OCI, and GHG. It is important to mention, however, that while the proposed framework is sound in theory, it lacks practical real-world case studies. X. Flores-Alsina et al. [11] found that aerobic tank dissolved oxygen, primary sedimentation tank suspended solid (SS) removal, anaerobic digester temperature, and reflux strategy had effects on the three objectives considered (GHG, EQI, and OCI). But these objectives could not be optimized for all of them. Similar conclusions were obtained in the study of C. Sweetapple et al. [12], where none of the 315 aeration strategies set up could simultaneously achieve the co-optimization of the three objectives. The operation energy consumption, water quality, and carbon emissions of WWTP based on continuous batch reactor were studied, but no collaborative optimization was achieved [13].

Since EQI, OCI, and GHG are three mutually constrained optimization objectives, it is difficult to achieve the optimal solution of the three objectives at the same time. Hence, it is necessary to adopt a nonlinear multi-objective optimization method to determine the overall optimal solution [8]. For this purpose, the study employs Pareto optimization principles and introduces the upstream logic in the NSGA-II algorithm—the Non-dominated Sorting Genetic Algorithms, which has proven effective in optimizing control strategies for WWTPs from prior studies. Chen et al. [14] used NSGA-II to achieve multi-objective optimization of operational energy consumption, effluent quality, total volume of structures, and SS of structures based on the activated sludge method. Beraud et al. [15] coupled NSGA-II with a common wastewater treatment plant model to illustrate how the algorithm can be used to determine the feasibility of Pareto optimality. It is worth noting that researchers use mathematical models to simulate complex process conditions and obtain basic data for optimization evaluation commonly. In this way, as the high input of time and labor decreases apparently, so does the cost of trial and error. Moreover, the issues caused by uncertainty and inefficiency in relying on traditional methods to optimize operational parameters can be avoided efficiently, as well [16].

At present, there are hundreds of domestic and foreign wastewater treatment processes. Among them, the combination of anaerobic-anoxic-oxic (AAO) and membrane bioreactor (MBR) processes (named AAO-MBR) has aroused general attention, because it can achieve high and stable effluent quality, short hydraulic retention time, and low residual sludge volume [17,18]. Due to its high effluent quality, the resulting wastewater can not only be discharged directly into the environment, but also reused for non-potable water applications [19]. Considering the increasing use of AAO-MBR in upgrading projects and underground WWTPs, more than 25% of underground WWTPs use AAO-MBR as the primary process [20], the typical AAO-MBR process was selected as the research object in this study. The nonlinear multi-objective optimization method was explored to achieve low-carbon emission taking into account the effluent quality and operation cost. We established a process model with the help of GPS-X simulation and modeling software to analyze the effects of six typical operational parameters on EQI, OCI, and GHG. A non-dominated sorting method was adopted to search for the Pareto-optimal set of solutions and screen the optimal solution from it. Through the above process, the necessary trade-offs between conflicting control objectives were made, which provided support for enhancing the sustainability of the wastewater treatment system.

2. Materials and Methods

2.1. A Multi-Objective Optimization Framework

The research framework of the multi-objective optimization comparison method used in this paper is visually depicted in Figure 1. Different from the traditional multi-objective weighted evaluation optimization method, this paper searches for the optimal combination of operational parameters based on the three mutual restraint objectives of EQI, OCI, and GHG by means of Pareto optimal solution set. The optimization process entails three major parts: model construction, multi-parameter evaluation, and optimization comparison.

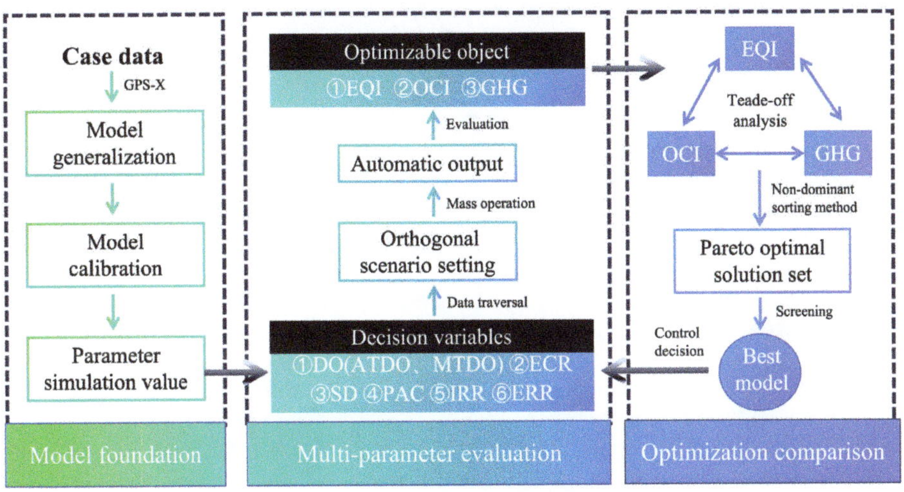

Figure 1. The research framework of the multi-objective optimization.

Firstly, the process model was built with the help of GPS-X (Hydromantics 8.0.1) simulation and modeling software. The key operating conditions included six operational parameters, including dissolved oxygen (DO) concentration, external carbon resource (ECR), phosphorus removal agent poly aluminum chloride (PAC), internal reflux ratio (IRR), external reflux ratio (ERR), and sludge discharge (SD). Among these, the DO concentration included the aerobic tank dissolved oxygen (ATDO) concentration and the membrane tank dissolved oxygen (MTDO) concentration. Based on the baseline scenario, the simulated value points of six operational parameters were set.

Secondly, the orthogonal combination of parameter simulation values was carried out through data traversal to form an orthogonal simulation scenario. GPS-X software was used to realize mass simulation calculation and automatic output results.

Finally, based on the simulation results of the orthogonal parameter set, the evaluation values of each scenario were solved by the equal proportion weighting calculation. The optimal scenario was selected to determine the optimal operation parameter combination.

2.2. *Technological Process Modeling and Scenario Analysis*

2.2.1. Construction of Process Model

The case plant is located in Luohu District, Shenzhen, China. It adopts the AAO-MBR treatment process, and the actual treatment scale is 33,500 m^3/d, with hydraulic retention time (HRT) of 17.33 h and solid retention time (SRT) of 10.94 d. The influent water quality is shown in Table S1, and the design effluent is required to meet the Class A standard of the China Urban Wastewater Treatment Plant Emission Standard (GB18918-2022) [21].

Based on the actual process and operational parameters of the case WWTP, the model is generalized and the parameters are calibrated according to the standardized procedure [22]. Its actual operational parameters and unit structures are shown in Tables S2 and S3. The generalized model of the case plant is depicted in Figure 2.

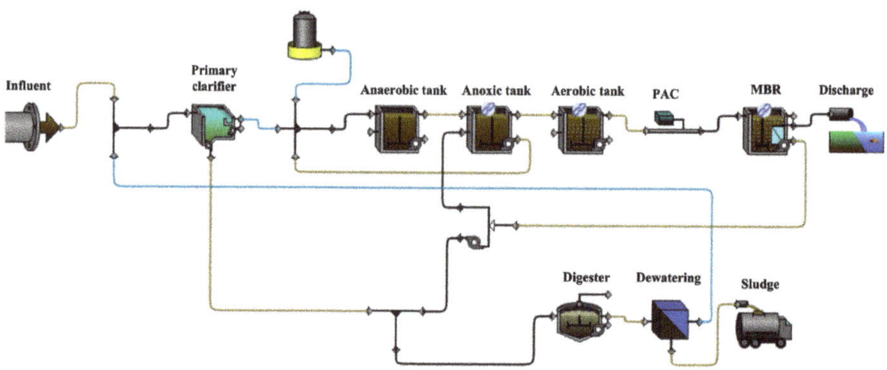

Figure 2. Generalized model of the case WWTP established in GPX-S software.

The daily water quality data in 2022 is collected for model calibration of the baseline scenario. Then, the relevant stoichiometric and dynamic parameters in the structure are adjusted to make the simulated effluent water quality the same or close to the actual effluent water quality, with reference to the previous studies [6,23]. The results of the calibration and the baseline simulation scenario of the effluent quality are shown in Table S1. After calibration, the average errors of BOD_5, COD, SS, TN, TP, and NH_4^+-N were 7.0%, 10%, 8.0%, 0.3%, 18%, and 25%, respectively, and the maximum absolute errors were 0.15, 1.23, 0.02, 0.03, 0.03, and 0.02 mg/L. Almost all the indicators' errors were within acceptable limits [24].

The Python Script Manager attached to the software is used to achieve external control. A large number of exhaustive simulations run within a certain range are completed under the premise of fixed process flow and structure. At the same time, the script is written in Python environment to realize automatic data import and export. The automated operation is implemented through the loop statements of the upper logic. The logical structure of which is shown in Figure S1.

2.2.2. Orthogonal Scenario Settings of Operational Variable

Referring to the basic operational parameters presented in Table S2, we set the optimized values of the six operational parameters, DO (ATDO, MTDO), ECR, PAC, IRR, ERR, and SD of the case WWTP, respectively. These six were chosen from a larger set of operational and design parameters as the set of most likely to be significant to the objective functions [25,26]. These values are combined with the actual operating data of WWTP and the simulative scenario values are shown in Table S4.

Through the way of data traversal, the value points of the operational parameters (Table S4) are arranged and combined successively, forming a total of 80,000 ($4 \times 4 \times 4 \times 5 \times 5 \times 5 \times 10$) groups of parameter combinations, forming orthogonal scenarios. These scenarios are simulated in the construction of the case WWTP model, and the results obtained are used for subsequent analysis. The values of MTDO had no significant effect on the three objectives (Figure S2). Therefore, the effects of six operational parameters, ATDO, ECR, SD, PAC, IRR, and ERR, on EQI, OCI, and GHG were examined separately.

2.3. Calculation Method of Three Optimizable Objectives

2.3.1. Effluent Quality Index (EQI)

The effluent quality is characterized by EQI with a unit of kg/m^3, which represents the total amount of pollutants discharged per ton of wastewater treated by the study object. In this study, EQI is characterized by COD, BOD_5, TSS, Soluble PO_4-P, NO_x (NO_2 + NO_3), and TKN. Their weighting factors are used with default values of 1, 2, 2, 2, 2, 1, and 20, respectively. The values are obtained directly from the software, and the reference formula in the software is shown in Equation (1):

$$EQI = Q\sum_{i=1}^{6} w_i S_i \quad (1)$$

where Q is the effluent flow of WWTP, m^3/d, w_i is the weight of the ith pollutant in the EQI. S_i is the concentration of the ith pollutant in the EQI, mg/L.

2.3.2. Operating Costs Index (OCI)

The operating cost is characterized by OCI with a unit of $\$/m^3$. The OCI includes energy costs, pharmaceutical costs, and sludge disposal costs. The energy costs include aeration energy costs, pumping energy costs, mixing energy costs, heating energy costs, and other energy costs. Similarly, the OCI value can be directly output by GPS-X software, and the OCI calculation formula is shown in Equation (2):

$$OCI = C_{are} + C_{che} + C_{disp} \quad (2)$$

where C_{are} is electronic energy cost, 0.1 \$/kWh. C_{che} is chemical cost, including the carbon source cost of 2.0 \$/kg, PAC cost of 0.5 \$/kg, and sludge pretreatment chemical cost of 1.0 \$/kg. C_{disp} is sludge disposal cost, including the sludge transportation cost of 80 \$/t.

Since the associated economic indicators in the software use USD$ as the unit, we convert them to USD$ in the analysis with reference to the USD-RMB exchange rate at the beginning of May 2023, based on 1 USD equals 6.91 RMB.

2.3.3. Greenhouse Gas (GHG)

Operational greenhouse gas emissions are characterized using GHG in units of tCO_2 eq/m^3. Among the widely used carbon emission accounting methods, the actual-measurement method is not applicable in the simulation and calculation scenarios. This study combines the emission factor method with the mass balance method to slice the carbon emissions of the wastewater treatment plant into multiple parts for separate calculations. The accounting equations are shown in Equations (3) and (4):

$$GHG = Q(C_{in} - C_{out}) \times EF \times GWP \tag{3}$$

$$GHG = AD \times EF \times GWP \tag{4}$$

where AD is the activity data for the emission source and the unit depends on the calculated emission source. EF is the emission factor and the unit depends on the activity data. Q is the effluent flow of WWTP, m^3/d. C_{in} is the concentration of the corresponding pollutant in the influent water, mg/L. C_{out} is the concentration of the corresponding pollutant in the effluent water, mg/L. GWP is Global Warming Potential. $GWPCH_4$ is 25 tCO_2 eq/tCH_4 and $GWPN_2O$ is 298 tCO_2 eq/tN_2O.

The total carbon emissions are composed of 10 components, which are presented in Table S5. For specific calculations of biochemical processes, power, and chemicals carbon emissions for each structure, refer to Equations (S1)–(S10). Three of the direct emissions (S1), (S2), and (S10) are calculated with the mass balance method (Equation (3)), while the rest of the emission processes are conveniently calculated with the emission factor method (Equation (4)).

2.4. Non-Dominated Sorting Method

Referring to the concept of Pareto optimum, the upstream logic-non-dominated sorting method was introduced in the NSGA-II algorithm to filter out the Pareto optimal solution from tens of thousands of simulation scenarios. It enables us to achieve multi-objective comprehensive optimization of the three objectives, EQI, OCI, and GHG. This method is ideal for optimizing control strategies in WWTPs as it can handle nonlinear optimization problems. Moreover, it is capable of evaluating the objective function with fewer conditions and achieving multi-objective optimization in a single simulation [27].

A cycle-based iterative ranking method is used to achieve the above optimization. For each round of the study objects within the study scope, the "dominance" relationship of each scenario point with other scenario points is examined separately. Then, the optimal scenario in a "non-dominated" state is selected in each round until all the remaining scenarios are not dominated by each other.

In general, the scenarios with Pareto rank 1 obtained by multi-objective sorting are not unique. In order to further complete the screening in "Pareto-optimal" scenarios, this study establishes a ranking method with constraints, focusing on the comprehensive level of objectives under certain weights. By setting the weights, we calculate the composite objective values to complete the final ranking. In this study, the three objectives are considered to be of the same importance, thus the weight factors $K_{i,j}$ of the three objectives were equal. MATLAB (2021B) programming is used to implement the non-dominated sorting and draw the Pareto frontier. The calculation of the integrated objective value for the i'th scenario is shown in Equation (5):

$$S_i = \sum_{j=1}^{3} K_{i,j} \cdot F_{i,j} / F_{0,j} \tag{5}$$

where S_i is the value of the composite objective for the ith scenario. j is the number of the objective. $K_{i,j}$ is the weight of the jth objective ($K_{i,1} = K_{i,2} = K_{i,3} = 1/3$). $F_{i,j}$ is the value of the jth objective for the ith scenario, with units referenced to the objectives. $F_{0,j}$ is the value of the jth objective for the based scenario, with units referenced to the objectives.

3. Results and Discussion

3.1. Effect of Operational Parameters on Effluent Quality Index (EQI)

The influence of the six operational parameters on the EQI values is shown in Figure 3. In general, the EQI values increase with the increase in ATDO and SD values, and decrease with the increase in PAC and ERR values. The change in ECR and IRR values does not have a significant effect on the EQI. Among these, the change in PAC values affects the most largely, indicating that PAC is the most influential operational parameter.

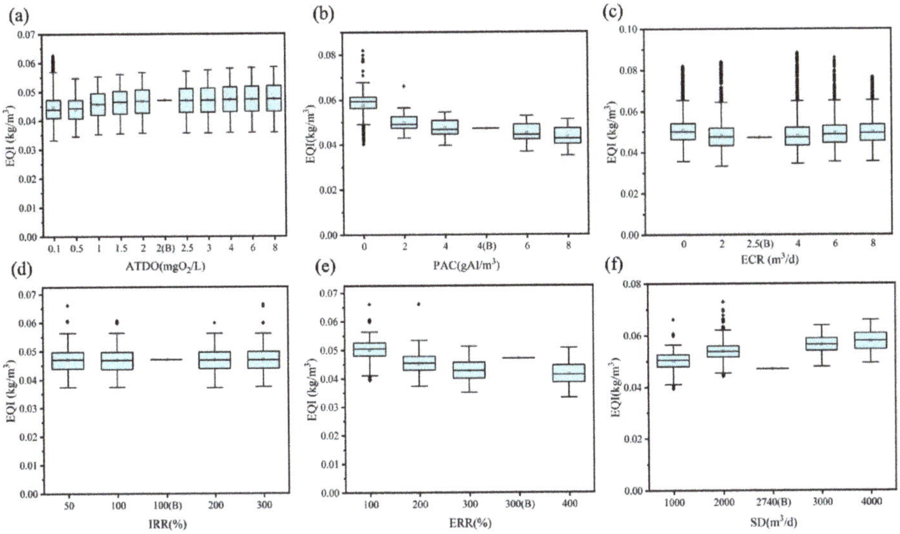

Figure 3. The effect of simulated value changes in each operational parameter on EQI for the case plant. (**a**) ATDO; (**b**) PAC; (**c**) ECR; (**d**) IRR; (**e**) ERR; (**f**) SD, and the values of the baseline scenario is denoted by (B) in the abscissa.

The EQI value of the case plant increases slightly with the increase in ATDO, which is shown in Figure 3a, indicating that high ATDO does not enhance the EQI, a similar conclusion was found in the study [28]. Since the subsequent membrane tank is able to share the aerobic tank pressure, the AAO-MBR process allows for extremely low ATDO values to be utilized, resulting in more stable EQI scenarios being met and retained. Jiang et al. [29] demonstrated that the AAO-MBR process improves water quality by efficiently removing nitrogen and phosphorus with limited aeration (DO = 0.5–1.0 mg/L), which is consistent with the simulation results. It is noteworthy that while reducing ATDO to 0.1 mg/L facilitates short-range denitrification, the actual engineering design parameter was set at 2 mg/L. This suggests that an empirical-based design can achieve local optimum values. However, due to insufficient empirical coverage of various scenarios, determining comprehensive and optimal solutions may not always be possible. The results depicted in Figure 3b reveal a noteworthy reduction in the EQI value, as the PAC value rises, which is especially prominent when the PAC value is low. It indicates the significance of PAC in decreasing the integrated water quality index. It also suggests that as the PAC dosage increases, its optimization effect on the EQI dwindles. These two points match with the findings [24,30]. By adding a small amount of ECR, the EQI values have a certain decrease, but with the increase in the dosage, the EQI values will turn to increase (Figure 3c). It indicates that ECR has a limited effect on the overall optimization of effluent water quality as the input of ECR to reduce effluent TN may lead to the increase in COD at the same time, which needs to be analyzed in conjunction with the actual influent C/N ratio. The increases in the C/N ratio in a certain range can enhance the removal of TN and COD [31]. IRR has little effect on the EQI, which is a relatively insensitive parameter, as presented in Figure 3d. Due to the fact that IRR mainly regulates the denitrification process while the carbon source is relatively low, increasing the IRR ratio does not contribute significantly to the effluent quality. The EQI value decreases significantly as the ERR rises (Figure 3e). This is because the rise in ERR increases the concentration of the bicycles, and further improves the EQI. Wang et al. [32] recorded that the NH_3-N removal was 98.1%, 98.5%, and 98.9% when the ERR was taken as 100%, 300%, and 500%, respectively. It indicated that the increase in ERR contributes to the water quality, which is in agreement with the simulation results. It has been noted that when the SD increases, the EQI value of the case plant also tends to increase, as depicted in Figure 3f. This is reason that higher SD results in a shorter SRT and lower concentration of mixed liquor suspended solids (MLSS), which are unfavorable conditions for biological treatment [33]. As a result, the effluent quality decreases. Furthermore, the study recommends that this process is better suited for a state of long sludge age.

3.2. *Effect of Operational Parameters on the Operation Cost Index (OCI)*

The influence of the six operational parameters on the OCI values is shown in Figure 4. It can be seen that IRR and SD do not have a significant effect on the OCI. An increase in the values of the remaining operational parameters causes an increase in the OCI, with a change in the value of ECR causing the largest change in the OCI. It suggests that ECR is the operational parameter with the greatest influence.

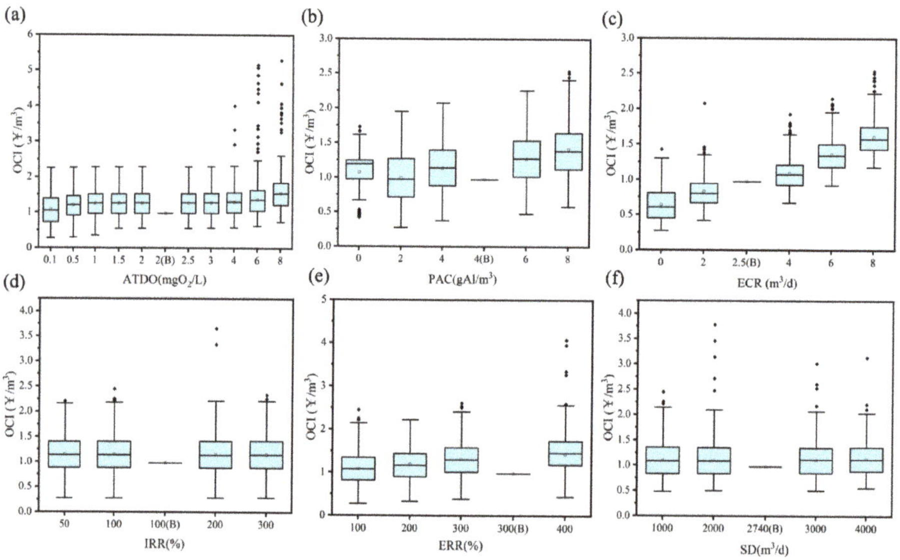

Figure 4. The effect of simulated value changes in each operational parameter on OCI for the case plant. (**a**) ATDO; (**b**) PAC; (**c**) ECR; (**d**) IRR; (**e**) ERR; (**f**) SD, and the values of the baseline scenario is denoted by (B) in the abscissa.

As Figure 4a shows, the OCI experiences gradual growth as the ATDO increases. This growth is especially pronounced when the ATDO concentrations are either lower than 0.5 mgO_2/L or higher than 4 mgO_2/L. This is the reason that low ATDO reduces energy consumption, and thus OCI in the AAO-MBR process [34,35]. Conversely, higher ATDO levels mean that the water is already saturated with dissolved oxygen, causing a decrease in oxygen transfer rate and utilization. This, in turn, results in an increase in energy consumption and OCI. Other studies have also arrived at similar conclusions [36,37]. Additionally, it has been shown that the aeration process accounts for around 50% of the total power consumption of a WWTP [38] and over 30% of the total OCI [39]. This highlights the crucial role of ATDO in optimizing OCI in wastewater treatment processes. A smooth increase in the OCI with the rise in PAC is presented in Figure 4b. And it matches the traditional concept of the impact of operation. The lower values of OCI for the base scenario demonstrate that the parameter values for the base scenario are relatively better. The OCI increases dramatically with the increase in ECR (Figure 4c). There is a positive correlation with OCI as the amount of added carbon source is linearly related to the cost of pharmaceuticals as the unit price remains constant. This finding is also well supported by a study that optimizing the dosage rate with a dosing control system can reduce OCI [40]. The effect of ERR is more pronounced on the case plant compared to IRR because the inner return pumps have a lower head than the outer return pumps, and therefore contribute relatively less to OCI, which are shown in Figure 4d,e. Kim et al. [41] analyzed the effect of six decision variables on OCI, in which the effect of IRR on OCI is smaller than ERR, which is consistent with the simulation scenario results. Meanwhile, due to the existence of a multi-stage sludge return line, the number of return pumps and energy consumption are greater compared to other wastewater treatment processes. The change in SD value does not contribute much to OCI (Figure 4f). Improving the OCI of the case plant by adjusting the SD poses a significant challenge. However, utilizing recycled biogas as an energy source can effectively lower electricity consumption. The amount of biogas produced is directly proportional to the SD; therefore, increasing biogas production by raising the SD can help reduce OCI [41]. Yet, this approach has its limitations. Heightening biogas production may increase carbon and nitrogen loads on the environment. Additionally, only the biodegradable portion of the sludge can be transformed into biogas, while the remaining part would require transportation, potentially increasing OCI from 25% to 65% [42]. As a result, further research is necessary to determine the feasibility of this pathway.

3.3. *Effect of Operational Parameters on Greenhouse Gas (GHG)*

The influence of the six operational parameters on the value of GHG is shown in Figure 5. It can be seen that GHG rises with the increase in the values of ECR and ATDO, and the effect of the remaining four operational parameters on GHG is not significant. Among them, the increase in the value of ECR caused a large increase in GHG, which is the same as OCI and is the most influential operational parameter.

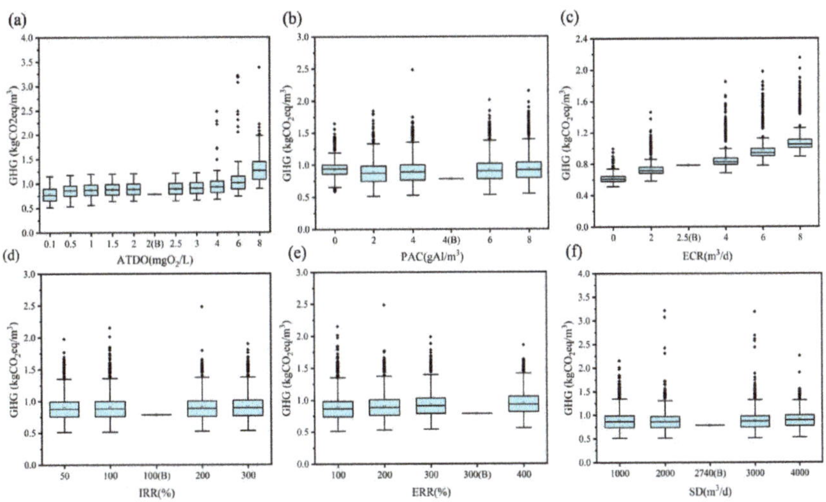

Figure 5. The effect of simulated value changes in each operational parameter on GHGI for the case plant. (**a**) ATDO; (**b**) PAC; (**c**) ECR; (**d**) IRR; (**e**) ERR; (**f**) SD, and the values of the baseline scenario is denoted by (B) in the abscissa.

GHG increases with an increase in ATDO which is presented in Figure 5a. This is because higher ATDO increases power consumption and GHG increases with power consumption [43]. A study showed a similar conclusion by demonstrating that electricity consumption is directly related to indirect GHG emissions [44]. The GHG corresponding to the baseline scenario values are lower than the remaining simulated scenario values, indicating that the ATDO design values have been taken to a more optimal level. Huang, F et al. [45] reduced power consumption and GHG emissions by optimizing DO settings through a DO control system using a conventional proportional-integral control strategy. Flores-Alsina et al. [46] investigated the control strategy of WWTP and activated sludge process modeling for GHG emission benchmarking. They used the Benchmark Simulation Model No 2 (BSM2G) to optimize the ATDO values to achieve carbon reduction. According to Figure 5b, GHG levels remain relatively stable as PAC increases. This suggests that carbon emissions from PAC have a minimal impact on the plant's overall GHG levels. However, as shown in Figure 5c, the GHG levels steadily increase with the rise in ECR, indicating a certain degree of linearity. Determining the emission factor reveals a linear relationship between the amount of additional carbon source injection and carbon emissions. When this component accounts for a significant proportion, an increase in added carbon sources can have a significant impact on the GHG emissions of the entire plant. It is worth noting that in practical situations, it is crucial to consider the influent of C/N ratio when making determinations. Kishida et al. [47] found that an appropriate C/N ratio can substantially decrease N_2O emissions during the denitrification stage. IRR and ERR have a small impact on the GHG and the trends are depicted in Figure 5d,e. The power consumption in the sewage treatment process contributes the most to GHG, while the power consumption in the whole process mainly lies in the energy consumption of sewage lifting pump and aeration energy consumption. The energy consumption of reflux pump accounts for a small proportion [48]; therefore, it contributes little to GHG. At the same time, ERR has a more significant influence on GHG than IRR. On the one hand, the internal reflux pump has a lower head than the external reflux pump. On the other hand, the internal reflux mainly controls the effect and process of denitrification. The denitrification process produces NO_2, and the emission of NO_2 is mainly affected by DO and C/N [49]. Thus, the contribution of IRR to GHG is relatively minor. Figure 5f shows that the change in the value of SD has almost no effect on the GHG of the case plant, which manifests that the optimization of the GHG by changing the value of SD alone is not very effective. In addition, CH_4 accounts for a relatively large proportion of the carbon emissions from wastewater treatment processes [50]; therefore, it is possible to generate electricity from the collection of discharged CH_4 and realize the recycling of resources, thus effectively reducing GHG [51].

3.4. Multi-Objective Optimization of the Anaerobic-Anoxic-Oxic and Membrane Bioreactor (AAO-MBR) Process

In order to synthesize the interrelationships among the three objectives and screen the strategies to optimize the operational results, a comprehensive comparison of the three objectives was performed using multi-objective optimization. The study involved analyzing 55,460 different optimization scenarios and comparing them to one set of baseline scenarios in the case plant. The relative relationship between these scenarios and their Pareto ratings was determined using a non-dominated ranking method. The data were compared to produce a total of 133 Pareto ratings (Table S6).

Scenarios with high Pareto ratings are better than scenarios with low Pareto ratings for all three objectives (here, lower values for all three metrics are preferred). If all scenarios were plotted in a three-dimensional coordinate system using the three objectives as coordinates, scenarios with high Pareto ratings would be at the more "frontier" position. Plots with Pareto ratings of 1, 50, and 127 in blue, orange, and yellow are depicted in Figure 6a, respectively, in the 3D coordinate system, interpolating the points to construct an envelope with extra data points.

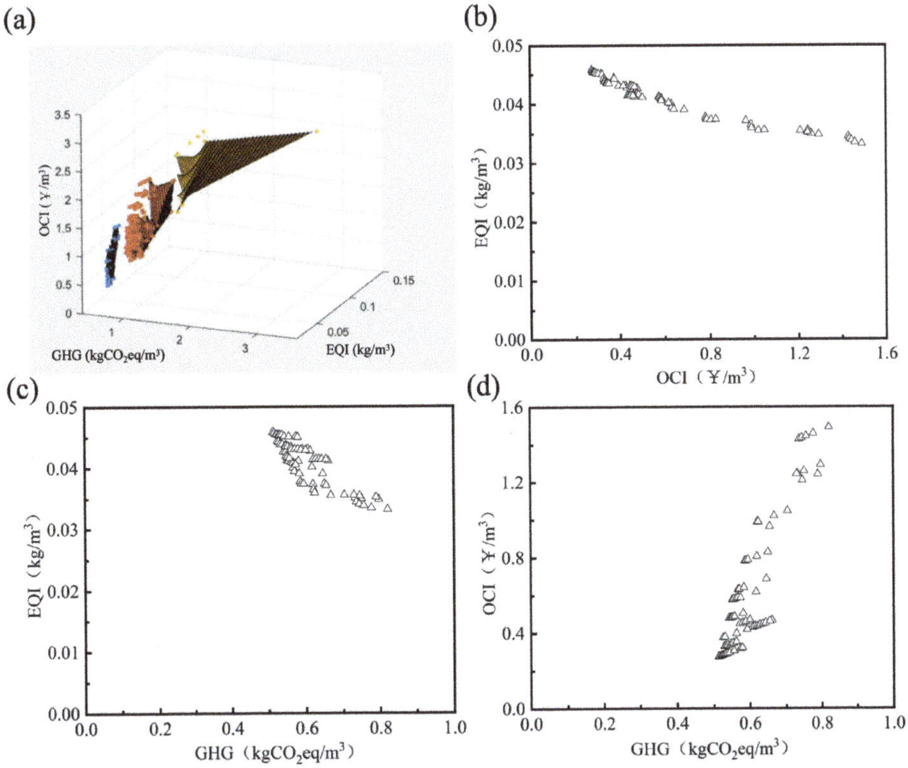

Figure 6. Scenarios with different Pareto levels and distribution of optimal scenario indicators. (**a**) Blue, orange, and yellow dots indicated the scenarios with Pareto levels of 1, 50, and 127, respectively; the relationship of OCI and EQI (**b**), GHG and EQI (**c**), and OCI and GHG (**d**) for the "Pareto-optimal" scenarios.

The 75 simulated scenarios with a Pareto rating of 1 can be considered "optimal". Although the data for these scenarios vary and there are even relatively discrete extremes in some of the scenarios, it is not possible to achieve full progress in all three objectives on the basis of these 75 sets of data. The distribution of the values of the three objectives for the "Pareto-optimal" scenario is shown in Figure 6b–d. There is a clear trade-off between the three objectives. In these scenarios, the OCI spans a wide range, indicating that there are still parameter combinations that have a large positive impact on the OCI, but a small negative impact on the other objectives, successfully realizing the "non-dominated" state. In the optimal scenario, with very few outliers removed, the EQI value is concentrated in the range of 0.033–0.046 kg/m^3, the OCI value is concentrated in the range of 0.27–1.49 ¥/m^3, and the GHG value is concentrated in the range of 0.51–0.82 kgCO_2/m^3.

After normalizing and weighting the EQI, OCI, and GHG objectives in equal proportions, the top 10 ranked scenarios were calculated as presented in Table 1, where the optimal scenario was 0.046 kg/m^3 for EQI, 0.27 ¥/m^3 for OCI, and 0.51 kgCO_2/m^3 for GHG. Compared to the baseline scenario (0.047 kg/m^3 for EQI, 0.97 ¥/m^3 for OCI, and GHG is 0.78 kgCO_2/m^3), the optimization is improved by 2.1%, 72.2%, and 34.6%, respectively, which significantly optimizes OCI and GHG under the premise of guaranteeing the EQI. Sweetapple et al. [52] used a non-dominated sorting genetic algorithm to find the Pareto-optimal solution for the three objectives. After the trade-off optimization, EQI is kept basically unchanged, while GHG and OCI are optimized to improve by 17.4% and 3.6%. Santín et al. [53] controlled the whole plant biological wastewater treatment process by the fuzzy controller, EQI, OCI, and GHG were optimized by 1.97%, 14.4%, and 8.24%, respectively. Although the differences in wastewater plant size and influent water quality may affect the optimization effect, it can reflect the advantage of the trade-off objective of the Pareto multi-objective optimization method to a certain extent.

Table 1. Multi-objective average weighted optimal ranking of EQI, OCI, and GHG for the case plant.

Serial Number	ATDO	MTDO	ECR	PAC	IRR	ERR	SD	EQI	OCI	GHG
	mgO_2/L	mgO_2/L	m^3/d	gAl/m^3	%	%	m^3/d	$kgCO_2/m^3$	¥/m^3	kg/m^3
251	0.1	4	0	2	50	100	1000	0.046	0.27	0.51
1501	0.1	4	0	2	100	100	1000	0.046	0.28	0.52
1511	0.1	5	0	2	100	100	1000	0.046	0.28	0.52
2751	0.1	4	0	2	200	100	1000	0.045	0.28	0.52
2761	0.1	5	0	2	200	100	1000	0.045	0.29	0.53
1521	0.1	6	0	2	100	100	1000	0.045	0.29	0.53
4001	0.1	4	0	2	300	100	1000	0.045	0.29	0.53
2771	0.1	6	0	2	200	100	1000	0.045	0.29	0.54
4011	0.1	5	0	2	300	100	1000	0.045	0.29	0.54
6501	0.1	4	0	2	100	200	1000	0.044	0.33	0.53

Focusing on the optimal combination of operational parameters, it can be seen that ATDO maintains a very low concentration of 0.1 mg/L, MTDO is also maintained by pulse aeration and other measures to maintain 4–6 mg/L, IRR is 50–300%, ERR is 100%, a small amount of PAC (2 gAl/m^3) is added, the SD is maintained at a low value of 1000 m^3/d, and the value of the ECR is taken to be 0, which means that the plant can meet the effluent standard without the addition of the carbon resource. The effluent standard can also be achieved. It can be seen that ERR, SD, and ATDO were taken to low values, which is the same as the results of univariate analysis. IRR and MTDO have a lower impact on the GHG, and therefore were taken to more dispersed values. PAC dosing was not taken to the lowest value, because 0 PAC would result in the effluent TP exceeding the standard, which would lead to a decrease in EQI [54].

To reduce pollution and carbon emissions in an environmentally-conscious and cost-effective manner, the sewage plant should prioritize reducing both ERR and SD. This approach will assist in diminishing the amount of sludge produced during the sewage treatment process and aid in lowering carbon emissions during sludge disposal. One effective way to control ERR is by adjusting the operating frequency of the return pump, while sludge settling time and organic loading rate can also be adjusted to minimize the solid content of the sludge [55]. In addition, the sludge production rate can be lowered by optimizing the operational parameters of the biochemical tank and accurately dosing phosphorus removal chemicals. By implementing these methods, it will further reduce GHG and OCI while still ensuring compliance with the EQI standard.

Based on the results of this study, some issues could be considered in the future. On the one hand, the main parameters for process simulation, such as structure size, stoichiometric parameters, and kinetic parameters, are mainly based on the default values in GPS-X software. It can be calibrated and localized with extensive data from actual wastewater plants if data are available in future studies. On the other hand, equal consideration is given to optimization objectives. When the weights of optimization objectives are not equal, such as carbon reduction as the goal, it is necessary to determine the index weight factor or adopt a fuzzy evaluation method to assist decision-making.

4. Conclusions

In this paper, three optimization objectives of WWTP were optimally weighed through a multi-objective optimization method. It demonstrated the potential to reduce GHG emissions cost-effectively while ensuring that the water quality meets the standards, and the following specific conclusions are drawn:

Starting from the idea of Pareto optimization, the GPS-X simulation and modeling software were used to screen 75 scenarios from tens of thousands of orthogonal simulation scenarios to reach the Pareto optimal level. Using the non-dominated sorting method, the optimal solutions were obtained by equal proportional weighting, which guided the attainment of the optimal combinations of operational parameters, as ATDO was kept at a very low concentration of 0.1 mg/L, MTDO was kept at 4 mg/L, IRR was 50%, ERR was 100%, a small amount of PAC (2 gAl/m^3) was added, SD was kept at a low value of 1000 m^3/d, and ECR was taken to be 0. This paper provides an optimal decision-making solution for this WWTP, which is conducive to the realization of sustainable development in the wastewater treatment industry.

Optimizing the performance of objectives requires a deep understanding of how it responds to operational parameters. Based on simulation results, there appears to be a significant linear relationship between ECR and the three key objectives that operators should prioritize. At the same time, the change in the C/N ratio due to the addition of ECR, thus affecting GHG emissions, is an essential direction for future research.

Supplementary Materials: The following supporting information can be downloaded at: https://www.mdpi.com/article/10.3390/w16070995/s1. Figure S1: Batch automation simulation script logic block diagram; Figure S2: Patterns of the effect of simulated value changes in MTDO on three objectives for the case plant; Table S1: The basic situation of influent water quality, actual effluent water quality, and simulated effluent water quality of the case plant; Table S2: Operational parameters of the case plant; Table S3: The parameters of each unit structure of the case plant; Table S4: Optimization scenario orthogonal take point setting of the case plant; Table S5: Summary of carbon emissions of each part; Table S6: Multi-objective optimization of Pareto rank distribution.

Author Contributions: Writing—original draft, data curation, investigation, formal analysis, visualization, J.L.; data curation, writing—original draft, visualization, S.L.; writing—original draft, data curation, visualization, Y.L.; methodology, investigation, writing—review and editing, S.M., T.T. and X.M.; conceptualization, formal analysis, writing—review and editing, B.L.; writing—review and editing, funding acquisition, Y.Q. All authors have read and agreed to the published version of the manuscript.

Funding: This study was support by the National Key Research and Development Program of China (2022YFC32031-04), Tsinghua-Toyota Joint Research Institute Inter-Disciplinary Program, and Tsinghua University INDITEX Sustainable Development Fund.

Data Availability Statement: Data are contained within the article and Supplementary Materials.

Conflicts of Interest: The authors declare no conflicts of interest. The funders had no role in the design of the study; in the collection, analyses, or interpretation of data; in the writing of the manuscript; or in the decision to publish the results.

References

1. Nguyen, D.H.; Chapman, A.; Tsuji, T. Assessing the Optimal Contributions of Renewables and Carbon Capture and Storage toward Carbon Neutrality by 2050. *Sustainability* **2023**, *15*, 13447. [CrossRef]
2. Zhang, Y.; Ge, T.; Liu, J.; Sun, Y.; Liu, Y.; Zhao, Q.; Tian, T. The Comprehensive Measurement Method of Energy Conservation and Emission Reduction in the Whole Process of Urban Sewage Treatment Based on Carbon Emission. *Environ. Sci. Pollut. Res.* **2021**, *28*, 56727–56740. [CrossRef] [PubMed]
3. Yao, H.; Gao, X.; Guo, J.; Wang, H.; Zhang, L.; Fan, L.; Jia, F.; Guo, J.; Peng, Y. Contribution of Nitrous Oxide to the Carbon Footprint of Full-Scale Wastewater Treatment Plants and Mitigation Strategies—A Critical Review. *Environ. Pollut.* **2022**, *314*, 120295. [CrossRef] [PubMed]
4. Ganora, D.; Hospido, A.; Husemann, J.; Krampe, J.; Loderer, C.; Longo, S.; Moragas Bouyat, L.; Obermaier, N.; Piraccini, E.; Stanev, S.; et al. Opportunities to Improve Energy Use in Urban Wastewater Treatment: A European-Scale Analysis. *Environ. Res. Lett.* **2019**, *14*, 044028. [CrossRef]
5. Wu, X.; Yang, Y.; Wu, G.; Mao, J.; Zhou, T. Simulation and Optimization of a Coking Wastewater Biological Treatment Process by Activated Sludge Models (ASM). *J. Environ. Manag.* **2016**, *165*, 235–242. [CrossRef] [PubMed]

6. Cao, J.; Yang, E.; Xu, C.; Zhang, T.; Xu, R.; Fu, B.; Feng, Q.; Fang, F.; Luo, J. Model-Based Strategy for Nitrogen Removal Enhancement in Full-Scale Wastewater Treatment Plants by GPS-X Integrated with Response Surface Methodology. *Sci. Total Environ.* **2021**, *769*, 144851. [CrossRef] [PubMed]
7. Vega, P.; Revollar, S.; Francisco, M.; Martín, J.M. Integration of Set Point Optimization Techniques into Nonlinear MPC for Improving the Operation of WWTPs. *Comput. Chem. Eng.* **2014**, *68*, 78–95. [CrossRef]
8. Zhang, R.; Xie, W.-M.; Yu, H.-Q.; Li, W.-W. Optimizing Municipal Wastewater Treatment Plants Using an Improved Multi-Objective Optimization Method. *Bioresour. Technol.* **2014**, *157*, 161–165. [CrossRef] [PubMed]
9. Guerrero, J.; Guisasola, A.; Comas, J.; Rodríguez-Roda, I.; Baeza, J.A. Multi-Criteria Selection of Optimum WWTP Control Setpoints Based on Microbiology-Related Failures, Effluent Quality and Operating Costs. *Chem. Eng. J.* **2012**, *188*, 23–29. [CrossRef]
10. Lu, H.; Wang, H.; Wu, Q.; Luo, H.; Zhao, Q.; Liu, B.; Si, Q.; Zheng, S.; Guo, W.; Ren, N. Automatic Control and Optimal Operation for Greenhouse Gas Mitigation in Sustainable Wastewater Treatment Plants: A Review. *Sci. Total Environ.* **2023**, *855*, 158849. [CrossRef] [PubMed]
11. Flores-Alsina, X.; Arnell, M.; Amerlinck, Y.; Corominas, L.; Gernaey, K.V.; Guo, L.; Lindblom, E.; Nopens, I.; Porro, J.; Shaw, A.; et al. Balancing Effluent Quality, Economic Cost and Greenhouse Gas Emissions during the Evaluation of (Plant-Wide) Control/Operational Strategies in WWTPs. *Sci. Total Environ.* **2014**, *466–467*, 616–624. [CrossRef] [PubMed]
12. Sweetapple, C.; Fu, G.; Butler, D. Does Carbon Reduction Increase Sustainability? A Study in Wastewater Treatment. *Water Res.* **2015**, *87*, 522–530. [CrossRef] [PubMed]
13. Sharawat, I.; Dahiya, R.; Dahiya, R.P. Analysis of a Wastewater Treatment Plant for Energy Consumption and Greenhouse Gas Emissions. *Int. J. Environ. Sci. Technol.* **2020**, *18*, 871–884. [CrossRef]
14. Chen, W.; Yao, C.; Lu, X. Optimal Design Activated Sludge Process by Means of Multi-Objective Optimization: Case Study in Benchmark Simulation Model 1 (BSM1). *Water Sci. Technol.* **2014**, *69*, 2052–2058. [CrossRef] [PubMed]
15. Béraud, B.; Steyer, J.P.; Lemoine, C.; Latrille, E.; Manic, G.; Printemps-Vacquier, C. Towards a Global Multi Objective Optimization of Wastewater Treatment Plant Based on Modeling and Genetic Algorithms. *Water Sci. Technol.* **2007**, *56*, 109–116. [CrossRef] [PubMed]
16. Jasim, N.A.; Aziz, H.A. The Design for Wastewater Treatment Plant (WWTP) with GPS-X Modelling. *Cogent Eng.* **2020**, *7*, 1723782. [CrossRef]
17. Pu, M.; Ailijiang, N.; Mamat, A.; Chang, J.; Zhang, Q.; Liu, Y.; Li, N. Occurrence of Antibiotics in the Different Biological Treatment Processes, Reclaimed Wastewater Treatment Plants and Effluent-Irrigated Soils. *J. Environ. Chem. Eng.* **2022**, *10*, 107715. [CrossRef]
18. Hu, Y.; Wang, X.C.; Zhang, Y.; Li, Y.; Chen, H.; Jin, P. Characteristics of an A^2O–MBR System for Reclaimed Water Production under Constant Flux at Low TMP. *J. Membr. Sci.* **2013**, *431*, 156–162. [CrossRef]
19. Le-Clech, P. Membrane Bioreactors and Their Uses in Wastewater Treatments. *Appl. Microbiol. Biotechnol.* **2010**, *88*, 1253–1260. [CrossRef] [PubMed]
20. Xia, Y.; Zhou, Y.; Wang, T.; Min, H.; Lu, Z.; Jiao, H.; Chen, Q. Application of AAO + MBR Processes in Underground Wastewater Treatment Plant (UWWTP). In *Water Purification Technology*; Shanghai Science and Technology Association: Shanghai, China, 2022; Volume 41, pp. 140–145. (In Chinese)
21. *GB 18918-2002*; Discharge Standard of Pollutants for Municipal Wastewater Treatment Plant. State Environmental Protection Administration: Beijing, China, 2002.
22. Ahmed, M.F.; Haider, M.Z.; Majid, H.; Hasan, S.M. Start-up and Operation of Novel EN-MBBR System for Sidestreams Treatment and Sensitivity Analysis Modeling Using GPS-X Simulation. *Alex. Eng. J.* **2022**, *61*, 10805–10818.
23. Liwarska-Bizukojc, E.; Biernacki, R. Identification of the Most Sensitive Parameters in the Activated Sludge Model Implemented in BioWin Software. *Bioresour. Technol.* **2010**, *101*, 7278–7285. [CrossRef] [PubMed]
24. Carnot, N.; Koen, V.; Tissot, B. *Economic Forecasting and Policy*; Palgrave Macmillan: London, UK, 2011.
25. Jia, Y.; Zeng, W.; Fan, Z.; Meng, Q.; Liu, H.; Peng, Y. An Effective Titanium Salt Dosing Strategy for Phosphorus Removal from Wastewater: Synergistic Enhancement of Chemical and Biological Treatment. *Sci. Total Environ.* **2022**, *842*, 156960. [CrossRef] [PubMed]
26. Kim, D.; Bowen, J.D.; Kinnear, D. Comprehensive Numerical Modeling of Greenhouse Gas Emissions from Water Resource Recovery Facilities. *Water Environ. Res.* **2015**, *87*, 1955–1969. [CrossRef] [PubMed]
27. Dai, H.; Chen, W.; Peng, L.; Wang, X.; Lu, X. Modeling and Performance Improvement of an Anaerobic–Anoxic/Nitrifying-Induced Crystallization Process via the Multi-Objective Optimization Method. *Environ. Sci. Pollut. Res.* **2019**, *26*, 5083–5093. [CrossRef] [PubMed]
28. Guo, X.; Kim, J.H.; Behera, S.K.; Park, H.S. Influence of Dissolved Oxygen Concentration and Aeration Time on Nitrite Accumulation in Partial Nitrification Process. *Int. J. Environ. Sci. Technol.* **2008**, *5*, 527–534. [CrossRef]
29. Jiang, L.; Liu, Y.; Guo, F.; Zhou, Z.; Jiang, J.; You, Z.; Wang, Q.; Wang, Z.; Wu, Z. Evaluation of Nutrient Removal Performance and Resource Recovery Potential of Anaerobic/Anoxic/Aerobic Membrane Bioreactor with Limited Aeration. *Bioresour. Technol.* **2021**, *340*, 125728. [CrossRef] [PubMed]

30. Nie, Z.; Huo, M.; Li, Q.; Shi, X.; Ma, Y.; Yu, G.; Zhu, S.; Ai, S.; Wang, F.; Bian, D. Sewage Treatment Effect of AOA-SBR under Different C/P Value and Its Mechanism of Nitrogen and Phosphorus Removal. *J. Water Process. Eng.* **2022**, *47*, 102774. [CrossRef]
31. Gou, J.; Hong, C.U.; Deng, M.; Chen, J.; Hou, J.; Li, D.; He, X. Effect of Carbon to Nitrogen Ratio on Water Quality and Community Structure Evolution in Suspended Growth Bioreactors through Biofloc Technology. *Water* **2019**, *11*, 1640. [CrossRef]
32. Wang, B.; Li, W.; Liu, L.; Huang, G.H. Influence of Sludge Reflux Ratios on Biodegradation Performance in a Coupled Landfill Leachate Treatment Process Based on UASB and Submerged MBR. *J. Environ. Sci. Health Part A* **2016**, *51*, 701–706. [CrossRef] [PubMed]
33. Irfan, M.; Waqas, S.; Khan, J.A.; Rahman, S.; Kruszelnicka, I.; Ginter-Kramarczyk, D.; Legutko, S.; Ochowiak, M.; Włodarczak, S.; Czernek, K. Effect of Operating Parameters and Energy Expenditure on the Biological Performance of Rotating Biological Contactor for Wastewater Treatment. *Energies* **2022**, *15*, 3523. [CrossRef]
34. Cao, Y.S.; Tang, J.G.; Henze, M.; Yang, X.P.; Gan, Y.P.; Li, J.; Kroiss, H.; van Loosdrecht, M.C.M.; Zhang, Y.; Daigger, G.T. The Leakage of Sewer Systems and the Impact on the 'Black and Odorous Water Bodies' and WWTPs in China. *Water Sci. Technol.* **2019**, *79*, 334–341. [CrossRef] [PubMed]
35. Liu, G.; Wang, J. Long-Term Low DO Enriches and Shifts Nitrifier Community in Activated Sludge. *Environ. Sci. Technol.* **2013**, *47*, 5109–5117. [CrossRef] [PubMed]
36. Sun, J.; Liang, P.; Yan, X.; Zuo, K.; Xiao, K.; Xia, J.; Qiu, Y.; Wu, Q.; Wu, S.; Huang, X.; et al. Reducing Aeration Energy Consumption in a Large-Scale Membrane Bioreactor: Process Simulation and Engineering Application. *Water Res.* **2016**, *93*, 205–213. [CrossRef] [PubMed]
37. Mallu, L.L.; Sean, W.-Y.; Pacheco, M. Optimization of Air Flowrate under Different Control Strategies Focus on Biological Process in Wastewater Treatment Plant. In Proceedings of the 2021 IEEE Symposium Series on Computational Intelligence (SSCI), Orlando, FL, USA, 5–7 December 2021.
38. Qambar, A.S.; Al Khalidy, M.M. Optimizing Dissolved Oxygen Requirement and Energy Consumption in Wastewater Treatment Plant Aeration Tanks Using Machine Learning. *J. Water Process. Eng.* **2022**, *50*, 103237. [CrossRef]
39. Zuo, R.; Ren, D.; Deng, Y.; Song, C.; Yu, Y.; Lu, X.; Zan, F.; Wu, X. Employing Low Dissolved Oxygen Strategy to Simultaneously Improve Nutrient Removal, Mitigate Membrane Fouling, and Reduce Energy Consumption in an AAO-MBR System: Fine Bubble or Coarse Bubble? *J. Water Process. Eng.* **2024**, *57*, 104602. [CrossRef]
40. Bagherzadeh, F.; Nouri, A.S.; Mehrani, M.-J.; Thennadil, S. Prediction of Energy Consumption and Evaluation of Affecting Factors in a Full-Scale WWTP Using a Machine Learning Approach. *Chem. Eng. Res. Des.* **2021**, *154*, 458–466. [CrossRef]
41. Kim, D.; Bowen, J.D.; Ozelkan, E.C. Optimization of Wastewater Treatment Plant Operation for Greenhouse Gas Mitigation. *J. Environ. Manag.* **2015**, *163*, 39–48. [CrossRef] [PubMed]
42. Lv, H.; Liu, D.; Xing, S.; Wu, D.; Wang, F.; Yang, J.; Wu, X.; Zhang, W.; Dai, X. The Effects of Aging for Improving Wastewater Sludge Electro-Dewatering Performances. *J. Ind. Eng. Chem.* **2019**, *80*, 647–655. [CrossRef]
43. Yapıcıoğlu, P.; Demir, Ö. Minimizing Greenhouse Gas Emissions of an Industrial Wastewater Treatment Plant in Terms of Water–Energy Nexus. *Appl. Water Sci.* **2021**, *11*, 180. [CrossRef]
44. Kyung, D.; Kim, M.; Chang, J.; Lee, W. Estimation of Greenhouse Gas Emissions from a Hybrid Wastewater Treatment Plant. *J. Clean. Prod.* **2015**, *95*, 117–123. [CrossRef]
45. Huang, F.; Shen, W.; Zhang, X.; Seferlis, P. Impacts of Dissolved Oxygen Control on Different Greenhouse Gas Emission Sources in Wastewater Treatment Process. *J. Clean. Prod.* **2020**, *274*, 123233. [CrossRef]
46. Flores-Alsina, X.; Corominas, L.; Snip, L.; Vanrolleghem, P.A. Including Greenhouse Gas Emissions during Benchmarking of Wastewater Treatment Plant Control Strategies. *Water Res.* **2011**, *45*, 4700–4710. [CrossRef] [PubMed]
47. Kishida, N.; Kim, J.H.; Kimochi, Y.; Nishimura, O.; Sasaki, H.; Sudo, R. Effect of C/N Ratio on Nitrous Oxide Emission from Swine Wastewater Treatment Process. *Water. Sci. Technol.* **2004**, *49*, 359–371. [CrossRef] [PubMed]
48. Li, Z.; Zou, Z.; Wang, L. Analysis and Forecasting of the Energy Consumption in Wastewater Treatment Plant. *Math. Probl. Eng.* **2019**, *2019*, 8690898. [CrossRef]
49. Hu, B.; Quan, J.; Huang, K.; Zhao, J.; Xing, G.; Wu, P.; Chen, Y.; Ding, X.; Hu, Y. Effects of C/N Ratio and Dissolved Oxygen on Aerobic Denitrification Process: A Mathematical Modeling Study. *Chemosphere.* **2021**, *272*, 129521. [CrossRef]
50. He, Y.; Li, Y.; Li, X.; Liu, Y.; Wang, Y.; Guo, H.; Hou, J.; Zhu, T.; Liu, Y. Net-Zero Greenhouse Gas Emission from Wastewater Treatment: Mechanisms, Opportunities and Perspectives. *Renew. Sustain. Energy Rev.* **2023**, *184*, 113547. [CrossRef]
51. Shrestha, A.; Bhattarai, T.N.; Ghimire, S.; Mainali, B.; Treichel, H.; Paudel, S.R. Estimation of Greenhouse Gases Emission from Domestic Wastewater in Nepal: A Scenario-Based Analysis Applicable for Developing Countries. *Chemosphere* **2022**, *300*, 134501. [CrossRef] [PubMed]
52. Sweetapple, C.; Fu, G.; Butler, D. Multi-Objective Optimisation of Wastewater Treatment Plant Control to Reduce Greenhouse Gas Emissions. *Water Res.* **2014**, *55*, 52–62. [CrossRef] [PubMed]
53. Santín, I.; Barbu, M.; Pedret, C.; Vilanova, R. Fuzzy Logic for Plant-Wide Control of Biological Wastewater Treatment Process Including Greenhouse Gas Emissions. *ISA Trans.* **2018**, *77*, 146–166. [CrossRef] [PubMed]

54. Huang, S.; Wang, R.; Yu, Y.; Zhang, W.; Tan, Z.; Zhou, H.; Wu, Y.; Chen, Y. Construction and Application of Effluent Quality Prediction Model with Insufficient Data Based on Transfer Learning Algorithm in Wastewater Treatment Plants. *Biochem. Eng. J.* **2023**, *191*, 108807. [CrossRef]
55. Heo, S.; Liu, Y.-Q. Dependence of Poly-β-Hydroxybutyrate Accumulation in Sludge on Biomass Concentration in SBRs. *Sci. Total Environ.* **2021**, *797*, 149138. [CrossRef]

water

MDPI

Article

A Modeling Analysis of Wastewater Heat Recovery Effects on Wastewater Treatment Plant Nitrification

Davide Mattioli [1], Gianpaolo Sabia [1], Luigi Petta [1], Margherita Altobelli [2], Margherita Evangelisti [2] and Marco Maglionico [2,*]

[1] ENEA Italian National Agency for New Technologies, Department for Sustainability, Division Resource Efficiency, Research Centre of Bologna, 40129 Bologna, BO, Italy; davide.mattioli@enea.it (D.M.); gianpaolo.sabia@enea.it (G.S.); luigi.petta@enea.it (L.P.)

[2] DICAM—Department of Civil, Chemical, Environmental and Materials Engineering, University of Bologna, Viale del Risorgimento 2, 40136 Bologna, BO, Italy; margherita.altobell3@unibo.it (M.A.); margherita.evangelisti@unibo.it (M.E.)

* Correspondence: marco.maglionico@unibo.it

Abstract: A global shift towards renewable energy production, driven primarily by the challenges posed by climate change, is currently underway. In this context, the utilization of heat recovery from municipal wastewater emerges as a promising green technology. Notably, the advantage of implementing energy recovery in sewers, as opposed to wastewater treatment plants (WWTPs), lies in the higher temperature of the wastewater and its proximity to potential heat users. Despite these benefits, concerns arise regarding the potential adverse effects on biological wastewater treatment processes downstream of the heat recovery section, particularly during colder seasons. This paper seeks to assess the impact of a heat recovery system along the sewer network on the efficiency of biological wastewater treatment processes. The methodology involves a modeling analysis of a real sewage network in Italy. Under typical northern Italy climate conditions, the results demonstrate the feasibility of heat recovery in sewers for WWTPs designed with a sludge residence time under aerobic conditions (SRT_{aer}) greater than 13 days. In such cases, the nitrification process remains relatively unaffected. However, for lower SRT_{aer} values, a case-specific feasibility assessment is recommended to evaluate the overall process efficiency comprehensively.

Keywords: heat recovery; sewer system; wastewater; biological nitrification; nitrogen removal rate

1. Introduction

Warm water production accounts for a significant share of the energy used in buildings; nearly 20 percent of the energy consumption in the domestic sector is associated with water heating for various purposes (e.g., shower, bath, dishwasher, washing machine, kitchen, etc.) [1]. A significant portion of this energy is unexploited since wastewater produced by residential, industrial, or commercial buildings and discharged after related uses still contains a significant amount of thermal energy. For example, in Switzerland, the loss of thermal energy in sewage has been estimated at around 6000 GWh per year, accounting for seven percent of the country's total heat needs [2]. Wastewater from domestic sources contains 2.5 times more recoverable thermal energy than the theoretical maximum chemical energy stored as COD (in the assumption of an energy recovery corresponding to a temperature effluent drop of 6 °C). According to an assessment carried out on sewage treatment plants in the Netherlands, assuming a thermal energy recovery from municipal effluents with a temperature drop of 5 °C, 24 h a day, 365 days a year, the total recovered energy would amount to about 40 PJ, equivalent to more than 40 percent of the total thermal energy produced by gas, coal, or biomass combustion processes [3]. As for the Italian potential, starting from the annual figure of 9.8 billion m^3 of wastewater treated by sewage treatment plants [4] and assuming a temperature drop of 5 °C, 24 h a day, 365 days a year, the calculated value for theoretically available thermal energy is about 200 PJ. Such a value corresponds to about 15 percent of the total thermal energy market in Italy for residential buildings and services [5].

The scientific community is now fully aware of the great potential of wastewater as an energy resource to be exploited to reduce fossil energy consumption. Recent European policies go in this direction; an example for all is the recently proposed revision of the Urban Waste Water Treatment Directive 91/271/EEC (UWWTD) (COM (2022) 541), which fosters the energy efficiency of the sector. Wastewater treatment needs are estimated at around 0.8 percent of the total energy consumption in the EU. The proposed revision would require sewage treatment plants larger than 10,000 P.E. to achieve energy neutrality by 2045 using renewable energy sources (of which no more than 30% can be purchased from external sources). Such an objective can be achieved through the implementation of efficiency measures, as well as by increasing energy production from renewable sources.

However, the implementation of heat recovery from sewage effluents requires careful evaluation, considering the effects on the entire drainage and treatment systems [6]. Heat recovery options for residential, commercial, and industrial facilities include the implementation of heat exchange technologies in sewer networks, near the wastewater discharge from the buildings, or at the centralized WWTP facilities. At present, there are several waste heat recovery technologies capable of producing thermal energy at different temperature levels, such as heat pumps and absorption refrigeration machines, systems already commercialized in many power ranges.

Several case studies show the achievable heat recovery potential in sewage networks or WWTPs [7]. Worldwide, at least 500 applications exist today [2], and many of them are implemented in Switzerland, Germany, and Scandinavia. Lucerne (Switzerland) and Oslo are examples of heat recovery in the sewer system, while in Berne (Switzerland), heat recovery is implemented within a WWTP achieving the potential for a heat recovery of more than 30 MW. Other practical applications of the thermal energy recovered from the effluent of WWTPs around the world are reported in [8].

Compared to WWTPs, the main advantage of providing energy recovery in sewer networks lies in the higher effluent temperature and the proximity to the potential users of the recovered heat [9]. On the other hand, sewer thermal energy recovery could result in appreciable changes in wastewater temperature and affect the WWTP performance [10,11]. In cold seasons, temperature drops can affect the kinetic rates of biological processes. Nitrification processes are particularly sensitive to temperature changes, as the relative maximum specific growth rate value μA of 0.8 d^{-1} at 20 °C slows down to 0.3 d^{-1} for a 10-degree decrease (ASM1) [12]. In activated sludge processes, the maintenance of microbial populations in reactors requires a proper balance between the new biomass growing and the excess sludge extraction for both autotrophic and heterotrophic bacteria populations [13]. The sludge retention time of autotrophic bacteria is carefully designed to prevent the washout of nitrifying bacteria as well as to avoid the resulting surge of ammonia nitrogen concentration in the effluent. As a rule of thumb, a sludge aerobic dilution rate (D_{aer}) significantly lower than the maximum net specific growth rate of nitrifying bacteria should be ensured, considering the temperature T of the process units. Wanner et al. [10] defined a nitrification safety factor (SF) as the ratio of the two quantities. Since the aerobic dilution rate is the reciprocal of the average sludge residence time in aerobic tanks (STR_{aer}), it follows that higher SRTs in the treatment plant are required to maintain a sufficient safety factor at low nitrification rates. Generally, nitrification requires a minimum SRT of about 10 days.

Given these drawbacks of heat recovery systems in sewers, an alternative and attractive option for energy recovery is represented by the application of such technologies directly at centralized WWTPs. In such plants, substantial quantities of wastewater from various sewer systems are collected, characterized by higher flow rates and less flow variability. The average wastewater temperature in WWTPs is typically lower than in the sewer network, attributed to the greater distances from production points; however, it also exhibits reduced susceptibility to extreme variability.

The present study examined the impacts of heat recovery in sewer networks on WWTP performances in terms of nitrogen removal rates. For this purpose, all significant contributions of heat transfer from the sewer to the WWTP as well as the effects of different temperatures of the wastewater on biological processes were calculated using different modeling software. The study provided a means of assessing the feasibility of thermal recovery from municipal wastewater, identifying the suitable conditions capable of ensuring both the treatment process performances and compliance with discharge limits. The paper addresses a relevant topic that has already been investigated in other contests. However, the adopted approach, which integrates hydraulics, heat exchange in sewage pipes and WWTPs, and biological treatment process modeling, represents a novel and distinctive aspect of the methodology employed in this study.

2. Materials and Methods

In the present study, the sewer flow rate of the Fossolo urban basin located in the southeastern suburbs of Bologna was first modelled. The basin extension is around 48 hectares, and its mixed-type drainage network has the advantage of being isolated from the surrounding sewer networks. The heat exchange between the wastewater and the environment was modelled as well. Both of these analyses were carried out with the TEMPEST calculation code developed by the Eawag-Swiss Federal Institute of Aquatic Science and Technology [12]. Heat transfer in the WWTP was then modelled by applying the "Basin Temperature" calculation code developed by researchers at the University of California, the University of Florence, and the Politecnico di Milano [14]. To evaluate the effect of temperature changes on biological processes in the WWTP, a model was built by using the BioWin 6.2 software (EnviroSim Associates Ltd., Flamborough, ON, Canada), based on the Activated Sludge/Anaerobic Digestion model (ASDM). The effect of installing a heat recovery system was evaluated starting from a five-degree reduced temperature.

The following paragraphs present the methodological details of the different phases of the applied modeling process.

2.1. Hydraulic Modeling

Fossolo hydraulic network collects wastewater mainly from domestic origin for an average dry weather flow rate of about 65 m^3/h. The total catchment area is about 48 ha, and there are about 10,000 inhabitants. The WWTP is located at a distance of about 10 km.

The network was studied extensively from a hydraulic point of view by Altobelli et al. [15] through the running of SWMM 5.1 software [16] by soliciting the network with 15 years of rainfall data starting from 2009 [15]. The choice of the years was derived from a numerical analysis aimed at identifying the average year over the period between 2005 and 2016. The analysis evaluated the average daily rainfall over the entire period and identified the year with the smallest difference, in absolute value, between the daily data for the year under study and the average daily data for the entire period. The simulations allowed the calculation of the flow rate released from the flood overflows placed along the sewer line and the flow rate conveyed to the WWTP. The trend over time of the flow rate reaching the WWTP is shown in Figure 1. The total volume conveyed from the sewer during the modeling year is approximately 690,000 m^3. This figure encompasses both the wastewater treated by the WWTP (638,000 m^3) and the upstream overflow discharged into the environment.

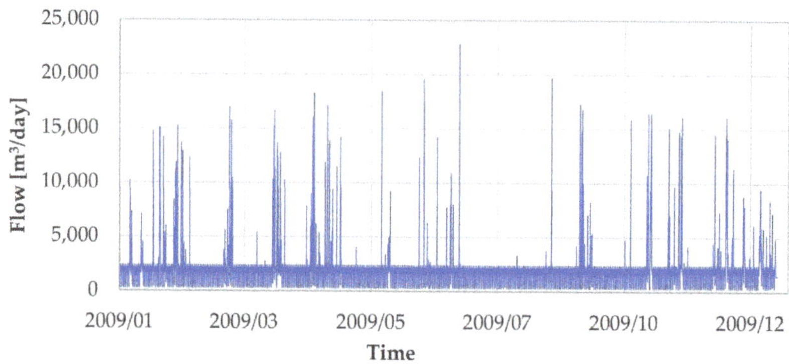

Figure 1. Flow rate collected by the sewer system.

2.2. *Thermal Exchange in Sewage Pipes*

Proper modeling of the effects of wastewater heat recovery is required, in addition to the flow rate, to estimate the temperature of the effluent within the sewage collectors. For the present study, flow rate and temperature monitoring data related to the Bologna sewer network were used as starting input. The dataset used refers to a measurement campaign from September 2006 to February 2007 [17], conducted on five measurement points at the sewer networks by the local water services provider, HERA S.p.A. The temperature values measured in the campaign are very similar among the various points. The temperature trend over time was interpolated by nonlinear regression by a periodic function of the equation: $y = y0 + a \sin(2\pi\, b\, x + c)$, with a period fixed at 365 days.

Once the temperature trend in the sewer network was defined, the heat transfer in the final sewer collector upstream to the WWTP was calculated for a total estimated distance of 10 km. To take into account heat exchanges between the network and the environment, the soil temperature at the depth of the pipeline was calculated using the method developed by Kusuda [18], which takes into account not only the depth with respect to the surface layer but also the characteristics of the soil. According to [17], where the sewer system of the city of Bologna was studied also in relation to the temperature of the wastewater and air in the various seasons, the soil temperature results as a function of the season and the depth below the surface layer.

One day in each month, from September 2006 through April 2007, was chosen to be reproduced, covering the entirety of the colder season this way.

The TEMPEST 1.02 software was employed to model heat exchange between the collector effluent and the soil using interpolated sewer temperature data and calculated soil temperature values. This tool facilitated the calculation of dynamic and longitudinal spatial profiles of wastewater temperature within the sewer collectors. Additionally, it allowed for simulating typical conditions on selected days during the colder season. One day was chosen each month, spanning from September 2006 through April 2007, to ensure coverage of the entirety of the colder season. The simulations were carried out considering constant soil and sewer pipe characteristics, as shown in Table 1.

For each day, two types of simulations were carried out:

- In the absence of heat recovery: the sewer water temperature at the upstream point (T_U) is actually that resulting from the interpolation;
- In the presence of heat recovery: the sewer water temperature value at the upstream point (T_U) is 5 °C lower because of the heat recovery operated in the sewer.

Thus, it was possible to define the temperature trend at the sewage inlet for the selected days. A sensitivity analysis of simulation data allowed for deriving the strong dependence of heat transfer on the difference between sewer and soil temperature.

Table 1. Soil and sewer pipe characteristics used in the model.

	Description	Unit of Measure	Value
PP	Pipe laying depth	m	4.6
L	Distance from the starting point	m	10,000
D	Pipe diameter	m	1.4
s	Thickness of pipe	m	0.2
S_0	Slope of pipe	m/m	0.003
r	COD degradation rate	mgCOD/(m^3 s)	2.8
Ksp	Roughness coefficient of pipe	$m^{1/3}$/s	83
C_P	Thermal conductivity of pipe	W/mK	0.7
T_P	Temperature diffusivity of pipe	m^2/s	0.49×10^{-6}
Q	Wastewater flow	m^3/s	0.02
	Air exchange coefficient	-	0.1
C_s	Thermal conductivity of soil	W/mK	1.88
T_s	Temperature diffusivity of soil	m^2/s	1.24×10^{-6}
ΔS	Penetration depth	m	0.13

Once the downstream temperature of the effluent entering the WWTP was denoted as T_D, the following parameters were calculated:

- $\Delta t = (T_D - T_U)$: temperature change of the effluent between the downstream and the upstream point;
- $(T_S - T_D)$: the difference between the soil temperature (Ts) and the temperature of the effluent at the downstream point.

The linear correlation between these two quantities was then checked, and the correlation line equation was used to estimate the temperature T_D for each day of the period.

2.3. Heat Exchange in the WWTP

In addition to the thermal exchanges within the sewer collector, thermal exchanges in the WWTP process sections need to be properly evaluated, and several contributions need to be considered. For each plant unit, the effects of energy exchanges across the external surface (i.e., the atmosphere and the walls), as well as of internal processes taking place (i.e., the effects of the mechanical aeration and the biochemical reactions), were assessed to evaluate the temperature variation of the mixed liquor. In the present study, the "Basin Temperature" calculation code was employed to calculate the net temperature change of wastewater in the WWTP [14]. The computational code allows for steady-state calculations. The model represents a development of that of Sedory and Stenstrom [19] and incorporates the heat transfer terms for biological processes of Makinia et al. [20], introducing new functions to account for surface convection and heat of evaporation. The model has been validated through several scenario studies [14].

The steady-state balance equation allows the calculation of the wastewater temperature in the plant units at equilibrium (T_w) as:

$$T_W = \frac{Q_t}{\rho_W q_W c_{pW}} + T_D \tag{1}$$

Q_t = net heat exchange
ρ_W = density of water
c_{pW} = specific heat of water
q_W = wastewater flow rate
T_w = tank equilibrium temperature
T_D = temperature of influent stream

Net heat transfer is the sum of various contributions according to the following equation:

$$Q_t = Q_{sr} + Q_p + Q_b + Q_{lr} + Q_e + Q_c + Q_a + Q_{tw} \tag{2}$$

Solar thermal radiation (Q_{sr}), biological reaction (Q_b), and power input (Q_p) terms are always positive and represent heat gains, while long wave thermal radiation (Q_{lr}), convection (Q_c), evaporation (Q_e), and tank wall conduction (Q_{tw}) terms can be gains or losses, with appropriate signs, depending on site-specific conditions. More information on the relationships used to estimate the quantities can be found in [14].

The input data to the Basin Temperature code for calculating the temperature in the WWTP units are summarized in Table 2.

Table 2. Basin Temperature input data used for heat transfer calculations in the WWTP units.

Months	Parameters					No Heat Recovery	With Heat Recovery
	T Air [°C]	T Soil [°C]	Wind [m/s]	Humidity [%]	Irradiance [W/m^2]	T Water T_i [°C]	T Water T_i [°C]
Sep	19.9	19.2	2.5	66.8	172.6	19.0	16.6
Oct	14.0	19.1	2.0	81.7	114.0	18.8	16.3
Nov	11.7	17.6	1.9	90.6	64.6	18.1	15.6
Dec	2.4	15.2	2.6	81.5	45.4	16.0	13.6
Gen	1.0	13.1	2.3	81.5	48.9	13.2	10.8
Feb	4.2	10.8	2.3	74.1	84.9	12.3	9.9
Mar	7.7	9.5	2.7	65.3	131.9	12.4	10.0
Apr	13.7	9.6	2.5	69.3	183.1	13.6	11.1

The modeling allowed the calculation of the temperature inside the WWTP sections by referring again to the two defined cases based on the presence or absence of heat recovery systems in the sewer.

2.4. *Effects on Biological Processes*

The biological nitrogen removal rates were simulated by running the ASDM model in the BioWin software [21]. The model accounts for 174 processes acting on 83 state variables, enabling the description of the typical chemical and biological processes occurring in WWTP processes [22].

The virtual WWTP considered for the simulations was purposely designed for the treatment of the wastewater produced by the Fossolo district, applying classic WWTP design criteria [13]. This choice is motivated by the need to fully calculate the effects of changes in sewer conditions on wastewater treatment processes. In centralized WWTPs, the mixing of several wastewater flows coming from other city districts would dilute such effects.

The main wastewater characteristics are shown in Table 3.

The layout of the WWTP, as depicted in Figure 2, comprises a single treatment line utilizing a conventional activated sludge treatment cycle, including pre-denitrification and nitrification units, followed by sedimentation and discharge processes. The oxidation phase takes place within a tank equipped with fine bubble diffusers, and the aeration rate is controlled based on the dissolved oxygen concentration.

The configuration of the sludge treatment line aligns with the typical characteristics found in small and medium-sized Italian plants. It involves aerobic digestion followed by a dehydration stage, with the recirculation of the water extracted from the sludge back to the pre-denitrification tank.

A combined sewer overflow discharger (CSO), designed for releasing excess wastewater from the sewer network, is situated upstream of the plant. The total annual volume of wastewater supplied to the WWTP, excluding the overflowed volume from the CSO upstream of the WWTP, totals approximately 638,000 m^3.

Table 3. Design data and main features of the WWTP.

Design Data	Value	Units
Flow rate	1530	m^3/d
BOD in	230	g/m^3
BOD out	10	g/m^3
TKN in	46	g/m^3
TKN out	1.5	g/m^3
SST in	230	g/m^3
SST out	25	g/m^3
NO_3-N out	10	g/m^3
T winter	12	°C
O_2 in aerated tanks	2	g/m^3
SST in recirculated sludge	8	kg/m^3
WWTP Main Features	**Value**	**Units**
SST in treatment tanks	3.5	kg/m^3
V nitrification tanks	840	m^3
V denitrification tanks	475	m^3
O_2 mass rate	38	kg/h
SRT	15–25	d
SRTaer	10–16	d

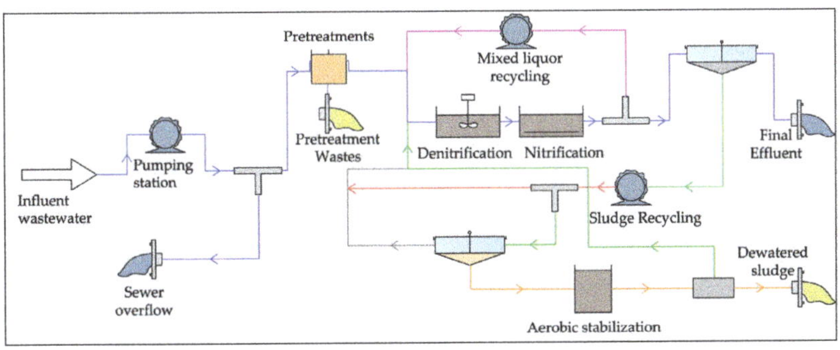

Figure 2. WWTP layout considered for simulations in the BioWin software.

The WWTP operation was simulated for two scenarios, each characterized by different operating conditions and varying cellular residence time (SRT) values. Consequently, different SRT under aerobic conditions (SRT_{aer}) were examined as follows:

- SRT = 15 days (SRT_{aer} = 10 days);
- SRT = 20 days (SRT_{aer} = 13 days);
- SRT = 25 days (SRT_{aer} = 16 days).

These SRT changes were achieved by acting on the setting of the biological sludge overflow rate.

The temperature effects on the nitrification processes of the WWTP were analyzed based on (i) time trends in the concentration of ammonia, nitrous, nitrate, and total nitrogen at the WWTP outlet; (ii) time trends in the concentration of nitrifying biomass in the WWTP units differentiated into ammonium (AOB) and nitrite-oxidizing (NOB) bacteria.

All simulations were conducted using the default values of the kinetic parameters in Biowin. The values of the maximum specific growth rate and decay rates and the related Arrhenius temperature constant of AOB and NOB are reported in Table 4.

Table 4. Maximum specific growth and decay rates of AOB and NOB as set in the model.

Kinetic Parameter	AOB	NOB
Maximum specific growth rate (μ_{max})	0.9 d^{-1}	0.7 d^{-1}
Arrhenius temperature constant for μ_{max}	1.072	1.060
Aerobic decay rate (b_{O2})	0.17 d^{-1}	0.17 d^{-1}
Arrhenius temperature constant for b_{O2}	1.029	1.029
Anoxic/anaerobic decay rate (b_{NOX})	0.08 d^{-1}	0.08 d^{-1}
Arrhenius temperature constant for b_{NOX}	1.029	1.029

3. Results and Discussion

Figure 3 shows the data recorded in the sewer during the monitoring campaign and the related curve obtained by nonlinear regression. The coefficient of determination is $R^2 = 0.8853$, while the resulting interpolating equation is shown in Figure 3.

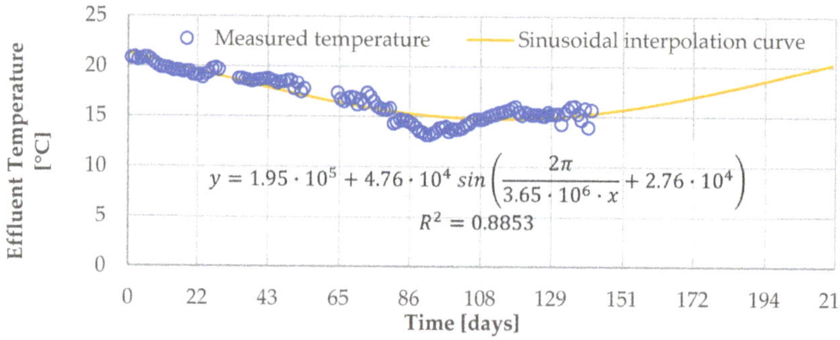

Figure 3. The trend of average daily temperature of the Bologna sewer from September 2006 to April 2007 interpolated by a periodic sinusoidal function. The y is the temperature of the effluent, and x is the number of days from 29 September 2006.

Figure 4 shows the annual trends of soil temperature (calculated by Kusuda's method) and air temperature at the sewer pipe depth of 4.6 m for the years 2006 and 2007.

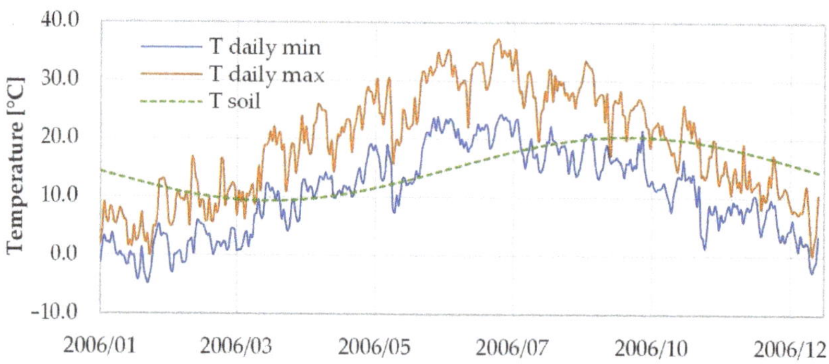

(a)

Figure 4. *Cont.*

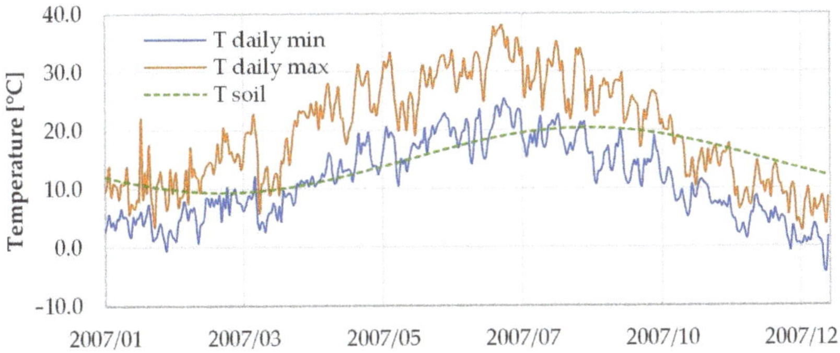

(**b**)

Figure 4. Trends of air temperature (daily minimum and maximum temperatures) and soil temperature, calculated by Kusuda's method, at the depth of the sewer pipe for the year 2006 (**a**) and for the year 2007 (**b**).

The maximum soil temperature value at the considered depth is reached in early autumn, while the lowest value falls in early spring.

Figure 5 depicts the interpolation of data obtained from the TEMPEST software, which simulated heat exchanges in the sewage collector. The graph illustrates the linear relationship between the upstream–downstream temperature drop in wastewater (*Y*-axis) and the difference between soil and sewage temperatures (*X*-axis). The linear interpolation of the points, supported by a high coefficient of determination, confirms the proportional relationship between these two quantities.

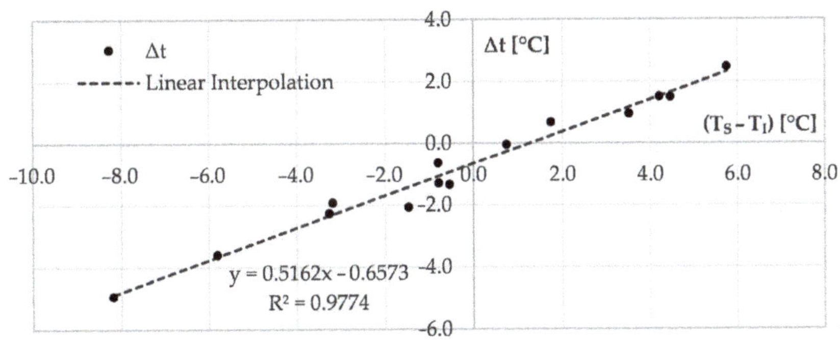

Figure 5. Linear interpolation of wastewater temperature difference as a function of the difference between soil and sewage temperature.

The temperature trend expected during the reference period, achieved through linear interpolation and related to the inlet section of the WWTP (T_d), is shown in Figure 6, together with the temperature expected in the sewer (T_u) and the temperature of the WWTP effluent (T_w) calculated from Tu via "Basin Temperature" code. As shown in Figure 6a, in the absence of heat recovery systems in the sewer, a significant variation in the wastewater temperature is expected, given the extended residence times available for heat exchange with the soil. In contrast, the heat exchange effect inside the WWTP treatment line is less significant. For almost the entire period, these exchanges result in a net wastewater cooling. Taking into account the T_u value in the first three months, substantially characterized by wastewater temperature drops, the mean value is 17.90 °C. In the same timeline, T_d and T_w assume an average value of 17.70 and 16.95 °C, respectively. In the following period when the heat exchanges with the surrounding structural assets cause a wastewater temperature rise, the related mean temperature values for T_u, T_d, and T_w stand at 16.46, 12.90, and 12.63 °C, respectively.

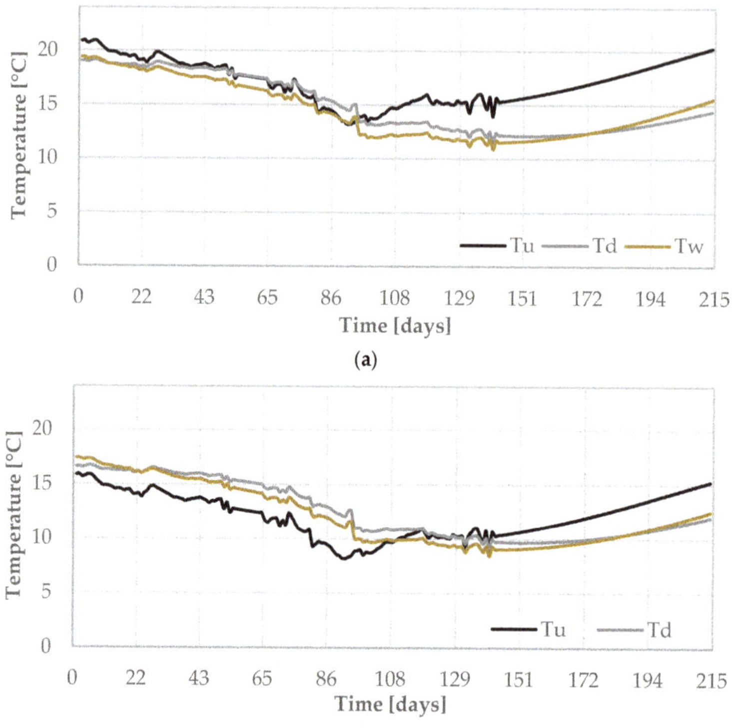

Figure 6. Modeled time trends of water temperature as discharged to the sewer (T_u), at the inlet of the WWTP (T_d), and in the WWTP (T_w), in the absence of heat recovery (**a**) and presence of heat recovery (**b**).

In Figure 6b, the expected effects of the presence of heat recovery in the sewer are represented. Following the wastewater temperature drop in the heat exchangers, throughout the first three months, significant wastewater warming was verified due to the heat exchanges with the soil. The mean temperature values for T_u, T_d, and T_w are 12.90, 15.28, and 14.84 °C, respectively. In the sewer system, the wastewater temperature is well lower than the WWTP inlet point, which shows a fluid temperature recovery along the path. During the following months, starting from January, the wastewater temperature values in the collector are significantly lower, with a mean value of 11.46 °C; T_D and T_w are 10.48 and 10.03 °C, respectively. Effectively, some heat is still lost, but the amount is lower with respect to case 1.

These temperature patterns cause an impact on the simulated WWTP nitrogen removal and on the AOB and NOB biomass concentration, as evidenced by Figures 7–9. In the absence of thermal recovery in the sewer and considering an SRT_{aer} of about 10 d (Figure 7a), the modeling output returns an outlet concentration of all nitrogen forms stably below the legal standards for discharge (i.e., 15 mg TN/l in surface water bodies in sensitive areas) during the whole experimental period. The simulated pollutant concentrations at WWTP discharge point are 9.43 ± 1.01 mg/L for TN, 2.98 ± 0.73 mg/L for TKN, and 6.04 ± 1.08 mg/L for nitrate and nitrite. However, the NOB biomass concentration in the aerobic reactor declines during the cold season (i.e., starting from January), and consequently, a slight increase in TKN concentration is expected.

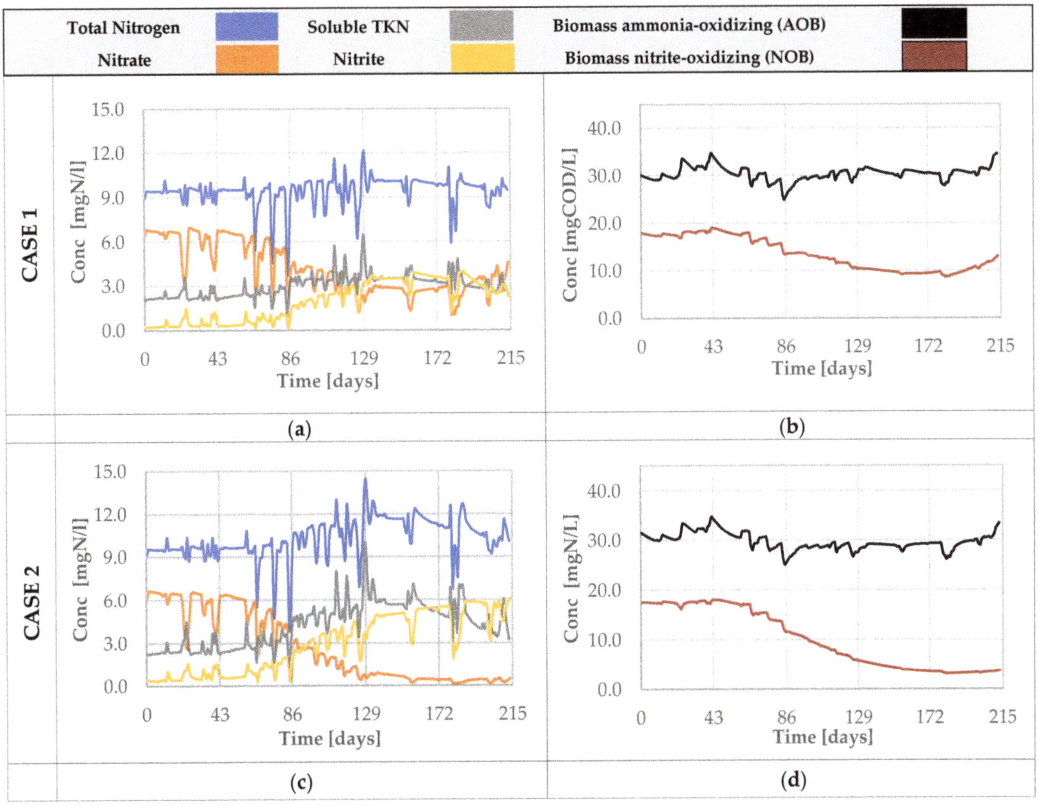

Figure 7. Simulated trends of the concentration of nitrogen compounds at the output of WWTP processes and the concentration of the main groups of nitrifying microorganisms calculated for SRT_{aer} of about 10 days and referred to: Case 1 (**a**,**b**); Case 2 (**c**,**d**).

Comparing this trend with the graph in Figure 7c, which refers to the case of thermal recovery in the sewer, it is possible to note how the effluent concentration of ammonia nitrogen is significantly affected by the reduced temperature, especially during the winter period.

The comparison of graphs in Figure 7b,d reveals that, in case 2, the model predicts a progressive decline for both AOB and NOB biomasses, more marked for the latter group. This observed difference is primarily influenced by the default values of the kinetic parameters of both biomasses (Table 4). Alternative assumptions regarding the Arrhenius temperature constants of the maximum specific growth rate could potentially result in a different pattern of AOB and NOB dynamics. For the purposes of this study, the main issue is the overall effect on the nitrification process while the specific impact of temperature on either group of nitrifiers is not so relevant and, thus, it was not further investigated.

As a direct consequence of the lower autotrophic biomass concentration and the decline of reaction kinetic rates due to the wastewater temperature drop, a higher increase in soluble TKN is expected (Figure 7c). At the same time, the greater drop in NOB concentration in the coldest months leads to a nitrite concentration increase, while nitrate declines. These trends imply a significant increase in TN discharge during winter months for case 2, with some daily values close to the allowed discharge limit. Average inorganic and organic nitrogen concentrations of the treated effluent are 10.29 ± 1.42, 5.65 ± 1.16, 4.23 ± 1.56 mg/L, and 2.98 ± 0.73 mg/L for TN, nitrate plus nitrite nitrogen, and TKN, respectively. Considering the inlet TKN concentration (i.e., 46.0 mg/L) and the averaged effluent concentration, the verified removal rates are 93.5% and 90.8% for Cases 1 and 2, respectively.

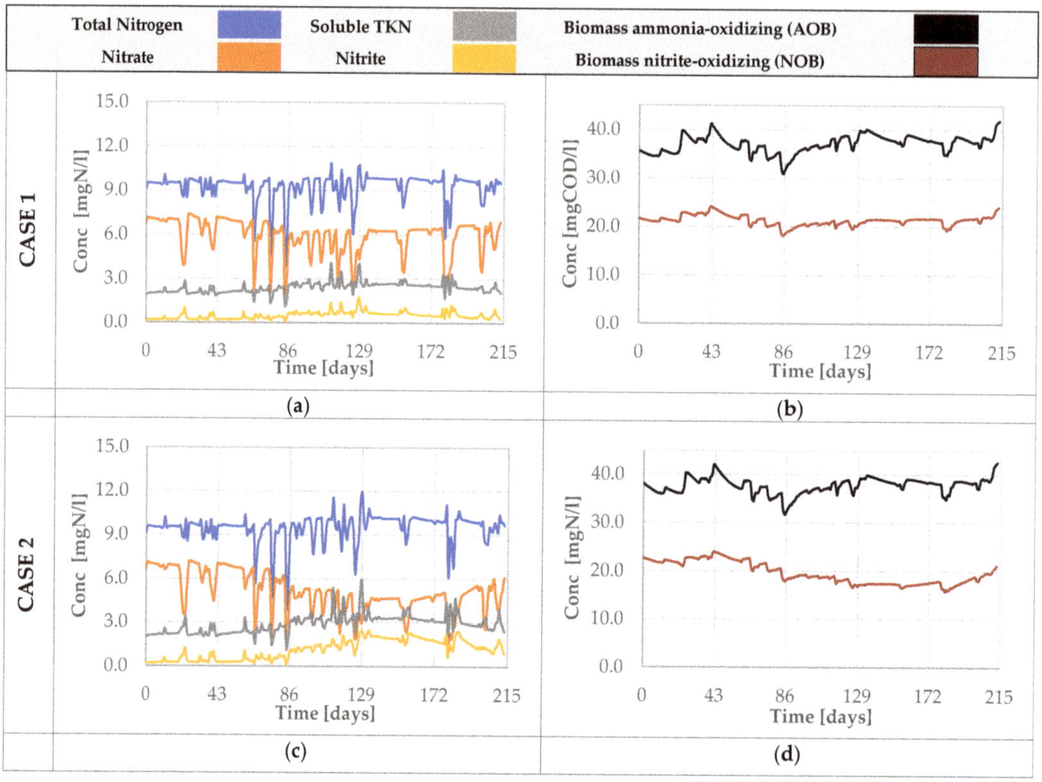

Figure 8. Simulated trends of the concentration of nitrogen compounds at the output of WWTP processes and the concentration of the main groups of nitrifying microorganisms calculated for SRT_{aer} of about 13 days and referred to: Case 1 (**a**,**b**); Case 2 (**c**,**d**).

Further WWTP simulations raising the SRT_{aer} to about 13 days again were performed to evaluate the effects of potential operational measures to be applied to face the effects of temperature variations. The results of such modifications are shown in Figure 8, revealing the same trends of the previous case, although significantly reduced. The comparison of Figure 8a,c shows only a slight increase in ammonia nitrogen concentration in the case of thermal recovery in the sewer. The time-averaged concentration of total, inorganic, and organic nitrogen are 9.25 ± 0.95 mg/L TN, 6.44 ± 1.07 mg/L nitrate plus nitrite nitrogen, and 2.39 ± 0.39 TKN in the option of no thermal recovery systems (Case 1), whereas for Case 2, the same chemical components amount to 9.50 ± 0.99 mg/L TN, 6.22 ± 1.09 mg/L nitrate plus nitrite nitrogen, and 2.85 ± 0.67 mg/L TKN. The comparison of the removal efficiency of soluble TKN shows a negligible drop in the removal rates from 99.2% to 98.6% in the presence of thermal recovery systems implemented into the sewer network. In both cases, such percentages are significantly higher at an SRT_{aer} of 13 days with respect to 10 days.

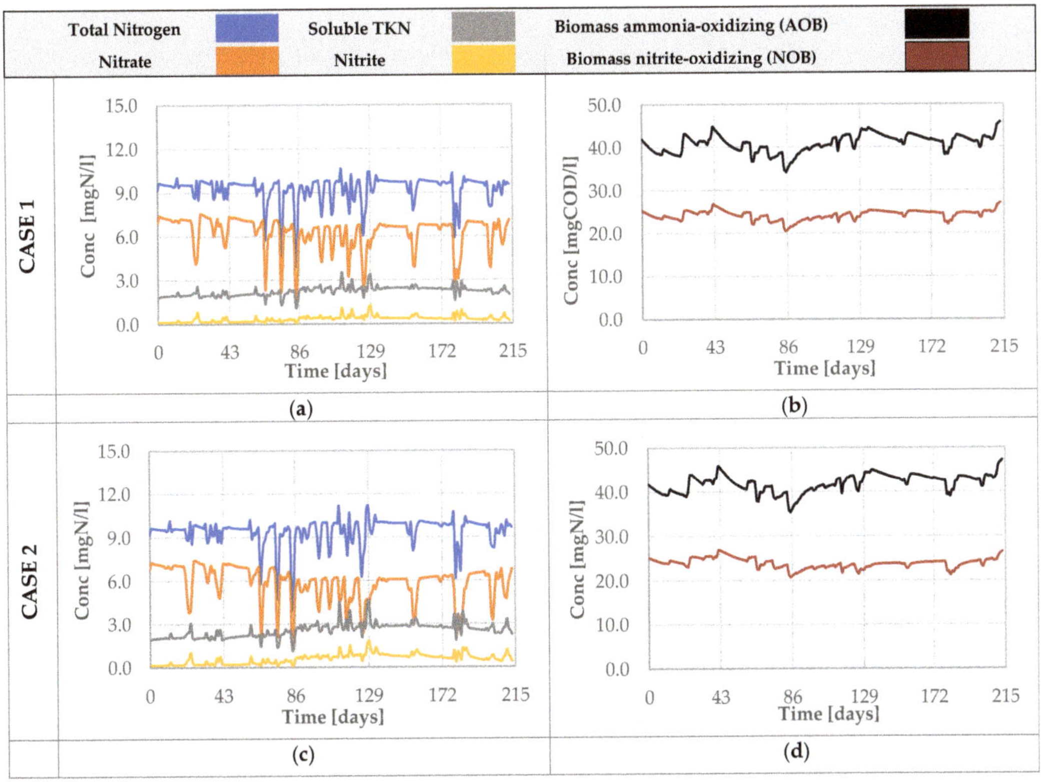

Figure 9. Simulated trends of the concentration of nitrogen compounds at the output of purification processes and the concentration of the main groups of nitrifying microorganisms calculated for SRT_{aer} of about 16 days and referred to: Case 1 (**a**,**b**); Case 2 (**c**,**d**).

Moreover, when considering an even higher SRT value, set at 16 days, no significant effect of the lower temperatures on nitrification processes is visible in Figure 9b,d.

The results indicate that with an SRT_{aer} of less than 13 days, commonly employed in denitrification–nitrification-based WWTPs, implementing a thermal recovery system in the sewer requires caution. Conversely, in the Italian climate, for extended or extensive aeration WWTPs with SRTs typically ranging between 20 and 30 days corresponding to SRT_{aer} values exceeding 13 days, heat recovery should pose fewer challenges for WWTP performance. The potential effects of a wastewater temperature drop on the nitrification process need to be evaluated on a case-by-case basis. If such effects are found to be significant, interventions to increase the SRT of the aerobic stage might be considered to be coupled to a heat recovery practice. This target can be achieved, for example, by increasing aeration volumes through the construction of new process tanks or by expanding the volume of existing ones. Alternatively, for the same total treatment volume, an increase in aerobic treatment volumes can be obtained at the expense of those reserved for the anoxic treatment phase. Such a modification, however, requires careful evaluation of the effects on the treatment process. Another way to increase the SRT_{aer}, avoiding any interventions on tank volumes, is to increase the biomass concentration by reducing the waste sludge flow. This option requires a preliminary evaluation to avoid the risk of significantly worsening sludge settling and overloading the secondary settlers, with possible biological sludge overflowing with the increase in the effluent flow rate, i.e., during rainy days.

4. Conclusions

This paper aimed to investigate the potential effects on the process performance of wastewater treatment plants (WWTPs) resulting from the implementation of heat recovery practices in sewer networks. Using a real sewer network servicing a residential area in Bologna, such as the Fossolo sewage network, and incorporating on-field monitoring data, we pursued this objective. Initially, hydraulic models were applied to estimate the incoming flow rates to the treatment plant facilities. Subsequent thermal modeling allowed for the calculation of temperatures in the final sewer section, representing the inlet section of the WWTP. With heat recovery systems in the sewer networks, the monthly average temperature reaches a minimum value of 10.8 °C in February, while without heat recovery, the minimum is 16.6 °C.

To assess the impact on treatment process units, a virtual WWTP was considered, and a dedicated model based on classic design criteria (including pre-denitrification and nitrification units followed by sedimentation, aerobic digestion of sludge, and a dehydration stage) was defined to simulate the processes in this plant. The temperature in the biological reactor and its influence on the nitrification process were then calculated. Our findings indicate that caution is advised when implementing thermal recovery systems in sewer networks for WWTPs with sludge residence times (SRT_{aer}) of less than 13 days, common in many denitrification–nitrification WWTPs in Italy. The study highlights that biological processes may be inhibited due to temperature drops, particularly during the cold season, leading to adverse effects on nitrogen compound removal rates.

Conversely, in the typical Italian climate, for WWTPs with SRT_{aer} values exceeding 13 days, no significant impacts on nitrification process performance are expected. This insight underscores the importance of considering specific operational parameters, such as SRT_{aer}, when evaluating the feasibility of heat recovery systems in sewer networks for preserving WWTP nitrogen removal efficiency.

Author Contributions: Conceptualization, D.M. and M.M.; methodology, D.M., M.M. and M.A.; investigation, D.M., G.S. and M.A.; writing—original draft preparation, D.M., G.S. and M.E.; writing—review and editing, all authors; funding acquisition, D.M. and L.P. All authors have read and agreed to the published version of the manuscript.

Funding: This research activity was funded by the project "Energy efficiency of industrial products and processes"—Research Program: "Piano Triennale della Ricerca del Sistema Elettrico Nazionale 2019–2021" funded by the Italian Ministry of Economic Development.

Data Availability Statement: The data presented in this study are available on request.

Conflicts of Interest: The authors declare no conflicts of interest.

References

1. Yao, R.; Steemers, K. A method of formulating energy load profile for domestic buildings in the UK. *Energy Build.* **2005**, *37*, 663–671. [CrossRef]
2. Schmid, F. Sewage Water: Interesting Heat Source for Heat Pumps and Chillers. In Proceedings of the 9th IEA Heat Pump Conference, Zürich, Switzerland, 19–23 May 2008.
3. Kehrein, P.; van Loosdrecht, M.; Osseweijer, P.; Garfí, M.; Dewulf, J.; Posada, J. A critical review of resource recovery from municipal wastewater treatment plants—Market supply potentials, technologies and bottlenecks. *Environ. Sci. Water Res. Technol.* **2020**, *6*, 877–910. [CrossRef]
4. Berardi, D.; Casarico, F.; Fatone, F.; Traini, S.; Zaramella, F. *Riuso Delle Acque Depurate in Agricoltura: Una Scelta Indifferibile*; Associazione Idrotecnica Italiana: Roma, Italy, 2020.
5. Persson, U.; Werner, S. *Quantifying the Heating and Cooling Demand in Europe*; European Commission: Brussels, Belgium, 2015.
6. Hadengue, B.; Joshi, P.; Figueroa, A.; Larsen, T.A.; Blumensaat, F. In-building heat recovery mitigates adverse temperature effects on biological wastewater treatment: A network-scale analysis of thermal-hydraulics in sewers. *Water Res.* **2021**, *204*, 117552. [CrossRef] [PubMed]
7. Nagpal, H.; Spriet, J.; Murali, M.K.; McNabola, A. Heat Recovery from Wastewater—A Review of Available Resource. *Water* **2021**, *13*, 1274. [CrossRef]
8. Hao, X.; Li, J.; van Loosdrecht, M.C.M.; Jiang, H.; Liu, R. Energy recovery from wastewater: Heat over organics. *Water Res.* **2019**, *161*, 74–77. [CrossRef] [PubMed]

9. Kretschmer, F.; Simperler, L.; Ertl, T. Analysing wastewater temperature development in a sewer system as a basis for the evaluation of wastewater heat recovery potentials. *Energy Build.* **2016**, *128*, 639–648. [CrossRef]
10. Wanner, O.; Panagiotidis, V.; Clavadetscher, P.; Siegrist, H. Effect of heat recovery from raw wastewater on nitrification and nitrogen removal in activated sludge plants. *Water Res.* **2005**, *39*, 4725–4734. [CrossRef] [PubMed]
11. Callegari, A.; Boguniewicz-Zablocka, J.; Capodaglio, A.G. Energy recovery and efficiency improvement for an activated sludge, agro-food WWTP upgrade. *Water Pract. Technol.* **2018**, *13*, 909–921. [CrossRef]
12. Henze, M.; Gujer, W.; Mino, T.; van Loosdrecht, M.C.M. *Activated Sludge Models ASM1, ASM2, ASM2d and ASM3*; IWA Publishing: London, UK, 2000; ISBN 978-1-900222-24-2.
13. Bonomo, L. *Trattamenti Delle Acque Reflue*; McGraw-Hill: New York, NY, USA, 2008.
14. Lippi, S.; Rosso, D.; Lubello, C.; Canziani, R.; Stenstrom, M.K. Temperature modelling and prediction for activated sludge systems. *Water Sci. Technol.* **2009**, *59*, 125–131. [CrossRef] [PubMed]
15. Altobelli, M.; Cipolla, S.S.; Maglionico, M. Combined Application of Real-Time Control and Green Technologies to Urban Drainage Systems. *Water* **2020**, *12*, 3432. [CrossRef]
16. Rossman, L. *Storm Water Management Model*; User's Manual; National Risk Management Research Laboratory, Office of Research and Development, US Environmental Protection Agency: Cincinnati, OH, USA, 2015.
17. Cipolla, S.S.; Maglionico, M. Heat recovery from urban wastewater: Analysis of the variability of flow rate and temperature. *Energy Build.* **2014**, *69*, 122–130. [CrossRef]
18. Kusuda, T.; Achenbach, P.R. *Earth Temperature and Thermal Diffusivity at Selected Stations in the United States*; National Bureau of Standards: Gaithersburg, MD, USA, 1965.
19. Sedory, P.E.; Stenstrom, M.K. Dynamic Prediction of Wastewater Aeration Basin Temperature. *J. Environ. Eng.* **1995**, *121*, 609–618. [CrossRef]
20. Makinia, J.; Wells, S.A.; Zima, P. Temperature Modeling in Activated Sludge Systems: A Case Study. *Water Environ. Res.* **2005**, *77*, 525–532. [CrossRef] [PubMed]
21. EnviroSim Associates Ltd. Available online: http://envirosim.com/products/biowin (accessed on 3 February 2020).
22. EnviroSim Associates Ltd. *BioWin Help Manual*; EnviroSim Associates Ltd.: Flamborough, ON, Canada, 2010.

processes

MDPI

Article

Innovative Plant-Derived Biomaterials for Sustainable and Effective Removal of Cationic and Anionic Dyes: Kinetic and Thermodynamic Study

El Mokhtar Saoudi Hassani [1,*], Dounia Azzouni [1], Mohammed M. Alanazi [2], Imane Mehdaoui [1], Rachid Mahmoud [1], Atul Kabra [3], Abdeslam Taleb [4], Mustapha Taleb [1] and Zakia Rais [1]

1 Laboratory of Engineering Electrochemistry, Modeling and Environment, Department of Chemistry, Faculty of Sciences Dhar Mahraz, Sidi Mohamed Ben Abdellah University, Fez 30000, Morocco; dounia.azzouni@usmba.ac.ma (D.A.); imane.mehdaoui@usmba.ac.ma (I.M.); rachid.mahmoud@usmba.ac.ma (R.M.); mustapha.taleb@usmba.ac.ma (M.T.); zakia.rais@usmba.ac.ma (Z.R.)
2 Department of Pharmaceutical Chemistry, College of Pharmacy, King Saud University, P.O. Box 2457, Riyadh 11451, Saudi Arabia; mmalanazi@ksu.edu.sa
3 University Institute of Pharma Sciences, Chandigarh University, Mohali 140413, Punjab, India; atul.e9963@cumail.in
4 Laboratory of Water and Environmental Engineering, Hassan II University of Casablanca, Casablanca 20150, Morocco; talebabdeslam1@gmail.com
* Correspondence: saoudimokhtar94@gmail.com

Abstract: The aim of this study is to purify industrial textile effluents by treating two types of commonly encountered dyes: blue maxilon (BM), of cationic nature, and black eriochrome (NE), of anionic nature. We intend to employ an innovative approach based on the adsorption of these dyes onto a novel vegetal biomaterial derived from Aleppo pine fibers (FPAs). A kinetic and thermodynamic study was conducted. The effect of some physicochemical parameters on both dye adsorption and FPAs was also evaluated. The modeling of the adsorption results was performed using Langmuir, Freundlich, Temkin, and Dubinin Radushkevich (D-R) isotherms. The results indicate that the equilibrium time strongly depends on the initial concentration of the two dyes, being 60 min with pseudo-second-order adsorption kinetics for both dyes. Adsorption isotherms under the optimal conditions of adsorbent mass, temperature, medium pH, and dye concentration were used to determine the maximum adsorption efficiency, which was close to 93% and 98% for BM and NE, respectively. The results also show that the adsorption of both dyes on FPAs fits well with Langmuir's model. The thermodynamic study indicates that the adsorption of both dyes on FPAs is spontaneous and exothermic in nature for BM and endothermic for NE.

Keywords: cationic dye; anionic dye; Aleppo pine fiber; adsorption; kinetics; thermodynamic parameters

1. Introduction

In most developing countries, the industrial revolution constitutes a real threat and affects health and the environment in an alarming way [1]. Numerous studies have shown that the industries generating significant pollution in wastewater highly charged with pollutants [2] are those of textiles [3], tanneries [4], brass factories [5], and paper mills [6]. Indeed, these effluents are often loaded with little or non-biodegradable pollutants, in particular, certain dyes that are harmful to health [7] and lead to a very high level of pollution of surface water and groundwater [7,8]. According to the United Nations World Water Development Report (2017), dyes are carcinogenic, mutagenic, and teratogenic chemicals, and also toxic to humans and some species of fish and microorganisms [9].

Despite the implementation of certain programs in well-advanced countries in terms of environmental protection, and the efforts made by the Moroccan government to establish laws requiring industries, according to their activities, to equip themselves with wastewater treatment units to reduce the pollutant load of effluents before they are discharged into watercourses via sewerage networks, the problem of surface water contamination by colored waters remains relevant and concerns the entire world [10].

Various methods have been developed for treating colored effluents, ranging from biological treatment [11,12] to coagulation/flocculation [13], chemical oxidation [11,14], membrane separation [15,16], and adsorption [17–38]. However, each method has potential drawbacks. For instance, biological treatment may be subject to specific constraints, such as the presence or absence of oxygen, and may be limited by the composition of the present dyes. In fact, coagulation/flocculation can lead to the formation of concentrated sludge, which may pose challenges in terms of disposal, particularly for high-flow effluents. In addition, chemical oxidation can be costly and energy-intensive, while membrane separation may require substantial investments in equipment and maintenance.

Adsorption is a widely used method in treating polluted waters due to its effectiveness and simplicity. It involves capturing dyes and pollutants with a solid material called an adsorbent. However, many traditional adsorbents, such as clays, zeolites, and activated carbons, may be costly to produce, originate from non-renewable natural deposits, and have limited regeneration capacity.

To address these challenges, many studies were focused on the development of alternative adsorbents that are biodegradable, affordable, and locally available from renewable natural resources or recyclable waste. These materials, such as biomass, agricultural residues, and industrial by-products, offer considerable potential for the sustainable treatment of colored effluents. So, by exploring these materials, we can discover multiple economical, ecological, and efficient solutions for industrial effluent treatment.

The objective of this study is to treat two dyes currently used in the textile industry, Maxilon Blue (BM) and Eriochrome Black T (NE), by adsorption on an innovative biomaterial derived from Aleppo pine fibers (FPAs). This biomaterial has never before been used to treat this type of pollution. The adsorption reaction time was determined by a kinetic study as a function of the BM and NE concentration of the dye solution, and pseudo-first- and pseudo-second-order models were used to correlate the kinetic data for dye adsorption on FPAs.

Operating conditions were optimized in terms of initial dye concentration, adsorbent mass, pH, and medium temperature for dye adsorption on the FPAs.

The results were modeled using Langmuir, Freundlich, Temkin, and D-R (Dubinin-Radushkevich) isotherms to understand the adsorption mechanism and determine thermodynamic parameters.

2. Materials and Methods

2.1. Adsorbent

The FPA powder was derived from the waste of Aleppo pine, supplied from the forest Ain chekaf (33°59′43.476″ N; 5°00′51.3″ W) in the city of Fez in Morocco, and treated physically by dry cleaning, then dried in the oven for 12 h, followed by grinding and sieving to a particle size less than 1 mm. The obtained powder, named FPAs, was analyzed by measuring physicochemical parameters and spectroscopic analysis.

2.2. Adsorbate

Two dyes were used in this study: Eriochrome Black T (NE) and Maxilon Blue (BM).

NE is an anionic dye with a gross chemical formula $C_{20}H_{12}N_3O_7SNa$ and a solubility in water exceeding 50 g/L. It is characterized by the presence of chromophore groups (Vinyl (-C=C-), azo (-N=N-), and Nitro ($-NO_2$), which is responsible for the colored aspect of the organic dyes due to the absorption of the ultraviolet rays, and of auxochrome groups, such as hydroxyl (-OH), which serve to enhance the effect of chromophore groups and ensure the solubility of dyes in water.

BM is a cationic dye with chemical formula $C_{19}H_{26}N_3O_6S_2$, characterized by the presence of chromophore groups Vinyl (-C=C-) and azo (-N=N-), and auxochromes groups Hydroxyl (-OH) (Table 1).

Table 1. Different characteristics of BM and NE.

Characteristics	Maxilon Blue	Eriochrome Black
Molecular weight (g/mole)	483	461.381
Chemical formula	$C_{19}H_{26}N_3O_6S_2$	$C_{20}H_{12}N_3O_7SNa$
Azo groups	Mono	Mono
Chromophore groups	(-N=N-) (-C=C-)	(-N=N-) (-C=C-) ($-NO_2$)
Auxochrome groups	(-OH)	(-OH)
Type	Cationic	Anionic
pH	5.6	5.6
λ_{max} (nm)	599	526
Chemical structure	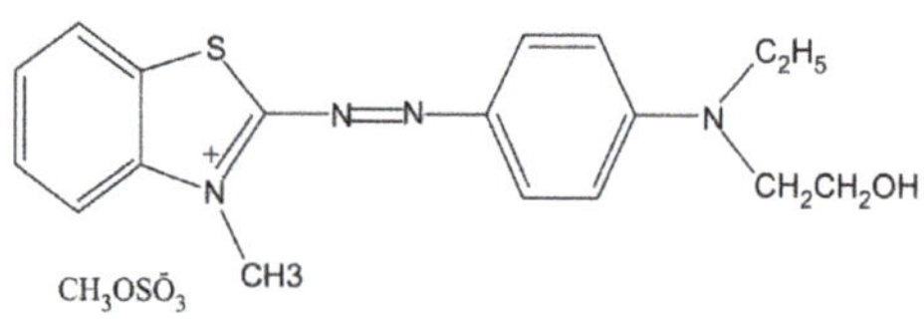	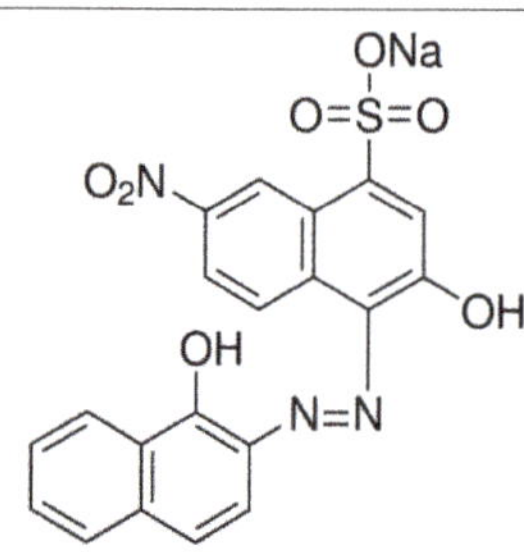

These dyes are widely used in laboratories as indicators for complexometric titrations and especially as dyeing dyes in textile industries because of their dye forms.

The 200 mg/L colored stock solution was prepared by dissolving 200 mg NE and BM in 1 L distilled water. The colored diluted solutions are produced by successive dilutions of the stock solution to obtain the solutions of different concentrations which will be used in the rest of this study.

Dye solutions before and after adsorption were analyzed by UV–Vis spectrophotometry of the UV-1800PC UV/VIS type, at wavelengths determined experimentally by scanning the wavelength spectrum between 400 nm and 900 nm for the different conditions studied (Figure 1).

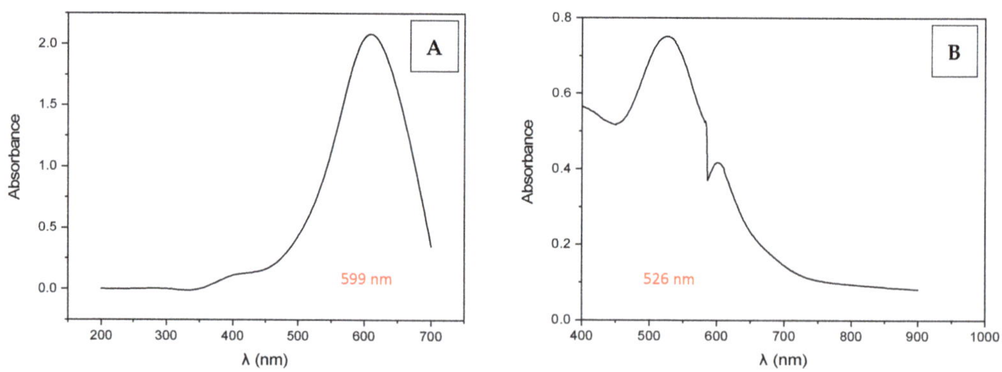

Figure 1. Adsorption maximum λ_{max} of BM (**A**) and NE (**B**) by UV/Visible spectroscopy.

2.3. Adsorption Process

2.3.1. Adsorption Kinetic

The kinetic study was performed on a suspension series of 0.05 g/L of FPAs in 100 mL of colored solutions with concentrations ranging from 5 to 120 mg/L. The mixtures were put under constant stirring of 300 rpm, at room temperature and pH of BM and NE solutions (5.8). Various samples were collected at various times, filtrated through a 0.45 µm PTFE syringe filter, and then analyzed by measuring the absorbance at the wavelength corresponding to the adsorption maximum, The adsorption capacity of the FPAs biomaterial was determined according to Equation (1):

$$Q_t = \frac{(C_i - C_t) \times V}{m} \tag{1}$$

where C_i (mg/L) is the initial dye concentration, C_t (mg/L) is the concentration of dye at time t, V (L) is the dye volume, and m (g) is the amount of adsorbent in the solution [19].

The results of the kinetic study were modeled according to the kinetic equations of pseudo-first-order (Equation (2)) or pseudo-second-order (Equation (3)) [20].

$$ln(q_e - q_t) = ln(q_e) - k_1 t \tag{2}$$

$$\frac{t}{q_t} = \frac{1}{k_2 {q_e}^2} + \frac{t}{q_t} \tag{3}$$

where q_e and q_t are, respectively, the amount of adsorbed material (mg/g) at equilibrium and at time t, k_1 (g/mg.h) is the first-order kinetic constant, and k_2 is the pseudo-second-order kinetic constant (g/mg.h).

The initial rate of adsorption, h, when t = 0 is defined by Equation (4):

$$h = k_2 \times q_e^2 \tag{4}$$

where q_e, h, and k_2 are calculated based on the slope and intercept of the t/q_t linear curve as a function of time.

The values of adsorption rate constants (k_2), initial velocities (h), and equilibrium adsorption capacities (q_e) are grouped in Table 4.

The results of the adsorption kinetics could be indicated on the identification of the diffusion mechanism by application of the intraparticle diffusion model given by Equation (5) after Webber and Morris [21].

$$Q_t = K_t.t^{(\frac{1}{2})} + C \tag{5}$$

where K_t is the velocity constant of the intraparticle diffusion. The value of C indicates the boundary layer thickness. It corresponds to the intercept value.

In this model, if the curve of q_t versus $t_{1/2}$ (Figure 5) presents a straight line passing through the origin, the speed of the adsorption phenomenon is considered limited by the process of intraparticle diffusion [22].

2.3.2. Adsorption Isotherms

Adsorption isotherms explain how the adsorbate molecules distribute between the liquid and solid phases when the adsorption process reaches a steady state. They were determined by adding a mass of 5 mg to a series of 250 mL vials containing 100 mL of solutions with the same concentrations of the 2 dyes as used in the adsorption tests. The solutions were stirred at 300 rpm for the equilibrium time determined by the reaction kinetics at room temperature (20 ± 2 °C), keeping the pH of the medium constant. The isotherm models considered were those of Langmuir [33], Freundlich [34,35], Temkin [36], and Dubinin Radushkevich (D-R) [37], as they better describe the adsorption mechanism of organic dyes.

- The Langmuir model is an adsorption model based on the formation of a single layer of adsorbate on the surface of the adsorbent. It is represented by Equation (6) [23]:

$$\frac{C_e}{q_e} = \frac{C_e}{q_m} + \frac{1}{K_L \times q_m} \quad (6)$$

where q_e (mg/g) is the quantity of dye adsorbed at equilibrium, K_L is the Langmuir constant related to the affinity of the binding sites and the adsorption energy, and C_e is the concentration of the dye solution at equilibrium (mg/L)

Plotting C_e/q_e in a function of C_e allows us to determine q_m and K_L.

The most important characteristics of the Langmuir isotherm can be represented by the separation factor R_L [23,24], which can be calculated by Equation (7):

$$R_L = \frac{1}{1 + K_L C_0} \quad (7)$$

where K_L is the constant of the Langmuir isotherm and C_0 is the largest starting concentration (mg/L).

The confirmation of the R_L separation factor is based on the following criteria (Table 2).

Table 2. Adsorption types according to R_L values.

R_L	$0 < R_L < 1$	$R_L > 0$	$R_L = 1$	$R_L = 0$
Adsorption type	Adsorption is favorable	Unfavorable adsorption	Linear adsorption	Irreversible adsorption

The isotherm of Freundlich is presented by the linearized form the defined Equation (8) [23]. It supposes that adsorption takes place on the heterogeneous surface according to a multi-layer adsorption process and that the quantity adsorbed increases according to the concentration.

$$log q_e = log K_F + \frac{1}{n} log C_e \quad (8)$$

where K_F and n are the constants of Freundlich, and q_e is the maximum saturation adsorption capacity.

The intensity parameter values $1/n$ indicate the variation of the adsorption isotherm from linearity:

1. When $1/n = 1$, the absorption process is linear, which indicates a homogeneity of the sites and the absence of any interaction between the adsorbed species.
2. When $1/n < 1$, the adsorption process is favorable, the retention capacity increases, and many new adsorption sites appear.
3. When $1/n > 1$, the adsorption process is not favorable and the adsorption chains become weaker, resulting in a decrease in the adsorption capacity.

- The Temkin isotherm is represented by the following equation:

$$q_e = B_T Ln(C_e) + B_T Ln(K_T) \quad (9)$$

where B_T and K_T are the Temkin isotherm constants.

The Temkin isotherm contains a factor that takes the interactions between the adsorbent and the adsorbate explicitly into account (Table 3).

- The Dubinin–Radushkevich (D-R) isotherm model has been studied to take into consideration the effect of the porous structure of the adsorbents as well as their free energy of adsorption. It is expressed by the following equation [25]:

$$log(q_e) = log(q_{DR}) - K_{DR}\varepsilon^2 \quad (10)$$

where q_e (mg/g) is the quantity adsorbed on the adsorbent at the equilibrium, q_{DR} (mg/g) is the capacity of adsorbate adsorbed on the surface of the solid, K_{DR} (mol^2 KJ^{-2}) is the Sorption Energy Constant, and ε is the potential of Polanyi, defined by Equation (11):

$$\varepsilon = RTln\left[1 + \frac{1}{C_e}\right] \tag{11}$$

Table 3. Correspondence of the values of the constant B_T of the Temkin isotherm.

B_T	$B_T < 8$	$8 < B_T < 16$	$B_T > 16$
Adsorption process	Physical adsorption	Ion exchange	Chemical adsorption

The average adsorption energy E (kJ/mol) is obtained using the following equation [26]:

$$E = \frac{1}{\sqrt{2K_{DR}}} \tag{12}$$

The value of the average adsorption energy E can give useful information about the adsorption process (physical or chemical) from the D-R isotherm, physical adsorption (E < 8 KJ.mol^{-1}), ion exchange (8 < E < 16 KJ.mol^{-1}), and chemical adsorption (E > 16 KJ.mol^{-1}) [26].

3. Results and Discussions

3.1. Main Characteristics of the FPA Adsorbent Material

The FPA powder is acidic in character, with a porous appearance that facilitates the adsorption and fixation of dyes. It also has a zero charge point equal to 5.8 and a specific surface of 384 m^2·g^{-1}. It is mainly composed of hydroxyl, carboxyl, amine functional groups, mineral compounds, and organic compounds, including cellulose and other mineral elements, which are Ca, Mg, Fe, Na, P, Al, K, Ni, and Mo.

3.2. Adsorption Kinetic

To fully understand the adsorption mechanisms of the two dyes BM and NE on FPAs, mixtures of different concentrations were stirred according to the contact time between adsorbent and adsorbate from 0 to 180 min with a mass of adsorbent of (0.05 g/L) at the residual dye pH. In fact, different samples were taken for analysis at regular times so that the decolorization yields of the solutions was determined in the function of the contact time (Figure 2).

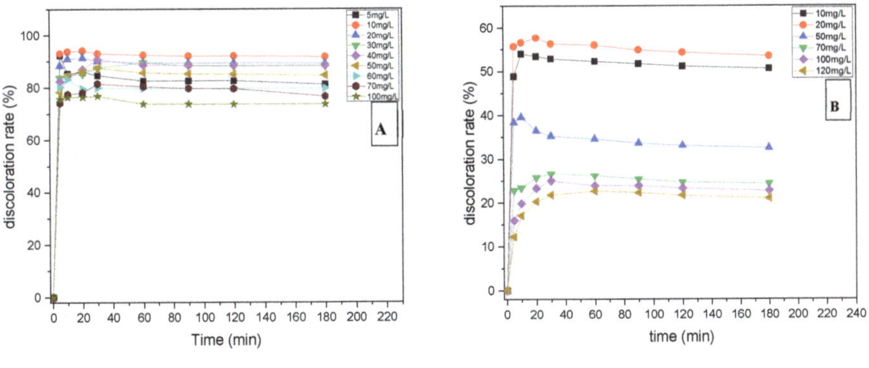

Figure 2. Adsorption kinetic of BM (**A**) and NE (**B**) on FPAs. Operating conditions: pH = 5.6, T = (20 ± 2) °C, adsorbent mass = 0.05 g/L.

For both dyes, the removal rate shows an increase with the reaction time following two different slopes. During the first one, a rapid increase of the decolorization yield is noticed at the first 10 min for all the studied concentrations of the 2 dyes. At this time, the dyes adsorb first on the easily accessible adsorbent sites. In contrast, the second stage is characterized by a slowing down of the adsorption efficiency to a maximum value corresponding to the equilibrium time of 60 min for both dyes. This can be expressed by the diffusion of the dye to the less accessible adsorption sites until an adsorption equilibrium is reached.

Modeling of Adsorption Kinetic

It took place through the pseudo-first-order (PPO) and pseudo-second-order (PSO) models.

a. Pseudo-first-order kinetics (PPO)

With this model, the rate of adsorption at time t is considered proportional to a difference between the equilibrium adsorbed amount Q_e and the amount Q_t adsorbed at that instant; the adsorption process is reversible [27].

The constants of the pseudo-first-order kinetics were determined by plotting the $\log(q_e\text{-}q_t)$ versus the timeline of the different studied concentrations (Equation (2)). The values of the first-order constants K_1, the theoretical and experimental quantities, and the coefficient of determination R^2 of the studied concentrations of BM and NE are presented in Table 4.

Table 4. Adsorption kinetic constants of BM and NE on FPAs.

	Kinetic Model	**Pseudo-First-Order**				**Pseudo-Second-Order**				
	Concentration (mg/L)	R^2	K_1 (min^{-1})	qe_{the} (mg/g)	qe_{exp} (mg/g)	R^2	K_2 (g/mg.min)	h (mg/g.min)	qe_{the} (mg/g)	qe_{exp} (mg/g)
BM	5	0.04	0.0003	1.314	8.25	0.99	−0.015	**−0.126**	8.123	8.25
	10	0.02	0.003	0.587	18.46	1	−0.188	**−3.472**	18.31	18.46
	20	0.05	0.005	1.33	35.80	1	−0.148	**−5.312**	35.71	35.80
	30	0.45	0.014	4.76	53.71	1	0.099	**5.322**	52.91	53.71
	40	0.25	0.011	6.436	70.84	1	0.221	**15.65**	70.92	70.84
	50	0.11	0.007	7.133	85.63	0.99	0.049	**4.192**	84.74	85.63
	60	0.03	0.004	6.099	95.61	1	−0.110	**−10.51**	95.23	95.61
	70	0.08	0.007	7.458	112.60	0.99	−0.011	**−1.261**	108.69	112.60
	100	0.02	−0.004	3.34	147.43	1	−1.492	**−219.96**	147.05	147.43
NE	10	0.03	−0.002	0.394	10.45	0.99	−0.125	**−1.306**	10.13	10.45
	20	0.17	−0.006	0.368	22.39	0.99	−0.044	**−0.985**	21.41	22.39
	50	0.61	−0.01	1.85	34.55	0.99	−0.017	**−0.587**	32.36	34.55
	70	0.07	0.003	1.719	36.59	0.99	−0.028	**−1.024**	34.24	36.59
	100	0.05	0.004	4.86	47.73	0.99	−0.074	**−3.532**	46.08	47.73
	120	0.20	0.009	6.35	53.32	0.99	0.042	**2.239**	51.51	53.32

The results of the R^2 values of both dyes are low and the theoretically adsorbed amounts (q_{the}) are also very low compared to the experimentally adsorbed amounts of both dyes. These observations indicate that the adsorption of BM and NE on FPAs does not correspond with a diffusion-controlled process since it does not satisfy the pseudo-first-order equation.

b. Pseudo-second-order kinetics (PSO)

The graph of t/Q_t versus t in Equation (3) is linear (Figure 3), implying that the adsorption of the two dyes on the FPAs has pseudo-second-order kinetics.

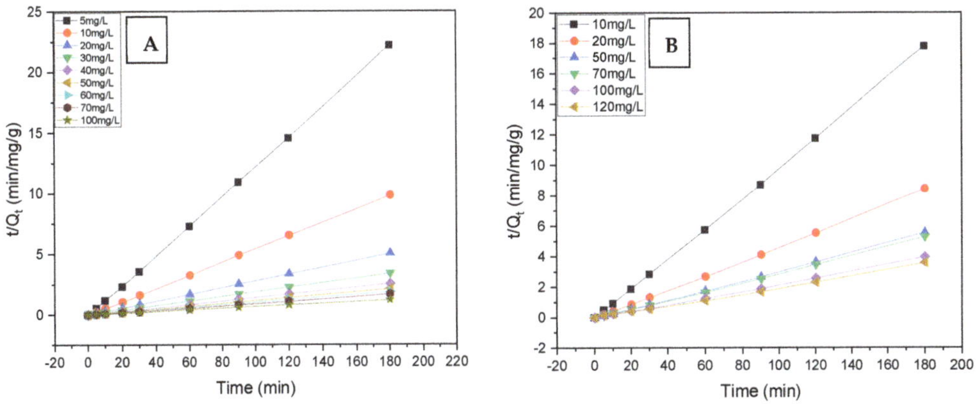

Figure 3. Application of the pseudo-second-order model for the adsorption of BM (**A**) and NE (**B**) on FPAs.

The adsorbed quantities q_e, the constants of this model K_2, and the regression coefficients R^2 of all employed concentrations are given in Table 4.

The results show that the equilibrium adsorbed amount q_e increases with the increasing initial concentration of both dyes. In addition, the values of the linear regression correlation coefficient R^2 for the second-order model are higher than those for the first-order model, and they are very close to 1 for all studied concentrations of both dyes. In addition, the theoretically adsorbed quantities (q_{the}) are very close to the values found experimentally and closer to those calculated using the second-order kinetics model.

All the results indicate that the adsorption kinetics of BM and NE on FPAs are perfectly defined by the pseudo-second-order kinetic model whatever the concentration of the dye solution is. This model considers chemisorption as the step that involves electron exchange and hydrogen bonding at the solid–liquid interface (Figure 4).

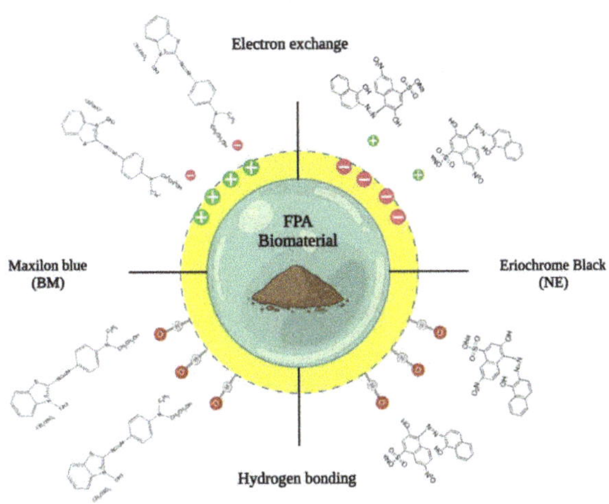

Figure 4. Chemical adsorption mechanism of BM and NE on FPA biomaterial.

When dyes come into contact with the biomaterial surface, electrostatic interactions occur, leading to electron transfer between dye molecules and active sites on the adsorbent surface. These electrostatic interactions can be enhanced by the presence of functional groups on the adsorbent surface, which can accept or donate electrons to form chemical bonds. At the same time, hydrogen bonds can form between the dye molecules and the adsorbent's functional groups. Hydrogen bonds are weak but significant interactions that occur between a partially positively charged hydrogen atom and a partially negatively charged oxygen, nitrogen, or sulfur atom in another molecule. These bonds can stabilize the dye molecules on the adsorbent surface and promote their adsorption. These electrostatic interactions and hydrogen bonds between the adsorbent and adsorbate contribute to the attachment of dyes to the biomaterial surface and play a decisive role in the adsorption kinetics described by the pseudo-second-order model.

The initial equilibrium adsorption rate calculation showed that the reaction rate is highly dependent on the dye concentrations (BM and NE).

These results imply that the BM and NE molecules adsorb onto the active sites of the FPAs with varying rates depending on the initial dye concentration.

3.3. Adsorption Mechanism

The adsorption of BM and NE dyes onto FPAs can be described in three successive steps. First, the dyes migrate through the solution to the outer surface of the FPA particles. Next, the dyes move inside the particles, and finally, they adsorb onto free positions on the inner surface of the adsorbent.

The representation of the adsorption capacities of the biomaterial as a function of the square root of t (Figure 5) is multilinear whatever the concentration of the solution and represents very low values of correlation coefficients R^2 and all lower than 0.59.

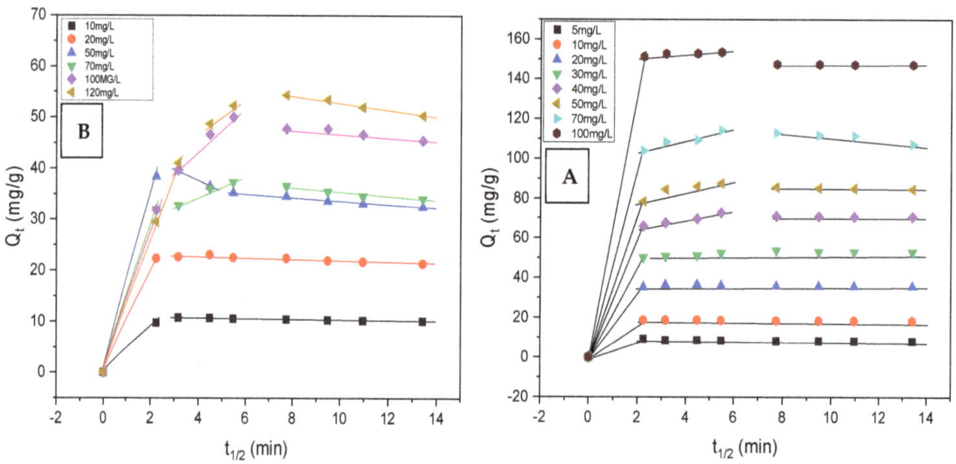

Figure 5. Intraparticle diffusion model for the adsorption of BM (**A**) and NE (**B**) on FPAs.

In the beginning, the dye adsorbs on the external surface of the FPA particles at a slightly faster speed for all the studied concentrations of the two dyes except for the first two concentrations of BM, which adsorb slowly. Until saturation of the external surface of the FPAs, at this time the dye molecules penetrate and adsorb on the internal surface of the FPAs. This increases the resistance to diffusion and subsequently leads to a decrease in the diffusion rate. As the dye concentration in the solution decreases, the rate of diffusion becomes lower and lower.

The intraparticle diffusion constants (K_t and C) results calculated from Equation (5) (Table 5) show that these constants (K_t and C) are different from zero and increase with dye concentration, confirming that intraparticle diffusion is not the only mechanism limiting the adsorption process of the two dyes BM and NE on FPAs.

Table 5. Parameters related to the intraparticle mechanism of the two dyes on FPAs.

	Concentrations Mg/L	Diffusion Constants		
		K_t	C	R^2
BM	5	0.4769	5.136	0.2116
	10	1.1935	10.513	0.2795
	20	2.366	20.074	0.293
	30	3.7704	27.481	0.3516
	40	5.0734	36.684	0.354
	50	5.9913	45.212	0.3361
	70	7.6828	59.136	0.3285
	100	9.3354	86.575	0.261
NE	10	0.678	5.827	0.2863
	20	1.3617	12.951	0.2521
	50	1.6705	23.138	0.1442
	70	2.5038	18.441	0.3411
	100	4.0568	19.223	0.4935
	120	4.9886	17.431	0.5922

3.4. Adsorption Isotherms

Figure 6 shows the curves of the quantity of adsorbate retained per gram of the adsorbent in the function of the equilibrium adsorbate concentrations of BM and NE. The shape of the BM curve indicates that as the initial concentration of dye increases, the quantity of the adsorbed dye on the adsorbent also increases to the steady state, reflecting that the interactions between the adsorbate and the adsorbent are greater than that between the adsorbate molecules. This minimizes adsorption site competition between the adsorbent and the dyes to promote monolayer adsorption.

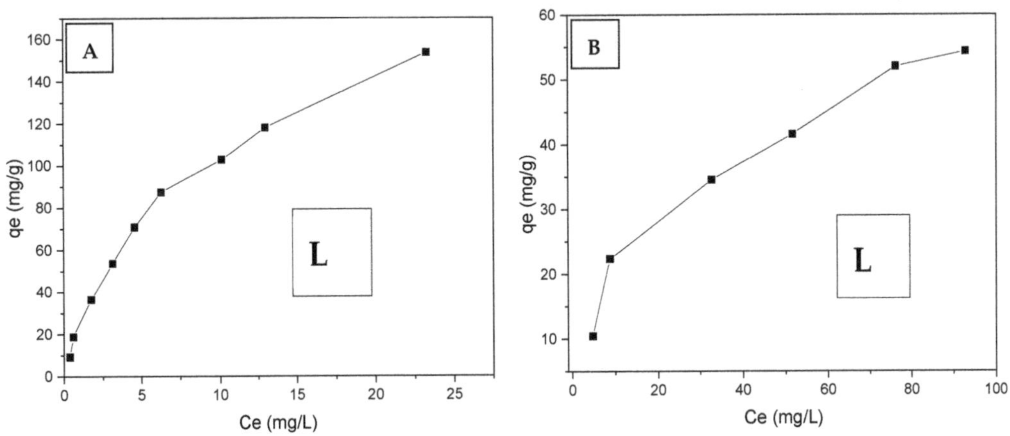

Figure 6. BM (**A**) and NE (**B**) adsorption isotherms on FPAs in aqueous solutions. Operating conditions: pH = 5.6, T = (20 ± 2) °C, adsorbent mass = 0.05 g/L.

Based on the Giles et al. [28] classification, the adsorption isotherms of BM and NE on FPAs are of type L.

The L-type isotherm for both BM and NE dyes indicates that the number of vacant sites decreases as the equilibrium concentration increases. This type of isotherm reflects the high affinity of the adsorbate/adsorbent. The solutes can be adsorbed horizontally on the surface of the adsorbent or vertically, which minimizes the competition of adsorption sites between the adsorbent and the dyes. The shape of the adsorption curve is characteristic of the chemical adsorption process, the isotherm can then be modeled by the Langmuir equation [29].

3.5. Optimization of Adsorption Parameters

The parameters influencing the adsorption phenomenon are the dye solution concentration, the pH, the temperature of the medium, and the mass of the adsorbent. Each optimized parameter is taken into consideration for the further optimization of the other parameters. The tests were performed on 100 mL solutions of the dyes at pH and temperature T. These mixtures were stirred with a speed of 300 rpm during the equilibrium time determined by the adsorption kinetics (60 min). The residual concentrations were determined by UV–Visible spectrophotometry and then exploited in order to follow the evolution of the adsorption yield of the dyes according to the values of the studied parameter.

3.5.1. Dye Solution Concentration Effect

The effect of the initial dye concentration was studied under the same conditions as before, this time changing the concentrations from 10 to 120 mg/L. The evolution of the decolorization yield according to the concentration of the dye (Figure 7) shows that for BM, the highest yield (93%) is reached for the concentration of 70 mg/L.

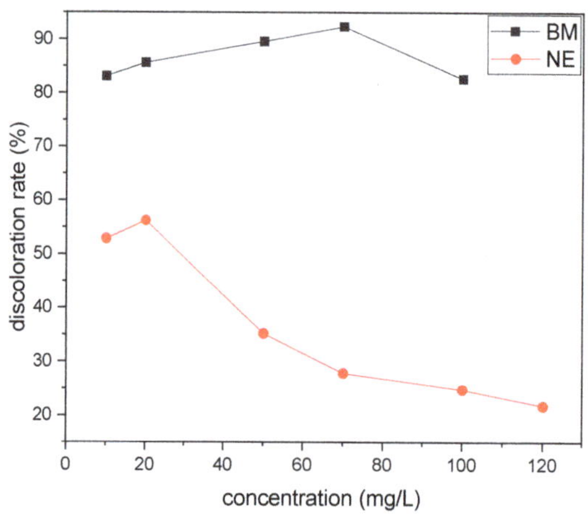

Figure 7. Initial dye concentration effect (BM and NE) on adsorption efficiency of FPAs. Operating conditions: pH = 5.6, T = (20 ± 2) °C, adsorbent mass = 0.05 g/L, (BM) = 10, 20, 50, 70 and 100 mg/L, (NE) = 10, 20, 50, 70, 100 and 120 mg/L, stirring time = 60 min.

For the NE, it is 57% for the concentration of 20 mg/L. Above these concentrations, these yields decrease because of the saturation of the sites active on the surface of the FPAs in the presence of a high dye content [30].

3.5.2. Adsorbent Mass Effect

The initial adsorbent mass effect was evaluated from 0.1–3 g/L and at optimum concentrations of 70 and 20 mg/L of BM and NE, respectively, while keeping other parameters constant.

The decolorization efficiency of both dyes changed with the increasing mass of the adsorbent used (Figure 8). Indeed, a mass of 1.5 g/L of FPAs is likely to adsorb a maximum of 94 and 69% of BM and NE, respectively. This phenomenon is the result of an increase in the number of sites available for dye binding, which consequently favors the phenomenon of discoloration [31–39]. Above this mass, the adsorption power decreases slightly, probably revealing the presence of other types of interactions between the dyes and the FPAs, or a competition between the sites retaining the dye molecules and the free sites of the adsorbent which attract them back into the solution.

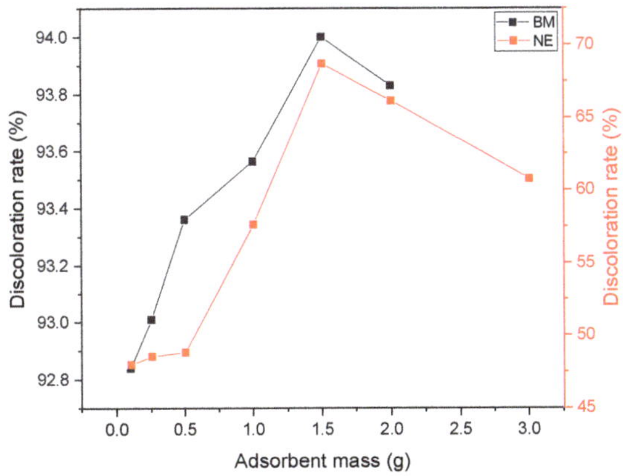

Figure 8. Effect of adsorbent mass on the adsorption efficiency of dyes (BM and NE) on FPAs. Operating conditions: pH = 5.6, T = (20 ± 2) °C, Adsorbent mass = 0.1 à 3 g/L, concentration NE = 20 mg/L, BM = 70 mg/L, stirring time: 60 min.

3.5.3. Medium pH Effect

We studied the impact of pH by setting the optimum concentrations and mass of the adsorbent and changing the solution pH from 2.5 to 10. Figure 9 shows that at an acid pH (3) below the FPAs' zero charge point value (pHzc = 5.9), the adsorption efficiency of NE reaches a maximum of 91% discoloration and a minimum of 54% for BM. At this acidic pH, the FPAs' surface is predominantly positively charged and can interact with negative species, favoring the adsorption of the anionic dye (NE) more than the cationic dye (BM). Consequently, at a solution pH above 5.9, the FPAs' surface is predominantly negatively charged and can interact with the cationic dye (BM), demonstrating the maximum decolorization efficiency of BM (93%) at pH = 7.5 and the decreased decolorization efficiency of NE (20%). It was highest (92%) at pH = 3.

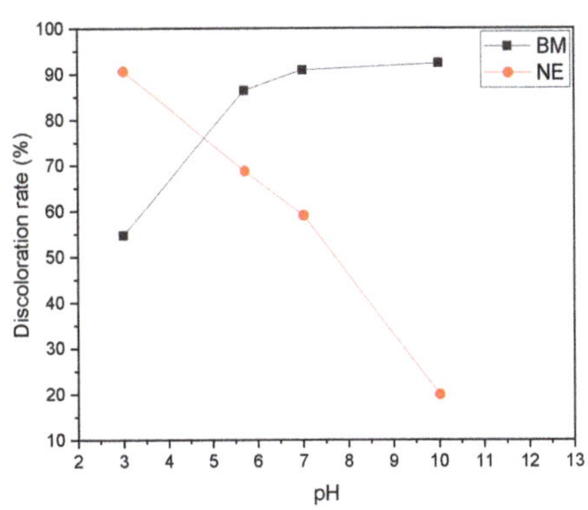

Figure 9. Effect of the pH solution on the adsorption efficiency of BM and NE on FPAs. Operating conditions: T = (20 ± 2) °C, adsorbent mass = 1.5 g/L, NE concentration = 20 mg/L, BM = 70, 3 ≤ pH ≤ 10, stirring time = 60 min.

3.5.4. Medium Temperature Effect

The effect of temperature on the rate of decolorization of the two dyes by FPAs was ensured by a series of experiments conducted at temperatures of 20, 30, 40, and 60 °C.

The results in Figure 10 show an increase of 3% for NE up to a temperature of 40 °C with a discoloration efficiency of 98%. For BM, the increase in temperature hinders its adsorption on the FPAs.

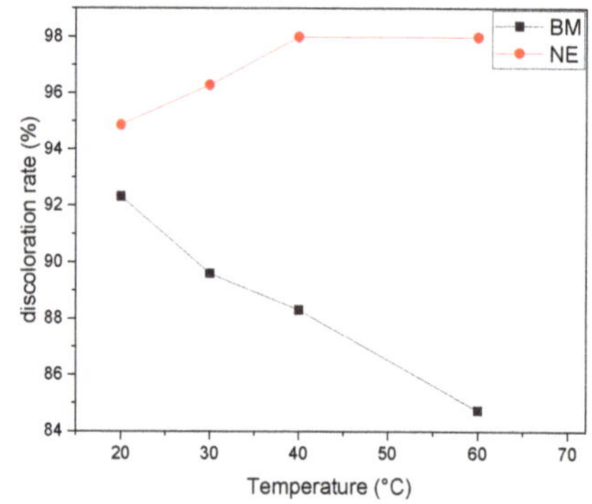

Figure 10. Medium temperature effect on the adsorption efficiency of BM and NE on FPAs. Operating conditions: pH (BM) = 10, pH (NE) = 3, adsorbent mass = 1.5 g/L, [NE] = 20 mg/L, [BM] = 70, $10 \leq T°C \leq 60$, stirring time = 60 min.

3.6. *Modeling of Adsorption Isotherms*

To optimize the process of adsorption, four models of adsorption isotherms are studied: Langmuir, Freundlich, Temkin, and Dubinin–Radushkevich in their linear form as shown in Figure 11.

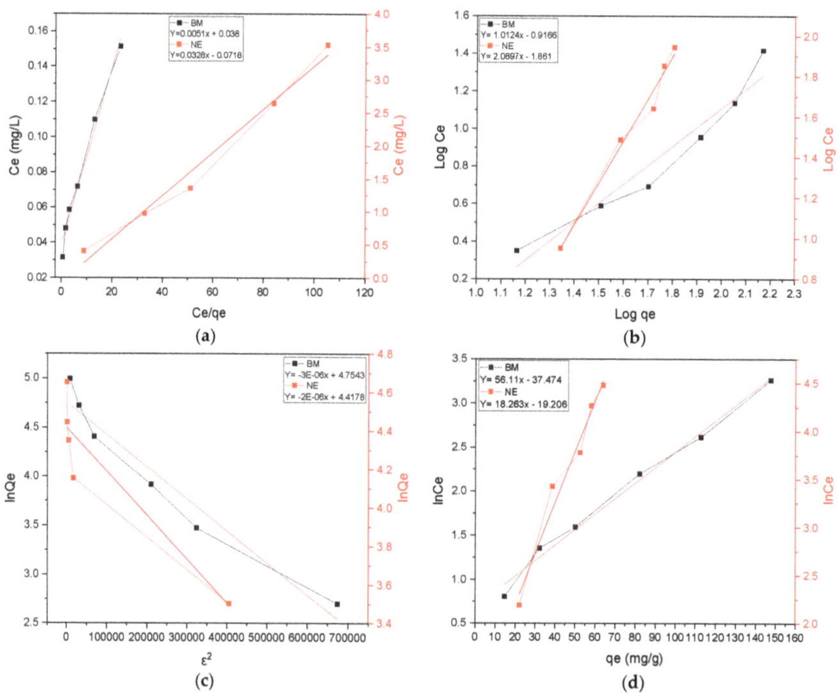

Figure 11. Linear representation of Langmuir (**a**), Freundlich (**b**), D-R (**c**), and Temkin (**d**) isotherms for the adsorption of BM and NE on FPAs.

The parameters of equilibrium and the constant values of the four isotherms were determined by linear regression (Table 6).

Table 6. Langmuir, Freundlich, Temkin, and Dubinin–Radushkevich parameters of BM and NE adsorption on FPAs.

Dyes	Langmuir			Freundlich			Temkin			D-R		
	(mg/g)		R^2	1/n		R^2	B_T	K_T	R^2	K_{DR}	E (Kj/mol)	R^2
BM	196.078	0.134	0.983	1.124	0.143	0.941	56.11	5.31×10^{-17}	0.991	3×10^{-6}	4.08×10^2	0.860
NE	30.488	−0.457	0.981	2.088	0.014	0.981	18.26	4.56×10^{-9}	0.970	2×10^{-6}	5×10^2	0.857

The results of modeling the adsorption isotherms of the two dyes BM and NE on FPAs record that the retention process of these dyes are both described by the Langmuir and Temkin models. Indeed, the coefficients of determination for the 2 isotherms are close to 1 for both dyes and the values of Langmuir's R_L separation factor for the different concentrations of BM and NE (Table 7) are between 0 and 1, which confirms the validity of this model. The numerical values of K_F and $1/n$ for both dyes (Table 6) do not meet the validation criteria of the Freundlich model.

Table 7. Separation factor for different concentrations of BM and NE dyes.

C (mg/L) / R_L	10	20	30	50	70	100	120
BM	0.427	0.271	0.199	0.131	0.097	0.07	-
NE	-	0.89	-	0.96	0.97	0.98	0.98

As for Temkin's model, it shows that there is no complex formation (K_T is very low) or interaction between adsorbed molecules, and the adsorption reaction is exothermic for the NE at the substitution of one or more water molecules by the dyes. This reinforces the Langmuir model that is based on the non-interaction between entities adsorbed on sites of the same nature.

These results can be explained by a monolayer adsorption characterizing the Langmuir isotherm, without any dye–dye interaction and without complex formation.

The average free energy E (kJ/mol) values for both dyes, provided by the K_{DR} constant calculation of the Dubin–Radushkevich isotherm (Table 6) are greater than 16 kJ/mol, revealing that the adsorption of both dyes is chemical in nature, which is also proven by the B_T values obtained from the Temkin isotherm (B_T > 16).

3.7. Thermodynamic Study

The nature of adsorption of BM and NE dyes on FPAs powder was evaluated by calculating the thermodynamics parameters standard free energy (ΔG°) (J/mol), standard enthalpy (ΔH°) (J.mol^{-1} K^{-1}), and standard entropy (ΔS°) (J/mol) (Table 8) according to the following equations [32]:

$$K_d = \frac{Q_e}{C_e} \quad (13)$$

$$\Delta G^\circ = -RT.lnK_d \quad (14)$$

$$lnK_d = \left(\frac{\Delta S^\circ}{R}\right) - \left(\frac{\Delta H^\circ}{R}\right)\frac{1}{T} \quad (15)$$

where K_d is the Distribution Constant deduced from the adsorption capacity Q_e (mg/g) of the FPA adsorbent, and C_e (mg/L) is the equilibrium concentration.

Table 8. Thermodynamic parameters of adsorption of BM and NE on FPAs.

Parameters	Temperatures	ΔH° (J.mol^{-1} K^{-1})	ΔS° (J/mol)	ΔG° (J/mol)
BM	20 °C 30 °C 40 °C 60 °C	−10,409	−16.85	−5460.406028 −5242.206332 −5261.755284 −4745.671639
NE	20 °C 30 °C 40 °C 60 °C	20,518.9	95.19	−7100.774357 −8195.002824 −10105.07668 −10750.76848

R: constant of perfect gases (J.mol^{-1} K^{-1}) (8.314 J mol^{-1} K^{-1})

T: temperature absolute (K)

The results shown in Figure 12 demonstrate that BM adsorption on FPAs is exothermic in nature (ΔH° < 0), while NE adsorption is endothermic (ΔH° > 0), meaning that increasing temperature increases the adsorbent's maximum carrying capacity at the same time. However, the negative value of ΔS° in the case of BM adsorption on FPAs, shows that the distribution order of the dye molecules on the adsorbent is important compared to that in the solution, while the positive value of ΔS° of NE suggests that some structural changes are taking place on the adsorbent and reveals the presence of random interferences at the solid-liquid interface. On the other hand, the negative values of ΔG° indicate that the adsorption process for both dyes is spontaneous throughout the temperature range studied. We also note that for BM, ΔG° increases with increasing solution temperature, which may be explained by the fact that adsorption becomes somewhat more difficult as temperature rises. As for NE, ΔG° values decrease with increasing solution temperature, which can be explained by the fact that NE adsorption is more favorable at high temperatures [33].

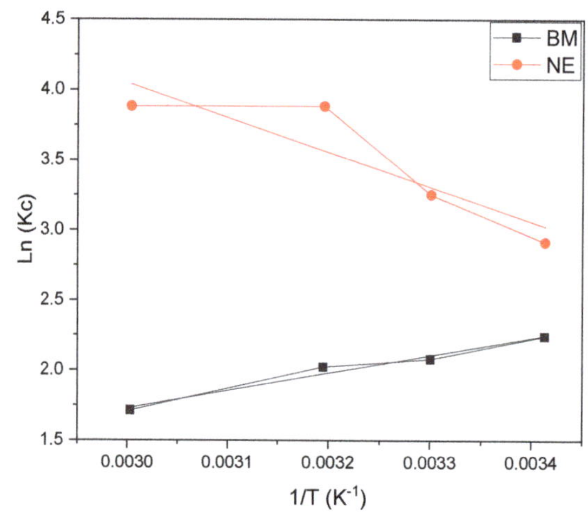

Figure 12. Linear representation of the variation of Ln (K_c) as a function of 1/T for the adsorption of BM and NE on FPAs.

This confirms the results obtained for the temperature effect on the adsorption of both dyes on FPAs (Figure 10).

4. Conclusions

This study focused on the adsorption of the dyes Maxilon Blue (BM) and Eriochrome Black (NE) on Aleppo pine fiber powder (FPAs). The results obtained concerning kinetics, adsorption isotherms, and thermodynamics were used to explain the way in which the two dyes attach to the adsorbent. A study of the influence of the initial concentration of the two dyes on the kinetics showed that the equilibrium time is 60 min and follows the pseudo-second-order model for both dyes. In addition, the adsorption parameter optimization study revealed that BM adsorption is more favored at basic pH with an adsorbent mass of 1.5 g/L and decreases with increasing temperature. NE adsorption, on the other hand, is favored at an acidic pH and a temperature of 40 °C. These two parameters give BM and NE removal rates of 93% and 98%, respectively. A study of the adsorption isotherms revealed that the binding of dye molecules to the active sites of the adsorbate was determined by monolayer adsorption characterizing the Langmuir isotherm, with no dye–dye interaction and no complex formation justified by the Temkin isotherm. The mean free energy (E) and constant (B_T) values determined from adsorption isotherms and D-R and Temkin, respectively, showed that the adsorption of both dyes on FPAs is chemical in nature. Thermodynamic parameters showed that the adsorption of BM and NE on FPAs is exothermic and endothermic in nature, respectively, and spontaneous throughout the temperature range studied.

In conclusion, the FPAs biomaterial recorded efficiency in the treatment and removal of cationic and anionic dyes typically used in textile industries with very interesting yields. These results recommend and suggest its application on an industrial scale.

Author Contributions: This article was a collaborative effort. E.M.S.H. supervised practical tasks, writing, and presentation of results, and handled the submission of the article. D.A., I.M. and R.M. contributed to practical tasks and results processing. M.M.A. provided funding, writing and reviewing. A.K. contributed to the interpretation of results. A.T. and M.T. supervised and verified the results. Z.R. guided supervised, and ensured the validity of results. All authors have read and agreed to the published version of the manuscript.

Funding: This research was supported by the Researchers Supporting Project number (RSPD2024R628), King Saud University, Riyadh, Saudi Arabia.

Data Availability Statement: Data are contained within the article.

Conflicts of Interest: The authors declare no conflicts of interest. The funders had no role in the design, execution, interpretation, or writing of the study.

References

1. Stearns, P.N. *The Industrial Revolution in World History*; Routledge: London, UK, 2020.
2. Kishor, R.; Purchase, D.; Saratale, G.D.; Saratale, R.G.; Ferreira, L.F.R.; Bilal, M.; Chandra, R.; Bharagava, R.N. Ecotoxicological and health concerns of persistent coloring pollutants of textile industry wastewater and treatment approaches for environmental safety. *J. Environ. Chem. Eng.* **2021**, *9*, 105012. [CrossRef]
3. Madhav, S.; Ahamad, A.; Singh, P.; Mishra, P.K. A review of textile industry: Wet processing, environmental impacts, and effluent treatment methods. *Environ. Qual. Manag.* **2018**, *27*, 31–41. [CrossRef]
4. Hansen, É.; de Aquim, P.M.; Hansen, A.W.; Cardoso, J.K.; Ziulkoski, A.L.; Gutterres, M. Impact of post-tanning chemicals on the pollution load of tannery wastewater. *J. Environ. Manag.* **2020**, *269*, 110787. [CrossRef] [PubMed]
5. Ndimele, P.E.; Pedro, M.O.; Agboola, J.I.; Chukwuka, K.S.; Ekwu, A.O. Heavy metal accumulation in organs of Oreochromis niloticus (Linnaeus, 1758) from industrial effluent-polluted aquatic ecosystem in Lagos, Nigeria. *Environ. Monit. Assess.* **2017**, *189*, 255. [CrossRef] [PubMed]
6. Ashrafi, O.; Yerushalmi, L.; Haghighat, F. Wastewater treatment in the pulp-and-paper industry: A review of treatment processes and the associated greenhouse gas emission. *J. Environ. Manag.* **2015**, *158*, 146–157. [CrossRef] [PubMed]
7. Hassaan, M.A.; El Nemr, A.; Hassaan, A. Health and environmental impacts of dyes: Mini review. *Am. J. Environ. Sci. Eng.* **2017**, *1*, 64–67.
8. Mashkoor, F.; Nasar, A. Polyaniline/Tectona grandis sawdust: A novel composite for efficient decontamination of synthetically polluted water containing crystal violet dye. *Groundw. Sustain. Dev.* **2019**, *8*, 390–401. [CrossRef]
9. Berradi, M.; Hsissou, R.; Khudhair, M.; Assouag, M.; Cherkaoui, O.; El Bachiri, A.; El Harfi, A. Textile finishing dyes and their impact on aquatic environs. *Heliyon* **2019**, *5*, e02711. [CrossRef] [PubMed]
10. Edokpayi, J.N.; Odiyo, J.O.; Durowoju, O.S. Impact of wastewater on surface water quality in developing countries: A case study of South Africa. *Water Qual.* **2017**, *10*, 66561. [CrossRef]
11. Collivignarelli, M.C.; Abbà, A.; Miino, M.C.; Damiani, S. Treatments for color removal from wastewater: State of the art. *J. Environ. Manag.* **2019**, *236*, 727–745. [CrossRef] [PubMed]

12. Bhatia, D.; Sharma, N.R.; Singh, J.; Kanwar, R.S. Biological methods for textile dye removal from wastewater: A review. *Crit. Rev. Environ. Sci. Technol.* **2017**, *47*, 1836–1876. [CrossRef]
13. Iwuozor, K.O. Prospects and challenges of using coagulation-flocculation method in the treatment of effluents. *Adv. J. Chem.-Sect. A* **2019**, *2*, 105–127. [CrossRef]
14. Särkkä, H.; Bhatnagar, A.; Sillanpää, M. Recent developments of electro-oxidation in water treatment—A review. *J. Electroanal. Chem.* **2015**, *754*, 46–56. [CrossRef]
15. Ma, W.; Zhang, M.; Liu, Z.; Kang, M.; Huang, C.; Fu, G. Fabrication of highly durable and robust superhydrophobic-superoleophilic nanofibrous membranes based on a fluorine-free system for efficient oil/water separation. *J. Membr. Sci.* **2019**, *570*, 303–313. [CrossRef]
16. Qing, W.; Shi, X.; Deng, Y.; Zhang, W.; Wang, J.; Tang, C.Y. Robust superhydrophobic-superoleophilic polytetrafluoroethylene nanofibrous membrane for oil/water separation. *J. Membr. Sci.* **2017**, *540*, 354–361. [CrossRef]
17. Zhu, Z.; Wu, P.; Liu, G.; He, X.; Qi, B.; Zeng, G.; Wang, W.; Sun, Y.; Cui, F. Ultrahigh adsorption capacity of anionic dyes with sharp selectivity through the cationic charged hybrid nanofibrous membranes. *Chem. Eng. J.* **2017**, *313*, 957–966. [CrossRef]
18. Mahmoud, M.E.; Nabil, G.M.; El-Mallah, N.M.; Bassiouny, H.I.; Kumar, S.; Abdel-Fattah, T.M. Kinetics, isotherm, and thermodynamic studies of the adsorption of reactive red 195 A dye from water by modified Switchgrass Biochar adsorbent. *J. Ind. Eng. Chem.* **2016**, *37*, 156–167. [CrossRef]
19. Wang, F.Y.; Wang, H.; Ma, J.W. Adsorption of cadmium (II) ions from aqueous solution by a new low-cost adsorbent—Bamboo charcoal. *J. Hazard. Mater.* **2010**, *177*, 300–306. [CrossRef]
20. Ho, Y.S.; McKay, G.A. comparison of chemisorption kinetic models applied to pollutant removal on various sorbents. *Process Saf. Environ. Prot.* **1998**, *76*, 332–340. [CrossRef]
21. Weber, W.J.; Morris, J.C. kinitics of adsorption on carbon from solution, J,of the sanitary Engineering division 1963, 89, 31–60.
22. Obradović, B. Guidelines for general adsorption kinetics modeling. *Hem. Ind.* **2020**, *74*, 65–70. [CrossRef]
23. Bouhamed, F.; Elouear, Z.; Bouzid, J. Adsorptive removal ofcopper(II) from aqueous solutions on activated carbon preparedfrom Tunisian date stones: Equilibrium, kinetics and thermody-namics. *J. Taiwan. Inst. Chem. Eng.* **2012**, *43*, 741–749. [CrossRef]
24. Wang, C.H.; Wen, W.C.; Hsu, H.C.; Yao, B.Y. High-capacitanceKOH-activated nitrogen-containing porous carbon material fromwaste coffee grounds in supercapacitor. *Adv. Powder Technol.* **2016**, *27*, 1387–1395. [CrossRef]
25. Dubinin, M.V.; Starinets, V.S.; Talanov, E.Y.; Mikheeva, I.B.; Belosludtseva, N.V.; Belosludtsev, K.N. Alisporivir improves mitochondrialfunction in skeletal muscle of mdx mice but suppresses mito-chondrial dynamics and biogenesis. *Int. J. Molec Sci.* **2021**, *22*, 9780. [CrossRef] [PubMed]
26. Torab-Mostaedi, M.; Ghaemi, A.; Ghassabzadeh, H.; Ghannadi-Maragheh, M. Removal of strontium and barium from aqueous solutions by adsorption onto expanded perlite. *Can. J. Chem. Eng.* **2011**, *89*, 1247–1254. [CrossRef]
27. Mohamad, A.D.M.; Abualreish, M.J.A.; Abu-Dief, A.M. Temperature and salt effects of the kinetic reactions of substituted 2-pyridylmethylene-8-quinolyl iron (II) complexes as antimicrobial, anti-cancer, and antioxidant agents with cyanide ions. *Can. J. Chem.* **2021**, *99*, 763–772. [CrossRef]
28. Owojori, O.J.; Reinecke, A.J.; Rozanov, A.B. Effects of salinity on partitioning, uptake and toxicity of zinc in the earthworm Eisenia fetida. *Soil. Biol. Biochem.* **2008**, *40*, 2385–2393. [CrossRef]
29. Seghier, A.; Habouchi, F.; Fortas, F.Z.; Cheklam, L.; Hadjel, M. Adsorption du bleu de sétapers issu de l'industrie de textile TAYAL de Relizane sur un biosorbant à faible coût préparé à base de feuilles d'artichaut. *Int. J. Nat. Resour. Environ.* **2021**, *3*, 61–72.
30. Kamran, U.; Bhatti, H.N.; Noreen, S.; Tahir, M.A.; Park, S.J. Chemically modified sugarcane bagasse-based biocomposites for efficient removal of acid red 1 dye: Kinetics, isotherms, thermodynamics, and desorption studies. *Chemosphere* **2022**, *291*, 132796. [CrossRef]
31. Makhlouf, M.; Hamacha, R.; Villiéras, F.; Bengueddach, A. Kinetics and thermodynamics adsorption of phenolic compounds on organic-inorganic hybrid mesoporous material. *Int. J. Innov. Appl. Stud.* **2013**, *3*, 1116–1124.
32. Bouberka, Z.; Kacha, S.; Kameche, M.; Elmaleh, S.; Derriche, Z. Sorption study of an acid dye from aqueous solutions using modified clays. *J. Hazard. Mater.* **2005**, *119*, 117–124. [CrossRef] [PubMed]
33. Abdellah, T. Etude de L'élimination du Chrome et de Bleu de Méthylène en Milieu Aqueux par Adsorption sur la Pyrophyllite Traitée et Non Traitée. Thèse Doctorale, Faculté des Sciences Rabat, Rabat, Morocco, 2006.
34. Temkin, M.J.; Pyzhev, V. Modifications récentes des isothermes de Langmuir. *Acta Phys. Chem.* **1940**, *12*, 217–222.
35. Freundlich, H. De l'adsorption des gaz. Section II. Cinétique et énergétique de l'adsorption des gaz. Article introductif à la section II. *Trans. Faraday Soc.* **1932**, *28*, 195–201. [CrossRef]
36. Chu, K.H. Revisiter l'isotherme de Temkin: Incohérence dimensionnelle et formes approximatives. *Rech. En. Chim. Ind. Et. Tech.* **2021**, *60*, 13140–13147.
37. Mahanty, B.; Behera, S.K.; Sahoo, N.K. Misinterpretation of Dubinin–Radushkevich isotherm and its implications on adsorption parameter estimates. *Sep. Sci. Technol.* **2023**, *58*, 1275–1282. [CrossRef]

38. Azzouni, D.; Baragh, F.; Mahmoud, A.M.; Alanazi, M.M.; Rais, Z.; Taleb, M. Optimization of methylene blue removal from aqueous solutions using activated carbon derived from coffee ground pyrolysis: A response surface methodology (RSM) approach for natural and cost-effective adsorption. *J. Saudi Chem. Soc.* **2023**, *27*, 101695. [CrossRef]
39. Azzouni, D.; Hassani, E.M.S.; Rais, Z.; Taleb, M. An Excellent Alternative to Industrial Activated Carbons for the Purification of Textile Water Elaborated from Waste Coffee Grounds. *Int. J. Environ. Res.* **2022**, *16*, 89. [CrossRef]

separations

MDPI

Article

Efficient Separation of Re (VII) and Mo (VI) by Extraction Using E-1006–Ammonium Sulfate Aqueous Two-Phase System

Linlin Fan, Wenhui Li, Zilong Dai, Min Zhou and Yunren Qiu *

School of Chemistry and Chemical Engineering, Central South University, Changsha 410083, China; fanlinlin0508@163.com (L.F.); liwenhui@163.com (W.L.); csudaizilong12152@163.com (Z.D.); zhouminnn@126.com (M.Z.)

* Correspondence: csu_tian@csu.edu.cn

Abstract: Aqueous two-phase extraction (APTE) stands out as an environmentally friendly technique for the separation of metal ions. The separation of Re (VII) and Mo (VI) in an aqueous solution was investigated using a novel aqueous two-phase system (ATPS) consisting of isodecanol polyoxyethylene ether (E-1006), ammonium sulfate, and water. A phase diagram of this system was developed, and the effects of pH, temperature, extraction time, the concentrations of E-1006 and $(NH_4)_2SO_4$, and metal ions on the separation of Re (VII) and Mo (VI) were examined. The results show that at pH 7.0, Mo (VI) had almost transformed into the $(NH_4)_2SO_4$-rich phase, while Re (VI) was extracted into the E-1006-rich phase. The increase in temperature induces a transition of Mo (VI) to the salt-rich phase, which is unfavorable for the extraction of Re (VII). The increase in the concentrations of E-1006 and $(NH_4)_2SO_4$ has a positive effect on the separation of rhenium and molybdenum. Overall, the ATPS consisting of 200 g/L of E-1006, 200 g/L of $(NH_4)_2SO_4$, and water yields an extraction efficiency of 97.2% for Re and a high separation factor of 2700 for Re (VII) and Mo (VI) from a mixture of 0.1 g/L of Re (VII) and 5 g/L of Mo (VI) at pH 7.0 and 323.15 K. Separation studies of the simulated leaching solution show that the extraction efficiency for Re (VI) is 99.1% and the separation factor of Re (VII) and Mo (VI) is 5100.

Keywords: aqueous two-phase system; phase equilibrium; rhenium-molybdenum separation; E-1006

1. Introduction

Molybdenum has excellent electrical and thermal conductivity and a melting point of up to 2620 °C [1], and it can be used as an important alloy additive [2], significantly improving the toughness and high-temperature strength of a material. Rhenium is an ultra-high-melting-point metal with a melting point of up to 3180 °C. Rhenium also has excellent stability, high-temperature elasticity, and ductility. The production of high-performance alloys, nanomaterials, and catalysts cannot be achieved without the addition of rhenium. In addition, molybdenum and rhenium are important raw materials for aerospace and national development [3].

Due to the similar ionic radii of Re^{4+} and Mo^{4+}, rhenium usually exists in the form of analogs in sulfide deposits such as molybdenite and copper ores. The rhenium content in molybdenite is relatively high and has a certain economic recovery value. Therefore, the separation and enrichment of molybdenum and rhenium in the leaching solution of molybdenite after ammonia leaching treatment is the key to obtaining rhenium and molybdenum resources. Currently, there are many methods for separating rhenium and molybdenum from a leaching solution, such as solvent extraction [4,5], ion exchange [6–8], and membrane separation [9–11]. Due to its advantages in terms of easily operable equipment and high selectivity, solvent extraction technology [2,5,12] stands out as the most widely used tool for this application. Commonly used extractants in the hydrometallurgical industry with respect to rhenium and molybdenum include organophosphorus extractants, as well as amine extractants, such as TBP, P204, Cyanex 27 [13], TOA, N263, etc. Aifei

Yi et al. [4] studied the extraction and separation of rhenium and molybdenum from the alkaline leaching solution of a waste-based high-temperature alloy using N263 as the extractant, and the extraction efficiency for Re reached 99% when O/A was 1/30 and under the first extraction condition. In the extraction of rhenium from acid chloride solutions using Cyanex 923, the extraction efficiency for Re can reach 99.2% [14]. Alamine 336 and TBP were used to synergistically extract perrhenate from an industrial leaching solution, in which the leaching solution was an acidic leaching solution of sulfuric acid and ammonium persulfate, and the separation efficiency for perrhenate reached more than 99.90% after a four-stage countercurrent extraction [15]. Solvent extraction has achieved good results in the separation and enrichment of rhenium and molybdenum. However, the traditionally used solvent extraction solution leads to the loss of organic solvents, environmental pollution, and adverse effects on human health.

Aqueous two-phase extraction (ATPE) is a promising alternative to hydrometallurgical processes for the selective extraction of metals since it does not involve organic solvents and is simple to perform. In ATPE, the differences in solute partition coefficients between the two phases caused by salting-out, intermolecular spatial hindrance, and external environments are used to achieve the selective separation of solutes. The hydrophobicity of polymers formed by an aqueous two-phase system (ATPS) enhances the interaction of the alkali component with the polymer phase. In the same way, the salting-out effect can further distinguish the phases, which may favor the distribution of the solute to the phase of greater affinity [16]. This method has been successfully studied for the extraction of Mo [17], the separation of molybdenum and vanadium [18], and the separation of other ions [19]. Li Ruisi et al. [20] used an ATPS consisting of the nonionic surfactant Triton X-100 and Na_2SO_4 to investigate the extraction separation effect of tungsten and molybdenum in an aqueous solution when no extractant was added, and the separation factor of tungsten and molybdenum was up to 328 at a pH 2.0, 343.15 K, and a mass fraction of sodium tartrate as a complexing agent of 3%. Daniela da Silveira Leited et al. [21] investigated the selective extraction of copper and cobalt from spent lithium-ion batteries via ATPS, and the optimal conditions resulted in a separation coefficient of 3.22×10^2 for copper and cobalt. However, little work has been reported for the separation of rhenium and molybdenum via ATPS.

A novel ATPS consisting of isodecanol polyoxyethylene ether (E-1006), ammonium sulfate, and water was constructed and applied for the separation of Re (VII) and Mo (VI). The effects of pH, temperature, extraction time, E-1006 concentration, ammonium sulfate concentration, and metal ion concentrations on the separation of Re (VII) and Mo (VI) were investigated to explore the optimal process conditions for the separation of Re (VII) and Mo (VI) from acidic solutions. This research can guide the separation of Re and Mo in the industry.

2. Materials and Methods

2.1. Materials and Apparatus

The commercial nonionic surfactant $(C_2H_4O)_nC_{10}H_{22}O$ (E-1006), with an average hydroxyl value of 132 (130–134), and ammonium perrhenate (NH_4ReO_4) with a purity of 0.99 (mass fraction, same as below) were purchased from Shanghai McLean Technology Biochemistry Co., Ltd. (Shanghai, China). The chemical structure of E-1006 is shown in Figure 1. Ammonium sulfate ($(NH_4)_2SO_4$) with a purity of 0.99 and ammonia solution (NH_4 OH) with a purity of 0.25–0.28 were supplied by China National Pharmaceutical Group Chemical Reagent Co., Ltd. (Shanghai, China). Ammonium molybdate tetrahydrate ($(NH_4)_6Mo_7O_{24}\cdot 4H_2O$) with a purity > 0.99 and sulfuric acid (H_2SO_4) with a purity of 0.95–0.97 were provided by Chengdu Kelong Chemical Co., Ltd. (Chengdu, China). Either NH_4OH and H_2SO_4 were used to adjust the pH of aqueous solutions. The inorganic chemicals used were of analytical grade. All the water used in the experiment was deionized water.

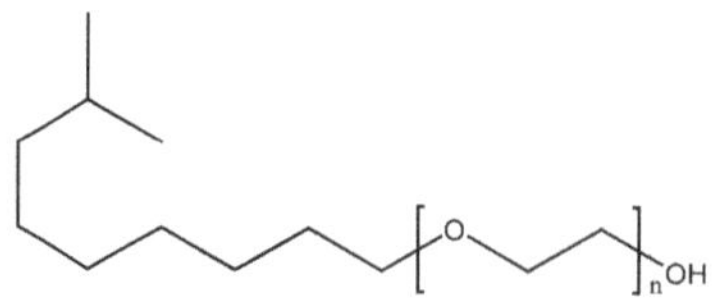

Figure 1. E-1006 chemical structural formula.

An analytical balance (AUY220, SHIMADZU, Kyoto, Japan) was used for weighting the reagents. The pH and conductivity measurements were carried out with a pH meter (PHS-3C, Shanghai Leici Instrument Co., Ltd., Shanghai, China) and a conductometer (DDS-307, Shanghai Leici Instrument Co., Ltd., Shanghai, China), respectively. A constant-temperature-heating magnetic stirrer (DF-101S, Gongyi Chuangyuan Instrument Manufacturing Co., Ltd., Zhengzhou, China) was used to stir the mixture of Re-Mo solution and ATPS. Then, phase separation was achieved via heating separation in a digital-display-equipped constant-temperature water bath pot (WB100-6, Qun'an Instrument Co., Ltd., Huzho, China). The concentrations of Re (VII) and Mo (VI) were quantified using Inductively Coupled Plasma Optical Emission Spectrometry (ICP-OES, ThermoFisher USA, Inc., Waltham, MA, USA, U.S.ICAP 6300 Radial, for which the RF power was 1300 W; the flux of auxiliary air was 0.2 $L{\cdot}min^{-1}$; the flux of carrying air was 0.7 $L{\cdot}min^{-1}$; the pump speed was 1.0 $r{\cdot}min^{-1}$; and the method of observing plasma was automatic). Fourier transform infrared spectrometry (FTIR, PerkinElmmer company, Waltham, MA, USA) was employed to analyze the changes in the chemical structure and functional groups of E-1006. The change in particle size of the E-1006-rich phase was determined using the Nano ZS90 dynamic light-scattering instrument (DLS, Malvern Instruments, Malvern, UK).

2.2. *Experimental Methods*

2.2.1. Phase Diagram of the ATPS

Liquid–liquid equilibrium data were obtained for an ATPS consisting of different concentrations of aqueous solutions of E-1006 and $(NH_4)_2SO_4$. The bimodal curve, which divides a region of component concentrations that will form two immiscible aqueous phases, was obtained using the gravimetric method. At a certain temperature, a 40 g solution mixture was vigorously stirred in a magnetic stirrer for 30 min and then deposited in a thermostatic bath with a graduated test tube for 10 h. Samples of the top and bottom phases were collected to determine the mass fractions of E-1006 and $(NH_4)_2SO_4$.

The concentration of $(NH_4)_2SO_4$ was determined by measuring conductivity within a linear range of $(NH_4)_2SO_4$ concentrations; further, we obtained the mass fraction of $(NH_4)_2SO_4$, for which the conductivity of the salt solutions was independent of the polymer composition. The mass fraction of water was calculated based on the water lost from the oven-dried specimen, and then the mass fraction of E-1006 was obtained. Utilizing the obtained data, a phase diagram was constructed, with the mass fraction of $(NH_4)_2SO_4$ on the horizontal axis and the mass fraction of E-1006 on the vertical axis.

2.2.2. Re-Mo Separation

The separation of Re (VII) and Mo (VI) in ATPS was investigated as a function of pH (1.0 to 7.0), temperature (303.15 K to 333.15 K), concentrations of E-1006 ($c_{E\text{-}1006}$) and $(NH_4)_2SO_4$ ($c_{(NH4)SO4}$) (50 g/L to 200 g/L), Re (VII) concentration (c_{Re}) (0.01 g/L to 0.1 g/L), and Mo (VI) concentration (c_{Mo}) (2 g/L to 10 g/L). Stock solutions of $(NH_4)_2SO_4$, Re (VII), and Mo (VI) were prepared using ammonium sulfate, ammonium perrhenate, and ammonium molybdate tetrahydrate, respectively.

For the extraction experiment, volumes of $(NH_4)_2SO_4$, Re (VII), and Mo (VI) stock solutions were mixed, adding a certain amount of deionized water and E-1006 to achieve a total volume of 40 mL. The pH of the solution was adjusted by adding the required amount of sulfuric acid and ammonia solution. At a certain temperature, the beaker containing the solution mixture was vigorously stirred in a magnetic stirrer for 20 min and placed in a thermostatic bath with a graduated test tube for 2 h. The scheme of the Re-Mo separation is shown in Figure 2. The volumes of the top and bottom phases were recorded, and samples of the $(NH_4)_2SO_4$-rich phase were collected. The samples were diluted for measuring the c_{Re} and c_{Mo} via ICP-OES.

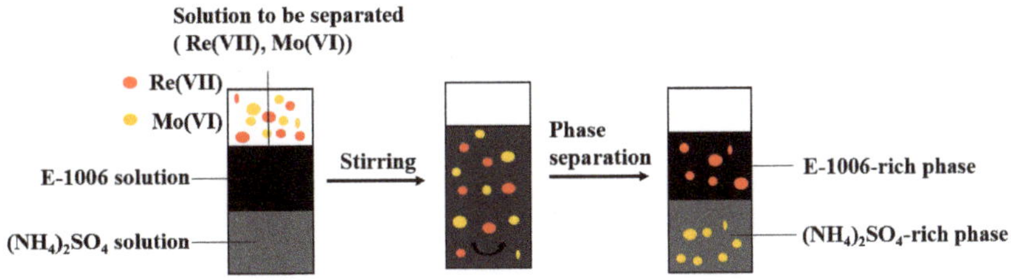

Figure 2. Scheme of Re (VII) and Mo (VI) separation.

The concentration of metal ions in the E-1006-rich phase, the distribution coefficient (D) and extraction efficiency (E) in the E-1006 phase, and the separation factor (S) of the aqueous two-phase extraction process are defined as follows:

$$c_1 = \frac{m - V_2 c_2}{V_1} \tag{1}$$

$$D = \frac{c_1}{c_2} \tag{2}$$

$$E = \frac{c_1 V_1}{m} \times 100\% \tag{3}$$

$$S_{Re/Mo} = \frac{D_{Mo}}{D_{Re}} \tag{4}$$

where c_1 and c_2 represent the concentrations of metal ions in the E-1006-rich aqueous phase and the $(NH_4)_2SO_4$-rich aqueous phase, respectively. V_1 and V_2 are the respective volumes of the E-1006-rich aqueous phase and the $(NH_4)_2SO_4$-rich aqueous phase. m represents the mass of metal ions added to the system.

3. Results and Discussion

3.1. Phase Diagrams

The ternary mixtures of surfactant, salt, and water, upon reaching thermodynamic equilibrium, cause exclusion among the components, resulting in the formation of two phases. Phase diagrams can explain the solute separation behavior of ATPS during the extraction process. A phase diagram of the ATPS formed by E-1006 and $(NH_4)_2SO_4$ at 313.15 K is shown in Figure 3. The good water solubility of E-1006 is mainly dependent on the hydrogen atoms in the water molecules connecting with the C–O–C groups on E-1006 to form hydrogen bonds and rapidly fuse with water [22]. The salting-out action of $(NH_4)_2SO_4$ breaks the above hydrogen bonds, leading to the dehydration of E-1006 [23]. In an ATPS, the content of E-1006 and $(NH_4)_2SO_4$ increases, that of the binodal curve is closer to the x-axis, and the tie-line length is longer, indicating a stronger ability for phase separation.

Temperature is an important factor affecting phase separation in ATPSs [24]. In this study, we investigated the phase equilibria of the ATPS in question at temperatures of 303.15 K, 313.15 K, and 323.15 K. The results are shown in Figure 4 and Supplementary, where it is shown that an increase in the temperature promotes an increase in larger biphasic regions [25]. This phenomenon can be explained by the enhanced irregular thermal motion of the molecules, the breakage of the hydrogen bonds between E-1006 and water molecules, and the decrease in the solubility of E-1006 in water, leading to an increase in the content of E-1006 in the upper phase and an increase in the content of salt in the lower phase. The area of the biphasic region increased, indicating that the increase in temperature promotes phase separation.

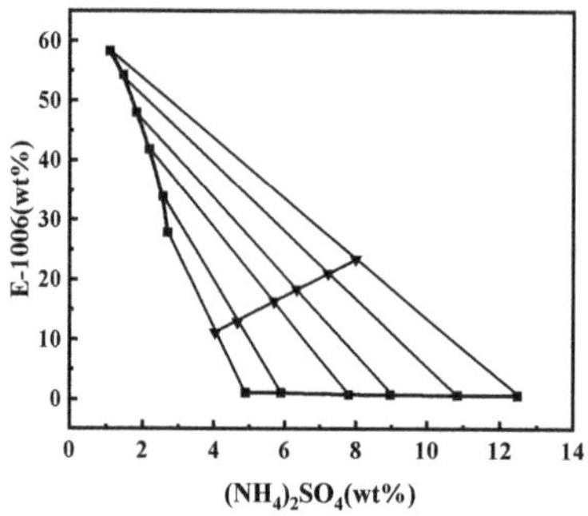

Figure 3. Phase diagram of ATPS composed of E-1006 and $(NH_4)_2SO_4$ at 313.15 K.

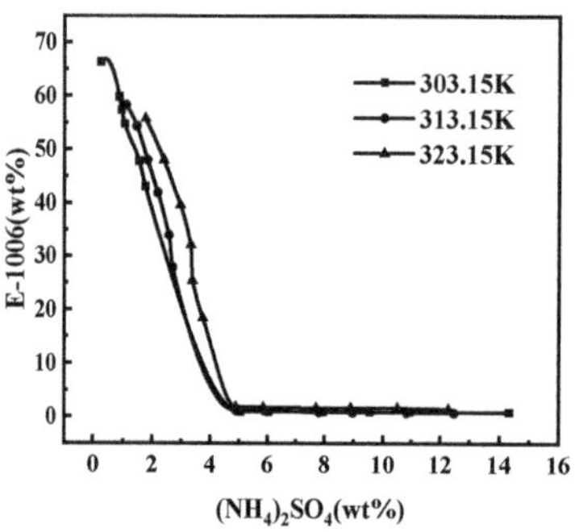

Figure 4. Phase diagram of ATPS composed of E-1006 and $(NH_4)_2SO_4$ at different temperatures.

3.2. The Separation of Re (VII) and Mo (VI)

3.2.1. Effect of pH

The morphology of Re (VII) and Mo (VI) present in solution varies considerably at different pH levels, as does the water solubility of the ions. The distribution of Re (VII) and Mo (VI) in this system is significantly influenced by pH. The separation of Re (VII) and Mo (VI) was studied at different pH levels using an ATPS consisting of 100 g/L of E-1006 and 100 g/L of $(NH_4)_2SO_4$, and the feed consisted of 0.1 g/L of Re (VII) and 5 g/L of Mo (VI). The temperature was maintained at 313.15 K. Figure 5 shows the variation in E_{Re}, E_{Mo}, and the $S_{Re/Mo}$ with respect to pH from 1.0 to 7.0. One can see that the E_{Re} increases from 71.44% to 83.60% as the pH increases from 1.0 to 2.0. A further increase in pH to 7.0 results in a slight increase in E_{Re}. On the other hand, the E_{Mo} reaches a maximum at pH 2.0, and most of the Mo (VI) is distributed in the $(NH_4)_2SO_4$-rich phase at pH 7.0. The $S_{Re/Mo}$ is very low at higher acidity levels, but it experiences a great increase as pH rises, reaching a maximum value of 129.67 at pH 7.0.

Accompanied by the formation of the ATPS, the system was divided into a hydrophobic phase, where E-1006 is located, and a hydrophilic phase, where $(NH_4)_2SO_4$ is located. The Fourier Transform Infrared Spectroscopy (FTIR) analysis results (Figure 6) show that the characteristic absorption peak at a wavenumber of 1100 cm^{-1} corresponds to the ether bond (C–O–C) in the molecular structure of E-1006 [26]. The position of this peak is 1100 cm^{-1} in pure E-1006 and 1099 cm^{-1} in E-1006 at pH 7.0. There is no obvious shift, which suggests that the addition of ammonia does not significantly affect the vibration of the ether bond. Noteworthily, the absorption peak of the ether bond is 1091 cm^{-1} after the separation of Re (VII) and Mo (VI). The slight redshift of the characteristic peak of C–O–C after extraction suggests the presence of a specific force between the E-1006 molecule and Re (VII) during the extraction process. The characteristic absorption peaks at 2860–3000 cm^{-1} were attributed to the C–H bond [27]. The significant change in the C–H characteristic absorption peaks after the extraction of Re (VII) can be attributed to the alteration of the environment surrounding the C–H bond in the E-1006 molecule, resulting in a shift in its vibrational state. In addition, the broad peak at 3000–3800 cm^{-1} represents the absorption peak of the hydroxyl group (–OH) [28]. The curves in Figure 6a,b demonstrate that the weak absorption peaks in this region for pure E-1006 and E-1006 at pH 7.0 correspond to the presence of hydroxyl groups in the molecular structure of E-1006. Corresponding to the period after the extraction of Re (VII), the curve in Figure 6c shows that the absorption peak of –OH in the E-1006-rich phase is significantly enhanced and encompasses lower wavenumbers. This indicates the presence of a hydrogen-bonding network, possibly due to the presence of specific water molecules in the E-1006-rich phase after the extraction. Meanwhile, the curve in Figure 6c shows a significant enhancement of the –OH bending vibrational peak at ~1600 cm^{-1} compared with that in Figure 6a,b, which also supports this conclusion.

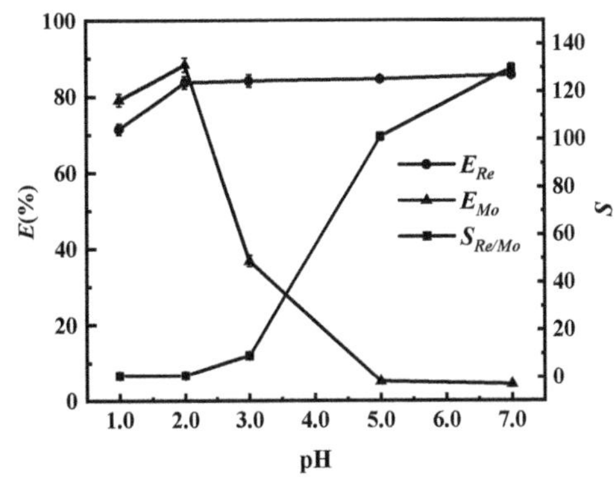

Figure 5. Effect of pH on the E_{Re}, E_{Mo}, and $S_{Re/Mo}$.

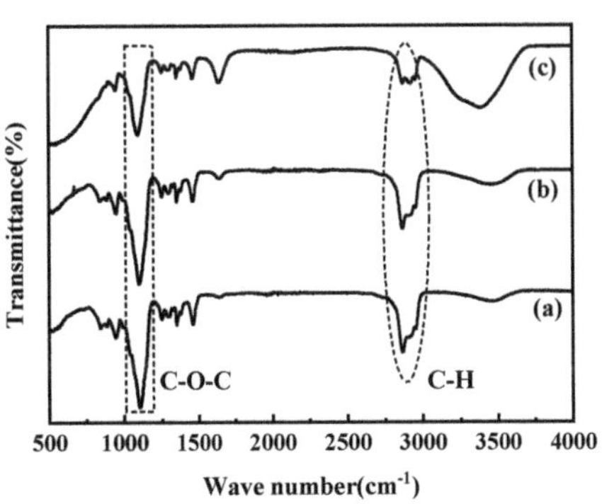

Figure 6. FTIR spectra of Re-Mo separated by the studied ATPS consisting of E-1006, $(NH_4)_2SO_4$, and H_2O. (**a**), pure E-1006; (**b**), E-1006 aqueous solution at pH 7.0; (**c**), E-1006-rich phase after Re-Mo separation at pH 7.0.

According to the distribution diagram of Re (VII) [14], Re (VII) primarily exists as ReO_4^- with strong hydrophobicity [29] in the pH range of 0–13. The forms of Mo (VI), along with the calculated ratios of charge, mass, and charge density at different pH levels, are shown in Table 1 [30]. $|z|$ denotes the absolute value of ion charge, n is the atomic number of an ion, and M is molar mass. $|z|/M$ is charge density. At pH $\geq$ 7.0, Mo exists mainly in the form of MoO_4^{2-}, which has a relatively large value of charge density, that is, 0.0125, and a higher hydrophilicity. Importantly, at pH 7.0, the hydrophobicity of ReO_4^- is responsible for the extraction into the E-1006 phase, and MoO_4^{2-} remains in the salt-rich phase due to its hydrophilicity, realizing an efficient separation of Re (VII) and Mo (VI).

Table 1. The ratio of charge, mass, and charge density of Mo (VI) at different pH levels.

pH	<1.8	<2.5	<4	<5	<7	>7
Species	MoO_2^{2+}	$H_2Mo_7O_{24}^{4-}$	$HMo_7O_{24}^{5-}$	$Mo_7O_{24}^{6-}$	$Mo_2O_7^{2-}$	MoO_4^{2-}
$\lvert z \rvert/n$	0.6667	0.1212	0.1562	0.1935	0.2222	0.4000
$\lvert z \rvert/M$	0.01562	0.003774	0.004726	0.005682	0.006579	0.01250

Therefore, the initial pH of the solution should be adjusted to 7.0 first for the APTE of Re (VII) and Mo (VI). The pH in the subsequent experiments was set to 7.0.

3.2.2. Effect of Temperature

In the exploration of ATPS phase equilibrium, temperature is closely related to phase separation. The effect of temperature, ranging from 303.15 to 333.15 K, on the E_{Re}, E_{Mo}, and $S_{Re/Mo}$ is shown in Figure 7.

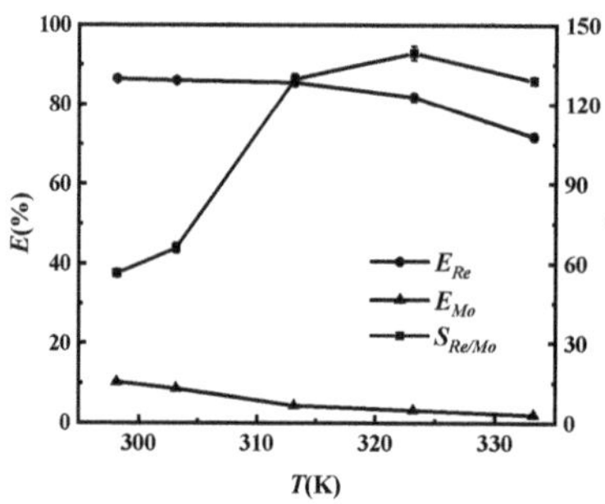

Figure 7. Effect of temperature on the E_{Re}, E_{Mo}, and $S_{Re/Mo}$.

Both E_{Re} and E_{Mo} decrease with temperature. The E_{Re} is favored at low temperatures, as demonstrated by the E_{Re} at 298.15 K that is close to 87%. The increase in temperature promotes the transfer of the more-hydrophilic MoO_4^{2-} to the $(NH_4)_2SO_4$-rich phase. The $S_{Re/Mo}$ shows a tendency to increase and then decrease, with a peak value of 139.37 at 323.15 K [31].

3.2.3. Effect of $c_{E\text{-}1006}$

$c_{E\text{-}1006}$ affects the number of micelles formed in an ATPS and plays a crucial role in the separation effect of metal ions. To study the effect of $c_{E\text{-}1006}$ on the separation of Re (VII) and Mo (VI), ATPSs composed of different concentrations of E-1006 in the range of 50 g/L to 200 g/L were prepared and mixed with 100 g/L of $(NH_4)_2SO_4$ at pH 7.0 and 313.15 K, as depicted in Figure 8.

With the increase in $c_{E\text{-}1006}$ from 50 g/L to 200 g/L, the E_{Re} increases from 60.71% to 93.66%, and the $S_{Re/Mo}$ increases from 25.03 to 514.24, which can be attributed to the greater number of micelles colliding with each other, prompting more ReO_4^- to be solubilized into the micelles of the E-1006 phase [30]. The increase in micelles makes the E-1006 phase more hydrophobic and promotes the transformation of MoO_4^{2-} into the salt-rich phase [32]. While an increase in $c_{E\text{-}1006}$ generally favors the separation of Re (VII) and Mo (VI), it is essential to consider the drawbacks of excessive concentrations. High concentrations can lead to emulsification of the solution, posing challenges for subsequent treatment processes.

The suitable $c_{E\text{-}1006}$ was determined to be 200 g/L, taking into account the favorable $S_{Re/Mo}$ and the economic cost of the agent.

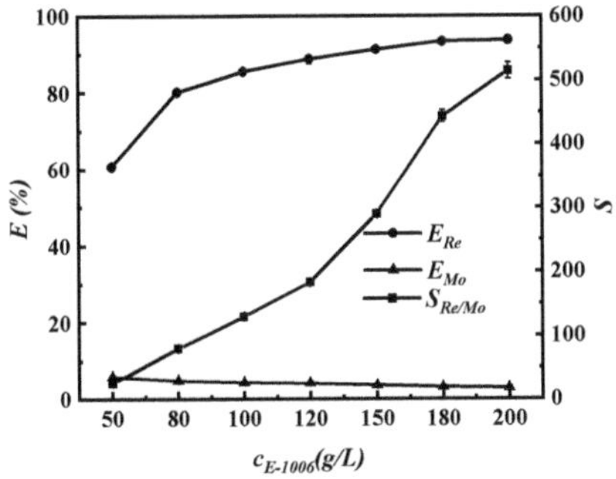

Figure 8. Effect of $c_{E\text{-}1006}$ on the E_{Re}, E_{Mo}, and $S_{Re/Mo}$.

The changes in particle size and zeta potential of the E-1006-rich phase were analyzed using dynamic light scattering (DLS) to speculate on the possible mechanism of the transition of Re (VII) into the E-1006-rich phase [33,34]. Figure 9 shows a micelle size analysis of the E-1006 phase, where the two lines are the micelle sizes in the aqueous solution of E-1006 and the aqueous solution of the E-1006 phase after Re (VII) extraction, respectively. The particles of the E-1006 micellar phase after Re (VII) extraction are larger. Table 2 shows the potential change of the E-1006 phase. Sample 1 is the E-1006 aqueous solution, sample 2 is the E-1006 aqueous solution at pH 7.0, and sample 3 is the E-1006 aqueous solution after Re-Mo separation at pH 7.0. The average zeta potential of the E-1006 micellar phase (0.110 mV) is slightly positive at pH 7.0 and becomes negative (−1.23 mV) after Re (VII) extraction, which can be attributed to ReO_4^- diffusing into the hydrophobic environment inside the E-1006 micelles due to its strong hydrophobicity [14,20].

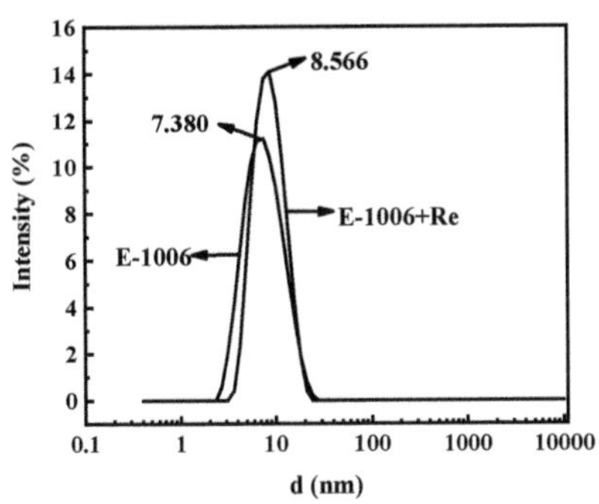

Figure 9. Particle size analysis of micelles in the E-1006 phase.

Table 2. Potential changes in the micellar phase of E-1006.

Sample Number	Zeta Potential (mV)
1	−0.642
2	0.110
3	−1.230

3.2.4. Effect of $c_{(NH4)SO4}$

Under specific conditions involving the addition of inorganic salts, the E-1006 aqueous solution undergoes phase separation [35]. However, at low salt concentrations, the experimental system fails to form distinct phases. On the premise of ensuring the formation of different phases, the addition of $(NH_4)_2SO_4$ was explored, ranging from 50 g/L to 200 g/L. Figure 10 presents the effect of $c_{(NH4)SO4}$ on the separation of Re (VII) and Mo (VI), indicating that $c_{(NH4)SO4}$ has an important influence on the separation of Re (VII) and Mo (VI).

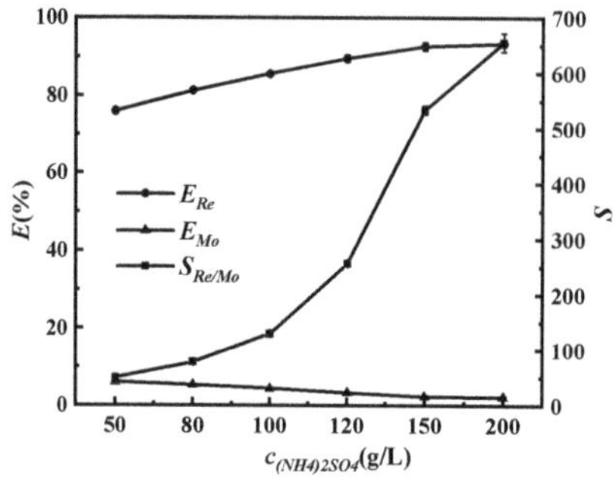

Figure 10. Effect of $c_{(NH4)SO4}$ on the E_{Re}, E_{Mo}, and $S_{Re/Mo}$.

As shown in Figure 10, as the amount of $c_{(NH4)SO4}$ increases from 50 g/L to 200 g/L, the E_{Re} increases from 75.92% to 93.39%, and the $S_{Re/Mo}$ rises from 49.35 to 655.90. $(NH_4)_2SO_4$ is a kosmotropic salt, and its free energy of hydration is large (in terms of the absolute value) [32]. Its salting-out ability is strong and becomes stronger with an increasing concentration, which results in a decreasing free water concentration, creating a more chaotropic polymolybdate anion dehydrate and increasing the number of micellar aggregates [36]. The difference in hydrophilicity between the two phases is larger, promoting the transition of ReO_4^- to the E-1006-rich phase, while MoO_4^{2-} is more prone to transition to the $(NH_4)_2SO_4$-rich phase. The selection of 200 g/L $c_{(NH4)SO4}$ is a suitable condition for the effective separation of Re (VII) and Mo (VI).

3.2.5. Effect of c_{Re} and c_{Mo}

The effects of c_{Re}, ranging from 0.01 g/L to 0.1 g/L, and c_{Mo}, ranging from 2 g/L to 10 g/L, on the separation of Re (VII) and Mo (VI) were investigated. To specifically assess the influence of the initial c_{Mo}, the c_{Re} was maintained at 0.1 g/L, while for studying the impact of the initial c_{Re}, c_{Mo} was held constant at 5 g/L. The obtained results are shown in Figure 11 and Figure 12, respectively.

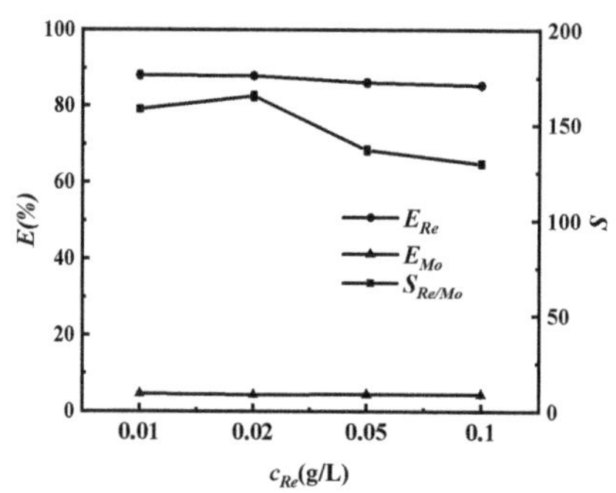

Figure 11. Effect of c_{Re} on the E_{Re}, E_{Mo}, and $S_{Re/Mo}$.

Figure 11 shows that an increase in c_{Re} has a negligible effect on E_{Re}, while E_{Re} tends to decrease. $S_{Re/Mo}$ decreases as c_{Re} increases. Figure 12 shows the effect of the initial c_{Mo} on the separation of Re (VII) and Mo (VI). It is clear that c_{Mo} has a negligible effect on E_{Mo}, but E_{Re} increases with an increasing Mo (VI) concentration. At pH 7.0, MoO_4^{2-} is highly hydrophilic, and the increase in the Mo (VI) concentration has a promoting effect on the transition of ReO_4^- to the hydrophobic phase.

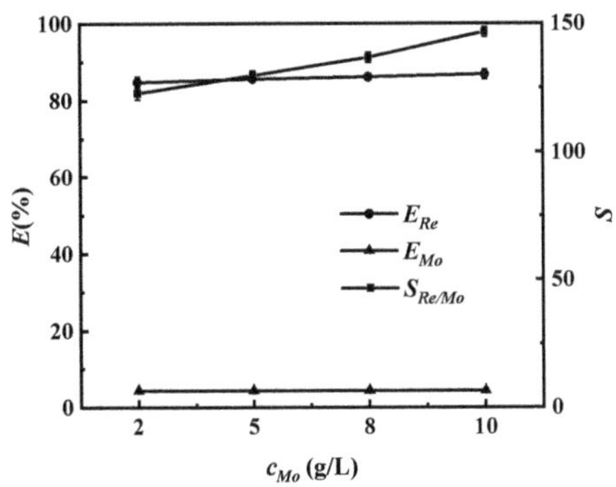

Figure 12. Effect of c_{Mo} on the E_{Re}, E_{Mo}, and $S_{Re/Mo}$.

3.2.6. Effect of Extraction Time

The extraction time greatly affects the extraction and separation of ions. In this section, the results of the separation of Re (VII) and Mo (VI) in the ATPS are investigated at an extraction time ranging from 15 min to 10 h. The results are shown in Figure 13.

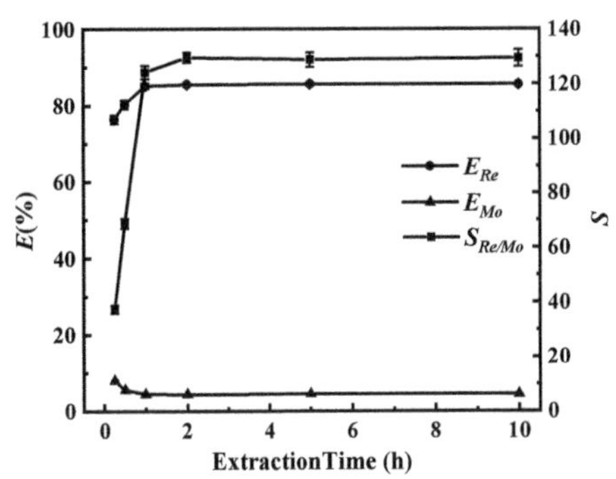

Figure 13. Effect of extraction time on the E_{Re}, E_{Mo}, and $S_{Re/Mo}$.

As can be seen in Figure 13, the extraction efficiencies and the separation factor of Re (VII) and Mo (VI) in the E-1006-rich phase increase significantly in the range of 15 min to 2 h and increase very little from 2 h to 10 h. The Re (VII) transitions to the E-1006-rich phase, and Mo (VI) transitions to the $(NH_4)_2SO_4$-rich phase. The extraction tends to equilibrate at 2 h, and the separation factor of rhenium and molybdenum tends to stabilize. Thus, 2 h is suitable for separation.

3.3. Separation of Re (VII) and Mo (VI) under Suitable Conditions

Based on the insights gained from the factors affecting the separation of Re (VII) and Mo (VI), the separation result of the solution containing 0.1 g/L of Re (VII) and 5 g/L of Mo (VI) under suitable conditions is outlined in Table 3. It indicated that E_{Re} reached 97.2%, and the $S_{Re/Mo}$ was 2700.

Table 3. Suitable conditions for separation behavior.

$c_{E\text{-}1006}$/ g L^{-1}	$c_{(NH4)2SO4}$/ g L^{-1}	pH	Temperature/ K	E_{Re}/%	E_{Mo}/%	$S_{Re/Mo}$
200	200	7.0	323.15	97.2 (±0.4)	1.3 (±0.1)	2700 (±200)

The leaching solution used is from a molybdenum mine. c_{Re} is 90–100 mg/L, c_{Mo} is 62–65 g/L, and the amount of free ammonia is 80–100 g/L. After the whole leaching solution was diluted tenfold, the separation of Re (VII) and Mo (VI) in a simulated solution with 0.01 g/L of Re (VII), 6 g/L of Mo (VI), and 8 g/L of free ammonia was investigated under the suitable conditions. As shown in Table 4, E_{Re} reached 99.1%, and the $S_{Re/Mo}$ was 5100, confirming the efficient separation of Re (VII) and Mo (VI). In comparison to the methods proposed in Salehi et al.'s study [12] and Aifei Yi et al.'s study [4], this method not only has a high extraction efficiency but also a much higher separation factor than those in the cited studies.

Table 4. Separation results for Re (VII) and Mo (VI) in simulated leaching solution.

$c_{E\text{-}1006}$/ g L^{-1}	$c_{(NH4)2SO4}$/ g L^{-1}	pH	Temperature/ K	E_{Re}/%	E_{Mo}/%	$S_{Re/Mo}$
200	200	7.0	323.15	99.1 (±0.1)	2.1 (±0.2)	5100 (±400)

4. Conclusions

(1) A phase equilibrium diagram of the ATPS, composed of E-1006, $(NH_4)_2SO_4$, and water, was developed.

(2) The effects of pH, temperature, concentrations of ATPS components, and metal ions on the separation of Re (VII) and Mo (VI) were investigated. The results show that pH plays an important role in the separation of Re (VII) and Mo (VI). At pH 7.0, Mo (VI) almost transitioned into the $(NH_4)_2SO_4$-rich phase, while Re (VII) was extracted into the E-1006-rich phase, and the separation factor of Re (VII) and Mo (VI) reached a maximum of 129.67. The extraction efficiency of Re (VII) is higher at low temperatures. The increase in temperature promotes the transition of Mo (VI) to the salt-rich phase, and the separation factor of Re (VII) and Mo (VI) reaches a maximum of 139.37 at 323.15 K. The separation of Re (VII) and Mo (VI) is favored by increasing the concentrations of E-1006 and $(NH_4)_2SO_4$. An increase in the Re (VII) concentration decreases the extraction efficiency of Re (VII). An increase in the Mo (VII) concentration promotes the extraction of Re (VII) and Re-Mo separation.

(3) The suitable conditions for the separation of Re (VII) and Mo (VI) were achieved using an ATPS composed of 200 g/L of E-1006, 200 g/L of $(NH_4)_2SO_4$, and water at a pH 7.0 heated to 323.15 K for 2 h. A mixed solution of 0.1 g/L of Re (VII) and 5 g/L of Mo (VI) was separated by the ATPS. The extraction efficiency of Re (VII) reached 97.2%, and the separation factor of Re (VII) and Mo (VI) reached 2700. A diluted simulated leaching solution consisting of 0.1 g/L of Re and 5 g/L of Mo was separated using the ATPS. The extraction efficiency of Re (VII) reached 99.1%, and the separation factor of Re (VII) and Mo (VI) reached 5100.

Supplementary Materials: The following supporting information can be downloaded at: https://www.mdpi.com/article/10.3390/separations11050142/s1, Table S1. Phase equilibrium data for the ATPS of E-1006 and $(NH_4)_2SO_4$ at different temperature.

Author Contributions: L.F.: methodology, formal analysis, investigation, data curation, and writing—original draft. W.L.: investigation and writing—review and editing. Z.D.: formal analysis and investigation. M.Z.: investigation and data curation. Y.Q.: conceptualization, writing—review and editing, funding acquisition, and supervision. All authors have read and agreed to the published version of the manuscript.

Funding: This research was funded by the National Key R&D Program of China, grant number 2022YFC2904602, and the Fundamental Research Funds for the Central Universities of Central South University, grant number 2023ZZTS0737).

Data Availability Statement: The data that support the findings of this study are available from the corresponding author upon reasonable request.

Conflicts of Interest: The authors declare no conflicts of interest.

References

1. Xu, W.; Wang, W.; Chen, S.; Zhang, R.; Wang, Y.; Zhang, Q.; Yuwen, L.; Yang, W.J.; Wang, L. Molybdenum disulfide (MoS_2) nanosheets-based hydrogels with light-triggered self-healing property for flexible sensors. *J. Colloid Interface Sci.* **2020**, *586*, 601–612. [CrossRef] [PubMed]
2. Liu, B.; Zhang, B.; Han, G.; Wang, M.; Huang, Y.; Su, S.; Xue, Y.; Wang, Y. Clean separation and purification for strategic metals of molybdenum and rhenium from minerals and waste alloy scraps—A review. *Resour. Conserv. Recycl.* **2022**, *181*, 106232. [CrossRef]
3. Hori, H.; Yonezato, Y.; Ito, K. Recovery of platinum and rhenium using selective precipitation induced by two-stage photochemical treatment. *Hydrometallurgy* **2022**, *211*, 105883. [CrossRef]
4. Yi, A.; Jiang, H. Rhenium-molybdenum separation in an alkaline leaching solution of a waste superalloy by N263 extraction. *Arab. J. Chem.* **2023**, *16*, 104516. [CrossRef]
5. Wang, Q.; Wang, S.; Ma, X.; Cao, Z.; Zhang, C.; Zhong, H. A green production process of electrolytic manganese metal based on solvent extraction. *Colloids Surf. A Physicochem. Eng. Asp.* **2023**, *670*, 131517. [CrossRef]
6. Joo, S.-H.; Kim, Y.-U.; Kang, J.-G.; Kumar, J.R.; Yoon, H.-S.; Parhi, P.K.; Shin, S.M. Recovery of Rhenium and Molybdenum from Molybdenite Roasting Dust Leaching Solution by Ion Exchange Resins. *Mater. Trans.* **2012**, *53*, 2034–2037. [CrossRef]
7. Feng, J.; Li, J.; Liao, Y.; Liu, F.; Li, H.; Jiang, Q.; Huang, B.; Wang, Y.; Xiao, L.; Liu, H.; et al. Rhenium recovery from roasting leachate of molybdenum concentrate by N-methylimidazole functionalized anion exchange resin. *J. Radioanal. Nucl. Chem.* **2023**, *332*, 747–760. [CrossRef]
8. Fu, Z.; Hou, Y.; Huang, J.; Cheng, D.; Li, G. Separation and recovery of rhenium(VII) from molybdenum concentrate oxygen pressure leach solution by modified D201 resin. *Conserv. Util. Miner. Resour.* **2022**, *42*, 115–122.
9. Wang, J.; Yu, Z.; Xiao, X. A novel hydroxyapatite super-hydrophilic membrane for efficient separation of oil-water emulsions, desalting and removal of metal ions. *Desalination Int. J. Sci. Technol. Desalt. Water Purif.* **2023**, *565*, 116864. [CrossRef]
10. Long, X.; Zhao, G.Q.; Zheng, Y.; Hu, J.; Zuo, Y.; Luo, W.; Jiao, F. A precise pyromellitic acid grafting prepared multifunctional MXene membranes for efficient oil-in-water emulsion separation and heavy metal ions removal. *Chem. Eng. J.* **2023**, *472*, 144904. [CrossRef]
11. Kang, S.; Qin, S.-J.; Wang, Q.; Hao, L.; Pang, W.; Li, S. Research Progress on Separation Enrichment and Extraction Techology of Rhenium. *Nonferrous Met. Smelt. Compon.* **2023**, 66–74.
12. Salehi, H.; Tavakoli, H.; Aboutalebi, M.R.; Samim, H.R. Recovery of molybdenum and rhenium in scrub liquors of fumes and dusts from roasting molybdenite concentrates. *Hydrometallurgy* **2019**, *185*, 142–148. [CrossRef]
13. Kim, H.S.; Park, J.S.; Seo, S.Y.; Tran, T.; Kim, M.J. Recovery of rhenium from a molybdenite roaster fume as high purity ammonium perrhenate. *Hydrometallurgy* **2015**, *156*, 158–164. [CrossRef]
14. Srivastava, R.R.; Kim, M.; Lee, J.; Ilyas, S. Liquid–liquid extraction of rhenium(VII) from an acidic chloride solution using Cyanex 923. *Hydrometallurgy* **2015**, *157*, 33–38. [CrossRef]
15. Hong, T.; Liu, M.; Ma, J.; Yang, G.; Li, L.; Mumford, K.A.; Stevens, G.W. Selective recovery of Rhenium from industrial leach solutions by synergistic solvent extraction. *Sep. Purif. Technol.* **2020**, *236*, 116281. [CrossRef]
16. Assis, R.C.; Mageste, A.B.; de Lemos, L.R.; Orlando, R.M.; Rodrigues, G.D. Application of aqueous two-phase system for selective extraction and clean-up of emerging contaminants from aqueous matrices. *Talanta* **2021**, *223*, 121697. [CrossRef] [PubMed]
17. Zhang, F.; Jia, A.; Guo, Q.; Sun, T.; Li, P.; Wang, P.; Pan, Y.; Zhang, Y. Mass transfer of molybdenum in L35 + sodium sulfate + H_2O aqueous two-phase system with a packed column. *Sep. Purif. Technol.* **2018**, *204*, 304–313. [CrossRef]
18. Pan, Y.; Sun, X.; Qi, M.; Qin, R.; Che, X.; Zhang, Y. A clean and efficient method for separation of vanadium and molybdenum by aqueous two-phase systems. *J. Mol. Liq.* **2020**, *313*, 113540. [CrossRef]
19. Huang, Y.; Chen, D.; Chen, S.; Su, M.; Chen, Y.; Yuvaraja, G. A green method for recovery of thallium and uranium from wastewater using polyethylene glycol and ammonium sulfate based on aqueous two-phase system. *J. Clean. Prod.* **2021**, *297*, 126452. [CrossRef]
20. Li, R.; Qin, R.; Li, Q.; Zhang, Y.; Zhang, Y. Study on extraction and separation of tungsten and molybdenum by nonionic surfactant Triton X-100/Na_2SO_4 two-phase aqueous system. *Chem. World* **2023**, *64*, 342–347. [CrossRef]
21. Leite, D.D.S.; Carvalho, P.L.G.; De Lemos, L.R.; Mageste, A.B.; Rodrigues, G.D. Hydrometallurgical separation of copper and cobalt from lithium-ion batteries using aqueous two-phase systems. *Hydrometallurgy* **2017**, *169*, 245–252. [CrossRef]
22. García-González, L.; Shirayama, S.; Morita, K. Cobalt and nickel separation in aqueous two-phase systems with polyethylene glycol 20,000 and sodium electrolytes. *Hydrometallurgy* **2023**, *222*, 106180. [CrossRef]
23. Hishida, M.; Kaneko, Y.; Yamamura, Y.; Saito, K. Salt Effects on Lamellar Structure of Nonionic Surfactants. *J. Solut. Chem.* **2016**, *45*, 1612–1619. [CrossRef]

24. Lindman, B.; Medronho, B.; Karlström, G. Clouding of nonionic surfactants. *Curr. Opin. Colloid Interface Sci.* **2016**, *22*, 23–29. [CrossRef]
25. Guedes De Carvalho, R.A.; Sampaio, M.N.M. Solvent extraction of tungsten by alkylamines hydrochloric acid and alkylamines-sulphuric acid systems. *Hydrometallurgy* **1991**, *26*, 137–150. [CrossRef]
26. Fei, Y.; Gou, S.; He, Y.; Zhou, L.; Peng, C.; Zhang, H.; Zhang, Q.; Wu, Y. The properties of polyoxyethylene polymers with temperature-sensitive and instant-solubility. *J. Mol. Liq.* **2019**, *275*, 146–156. [CrossRef]
27. Tang, Z.; Wang, H.; Wu, P.-F.; Zhou, S.-Y.; Huang, Y.-C.; Zhang, R.; Sun, D.; Tang, Y.-G.; Wang, H.-Y. Electrode–Electrolyte Interfacial Chemistry Modulation for Ultra-High Rate Sodium-Ion Batteries. *Angew. Chem. Int. Ed.* **2022**, *61*, e202200475. [CrossRef] [PubMed]
28. Vasylieva, A.; Doroshenko, I.; Vaskivskyi, Y.; Chernolevska, Y.; Pogorelov, V. FTIR study of condensed water structure. *J. Mol. Struct.* **2018**, *1167*, 232–238. [CrossRef]
29. Han, D.; Li, X.; Cui, Y.; Yang, X.; Chen, X.; Xu, L.; Peng, J.; Li, J.; Zhai, M. Polymeric ionic liquid gels composed of hydrophilic and hydrophobic units for high adsorption selectivity of perrhenate. *RSC Adv.* **2018**, *8*, 9011–9319. [CrossRef]
30. Zhang, Y.; Sun, T.; Hou, Q.; Guo, Q.; Lu, T.; Guo, Y.; Yan, C. A green method for extracting molybdenum (VI) from aqueous solution with aqueous two-phase system without any extractant. *Sep. Purif. Technol.* **2016**, *169*, 151–157. [CrossRef]
31. Ahsaie, F.G.; Pazuki, G. Effect of carbohydrates, choline chloride based deep eutectic solvents and salts on the phase behavior of PEG-PPG copolymer ATPSs and partitioning of penicillin G. *J. Mol. Liq.* **2021**, *339*, 117152. [CrossRef]
32. Sinoimeri, E.; Pescheux, A.-C.; Guillotte, I.; Cognard, J.; Svecova, L.; Billard, I. Fate of metal ions in PEG-400/Na_2SO_4/H_2O aqueous biphasic system: From eviction to extraction towards the upper polymer-rich phase. *Sep. Purif. Technol.* **2023**, *308*, 122854. [CrossRef]
33. Jin, L.; Jarand, C.W.; Brader, M.L.; Reed, W.F. Angle-dependent effects in DLS measurements of polydisperse particles. *Meas. Sci. Technol.* **2022**, *33*, 045202. [CrossRef]
34. Akach, J.; Kabuba, J.; Ochieng, A. Simulation of the Light Distribution in a Solar Photocatalytic Bubble Column Reactor Using the Monte Carlo Method. *Ind. Eng. Chem. Res.* **2020**, *59*, 17708–17719. [CrossRef]
35. Lou, Z.; Guo, C.; Feng, X.; Zhang, S.; Xing, Z.; Shan, W.; Xiong, Y. Selective extraction and separation of Re(VII) from Mo(VI) by TritonX-100/N235/iso-amyl alcohol/n-heptane/NaCl microemulsion system. *Hydrometallurgy* **2015**, *157*, 199–206. [CrossRef]
36. Boysen, R.I.; Wang, Y.; Keah, H.H.; Hearn, M.T. Observations on the origin of the non-linear van't Hoff behaviour of polypeptides in hydrophobic environments. *Biophys. Chem.* **1999**, *77*, 79–97. [CrossRef]

water

MDPI

Article

The Elimination of Levofloxacin from High-Salinity Wastewater via the Electrochlorination Process

Mingfei Wei, Jingyu Li, Bingqing Jing, Xuankun Li * and Guanghui Li *

School of Chemistry and Chemical Engineering, Shanghai University of Engineering Science, Shanghai 201620, China; wmf_0831@163.com (M.W.); ljy20001312022@163.com (J.L.); 19921296781@163.com (B.J.)
* Correspondence: li.xuankun@foxmail.com (X.L.); li_ripe@163.com (G.L.)

Abstract: The electrochlorination (E-Cl) process has attracted much attention as it is a highly efficient method for treating organic compounds in hypersaline wastewater. In this study, the E-Cl process was utilized for the removal of antibiotics. The optimal experimental conditions were determined to be a NaCl concentration of 100 mM, a current density of 1.5 mA/cm^2, a pH of 7.0, and a plate spacing of 1 cm, with a levofloxacin (LEV) degradation efficiency reaching as high as 99% using this setup. The effects of the presence of other ions and humic acid on the E-Cl process were investigated, and it was found that the degradation of LEV was not significantly affected by the presence of coexisting substances. In addition, free chlorine was identified as the primary active species for the degradation of LEV by means of a quenching experiment. It was demonstrated by 3D EEM and TOC that LEV was not completely mineralized and that intermediate products may be present. In order to reveal the degradation pathways of LEV, its degradation products were also analyzed via LC-MS, and some possible pathways of LEV degradation in this system were proposed. The successful degradation of LEV demonstrated that the E-Cl process is an efficient and promising technique for the treatment of organic pollutants in high-salinity wastewater.

Keywords: electrochlorination process; levofloxacin; high-salinity wastewater; active chlorine

1. Introduction

High-salinity wastewater (>1% salt), such as that generated by mining, oil production, and chemical manufacturing, poses significant challenges for the treatment and application of wastewater [1]. The volume of high-salinity wastewater produced has dramatically increased [2]. Pharmaceutical wastewater contains a high concentration of recalcitrant organic pollutants, a large amount of inorganic salts and different kinds of pharmaceutical residues. Therefore, the development of the pharmaceutical industry and the production of high-salinity wastewater are closely related. Jiang et al. [3] used collected pharmaceutical wastewater at a salinity of 2.5% from a pharmaceutical manufacturing plant in Singapore. People are becoming more dependent on the use of antibiotics, and the application of antibiotics in various fields, such as medical care and animal husbandry, among others, has become increasingly extensive. Fluoroquinolones (FQs), including levofloxacin (LEV), moxifloxacin, norfloxacin and ciprofloxacin [4], are highly concentrated in the environment [5] and pose a high risk to both ecosystems and public health [6]. Among them, levofloxacin (LEV) is widely overused as an anti-inflammatory agent and is discharged into the environment mainly through the effluent of wastewater treatment plants [7]. Hanna et al. [8] investigated the presence of antibiotic residues in twelve villages in Shandong Province, China. Their results showed that levofloxacin was detected at concentrations ranging between 0.3 and 3.9 ng/L in river water, 1.3 and 12.5 ng/L in wastewater, 0.5 and 21.4 ng/L in drinking water and 0.5 and 2.5 μg/kg in soil. The long-term existence and accumulation of LEV in water bodies pose a significant threat to human beings, aquatic organisms and the ecosystem [9]. Therefore, it is hoped that an efficient, convenient, environmentally friendly, and safe wastewater treatment method can be found to reduce the environmental and health hazards posed by LEV [10].

Many technologies have been developed to treat high-salinity organic wastewater, such as advanced oxidation processes [11], membrane separation [12], and osmosis technologies [13]. Among these technologies, advanced oxidation processes (AOPs) represent a promising method for the removal of emerging organic pollutants from water [14]. These processes involve the generation of highly reactive hydroxyl radicals (e.g., $^{\bullet}OH$, $Cl^{\bullet}$, and $SO_4^{\bullet-}$) through the use of various oxidizing agents, such as ozone [15], UV radiation [16], Fenton reactions [17], and electrochemical oxidation [18]. Wang et al. [19], using mesoporous $MnO@MnO_x$ microspheres with peroxymonosulfate (PMS), successfully achieved the degradation of LEV. In particular, the $MnO@MnO_x$ catalyst achieved 98.1% degradation and 81.4% mineralization of LEV after being irradiated for 30 min. In addition, the stability of the catalyst, its reaction kinetics and the degradation mechanism were also systematically studied. Li et al. [20] developed a novel catalyst (MS-N_3H) prepared directly from electrolytic manganese slag, and MS-N_3H could be applied for the efficient degradation of LEV with a broad pH range from 2.0 to 10. In the MS-N_3H/PMS system, after PMS (0.4 g/L) had been added, the concentration of LEV decreased obviously within 60 min with a degradation efficiency of 82.6%. It was demonstrated that while non-radical $^1O_2$ was the dominant contributor, the MS-N_3H/PMS was stable for the degradation of LEV in different water matrixes and maintained high recyclability even after four recycling processes. In addition, unlike membrane filtration technologies, AOPs can decompose hazardous substances directly rather than simply transferring the pollutants [21]. However, conventional AOPs may not be efficient at removing organic pollutants from high-salinity wastewater. Due to the presence of high concentrations of chlorides, sulfates, nitrates, phosphates, bicarbonates, carbonates, and so on in wastewater, the free radicals are largely scavenged [22].

The electrochemical oxidation (EO) process is able to degrade organic pollutants through the electrogenesis of reactive radicals and direct electron transfer [23]. In high-salinity wastewater, when a voltage is applied, the presence of large amounts of Cl^- can lead to the in situ generation of reactive chlorine species (RCS), which act as reactive oxidants to degrade organic pollutants [24]. Moreover, the in situ generation of RCS and free chlorine is less materially restricted and is more cost-effective than the generation of $^{\bullet}OH$ [25]. Kuang et al. [26] proposed a new bipolar-membrane-integrated electrochlorination (BPM-EC) process for removing biorefractory organic pollutants in landfill leachate. This BPM-EC process is efficient for the generation of $ClO^{\bullet}$, which significantly accelerates the oxidation of ammonium in landfill leachate and its selective transformation into nitrogen. Using this method, it was found that 100% of the NH_4^+-N present was removed within 3 h, demonstrating the superiority of this method in promoting ammonium oxidation. Huo et al. [27] coupled electroporation with electrochlorination to establish an efficient method of water disinfection. Electroporation is sub-lethal to microorganisms, resulting in limited efficiency of microbial inactivation. However, the electrogenerated active chlorine at the anode causes more lethal damage. This process completely disinfected water (>6.0-log) under a high water flux of 2.4×10^4 L/(m^2·h) and an applied voltage of 2.0 V, removing bacteria and viruses.

In this study, the electrochlorination (E-Cl) process was applied to treat high-salinity wastewater containing LEV. In the E-Cl system, a Ti/RuIr electrode was first selected as the anode, and a Ti/Pt electrode was selected as the cathode. Next, the effects of several experimental parameters, such as the NaCl concentration, current density, pH, and plate spacing between the electrodes, were investigated. Additionally, the effects of the presence of other substances on the system were explored, and the reactive chlorine species in the system were identified through quenching experiments. Lastly, the possible pathways and intermediates of LEV degradation were hypothesized. The aim of this study was to investigate the optimal degradation conditions and the degradation effect of the E-Cl process on antibiotic LEV and to analyze the degradation mechanism and degradation pathway. Another type of wastewater, high-salinity wastewater, was efficiently utilized to establish an efficient method for the removal of antibiotics from water, providing a new perspective on the treatment of organic pollutants in high-salinity wastewater.

2. Materials and Methods

2.1. Chemicals

Levofloxacin, sodium chloride, methanol, tert-butyl alcohol, sodium thiosulfate, nitrobenzene, sodium sulfate anhydrous, sodium bicarbonate, sodium nitrate, sodium phosphate dibasic (Sinopharm Chemical Reagent Co., Ltd., Shanghai, China), humic acid, triethylamine, and acetonitrile (Titan Scientific Co., Ltd., Shanghai, China). Ti/RuIr and Ti/Pt electrodes were purchased from Suzhou Schulte Industrial Technology Co., Ltd., Suzhou, China.

2.2. Electrochlorination Experiments

Experimental setup: A Ti/RuIr electrode was used as the anode and a Pt electrode was used as the cathode, both of which were 50 × 100 mm in size and 1 mm in thickness. A total of 300 mL of synthetic LEV wastewater (10 mg/L) was treated, and the required constant current density was provided using an IT6302 DC power supply (ITECH, Shanghai, China).

Experimental method: The experiment was carried out for 60 min, and samples were collected every 10 min. Magnetic stirring was performed at 300 rpm, and the pH of the solution was adjusted using HCl or NaOH.

2.3. Experimental Instrument

The concentration of LEV was measured using high-performance liquid chromatography (HPLC; Shimadzu, Kyoto, Japan). Total organic carbon (TOC) was used to detect the degree of mineralization of LEV, which was analyzed using a TOC-L analyzer (Shimadzu, Japan). The excitation–emission matrix (EEM) fluorescence spectra were measured using a fluorescence spectrophotometer (Hitachi, Japan). Liquid-phase chromatography/mass spectrometry (LC-MS; Agilent 1290 Infinity, Shanghai, China) was used to detect the intermediate products during the degradation of levofloxacin.

2.4. Analytical Methods

High-performance liquid chromatography (HPLC) with a symmetrical C18 column (150 mm × 4.6 mm i.d., 5 μm) was used to determine the concentration of LEV. The pH of 1% triethylamine was adjusted to 3.0 using phosphoric acid. A mixture of 1% triethylamine and acetonitrile at a volume ratio (86:14, *v*/*v*) was used as the mobile phase. The separation was set up at a column temperature of 40 °C. The detection wave lengths for LEV were set at 294 nm. The flow rate was set to 1.0 mL/min. The excitation–emission matrix (EEM) fluorescence spectra was used to investigate the degradation process. The LEV solution displayed fluorescence peaks in the Ex/Em = 200–300/400–550 nm range and the Ex/Em = 300–350/400–500 nm range. The width of the excitation and emission slits was fixed at 5 nm and the scan speed was 12,000 nm/min. The experimental conditions for LC-MS were as follows: a mixture of 0.1% formic acid and acetonitrile at a volume ratio of 2:3 was used as the mobile phase, and the column was a Waters BEH C18 (2.1 × 100 mm i.d., 1.7 μm). A volume of 5 μL of the reaction sample was extracted. The flow rate was set to 1.0 mL/min.

3. Results and Discussion

3.1. Effects of Key Operational Parameters on LEV Degradation

3.1.1. Effects of Cl^- Concentration

Cl^- is a common anion in water bodies, and the magnitude of its concentration has a direct effect on the generation of RCS in the E-Cl process. As shown in Figure 1, without the addition of NaCl, the degradation efficiency of LEV was only 62.92% after 60 min. With the addition of NaCl, the degradation efficiency increased significantly as the NaCl concentration increased. In the first 10 min, when the concentration of NaCl increased from 25 mM to 200 mM, the degradation efficiency of LEV increased from 41.96% to 87.53%, respectively. As the reaction time increased, the degradation efficiency of LEV further increased; when the reaction time reached 60 min, the degradation efficiency tended to be around 99% if the NaCl concentration was over 50 mM, which is approximately 36% higher compared to the degradation efficiency when no NaCl was added. The experimental results proved that the concentration of chloride is important for pollutant degradation in the E-Cl process; when the chloride concentration reached 100 mM, nearly complete LEV degradation was achieved.

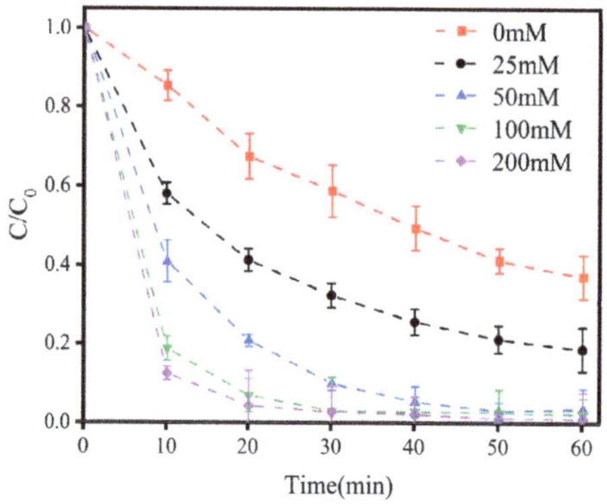

Figure 1. Variations in LEV degradation at different concentrations of NaCl. Experimental conditions: current density, 1.5 mA/cm^2; solution pH, 7.0; plate spacing, 1 cm. C/C_0 represents the degradation efficiency of LEV.

3.1.2. Effects of Current Density

In the E-Cl process, generally, as the current density increases, more reactive chlorine species will be produced, which can accelerate the reaction and improve the degradation efficiency [28]. Therefore, it is of great significance to study the effect of the current density on the degradation of LEV in the E-Cl process.

As depicted in Figure 2, generally, the LEV degradation efficiency increased with the increase in current density. In the first 10 min, the difference in the LEV degradation efficiency was obvious with the variation in current density. The LEV degradation efficiency was only 17.26% when the current density was 0.5 mA/cm^2; when the current density increased from 1 mA/cm^2 to 2.5 mA/cm^2, the LEV degradation efficiency increased from 34.76% to 92.95%, respectively. When the reaction time reached 20 min, the LEV degradation efficiency was close to 94% as the current density increased above 1.5 mA/cm^2. Therefore, considering the cost-effectiveness of LEV degradation, a current density of 1.5 mA/cm^2 was selected for further research.

3.1.3. Effects of pH

The pH level is also important to consider. It has been reported that pH determines the form of reactive chlorine species—Cl_2 (acidic), HOCl (neutral), or OCl^- (alkaline) [29]. Therefore, the effect of pH on the degradation of LEV via the E-Cl process was studied.

As depicted in Figure 3, within 10 min, as the pH value decreased, the efficiency of LEV degradation increased. However, the efficiency of LEV degradation at pH levels ranging from three to nine tended to be consistent when the reaction time reached 20 min, and the degradation efficiency was only obviously suppressed when the pH was eleven. The LEV degradation efficiency also decreased to 78.33% at a reaction time of 60 min and a pH of eleven. Although there was an obvious degradation effect, the degradation rate became slower. Therefore, a pH of seven was chosen as the optimal condition in this experiment.

Meanwhile, the experiment showed that the E-Cl process, which has a wide range of applicability in acidic and alkali conditions, had an inhibitory effect on the efficiency and rate of LEV degradation under a strong alkali pH of 11 or more, demonstrating good application prospects.

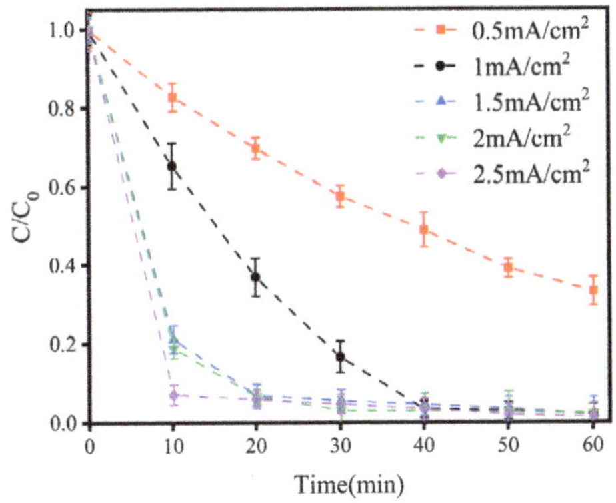

Figure 2. Variations in LEV degradation at different current densities. Experimental conditions: concentration of NaCl = 100 mM; solution pH = 7.0; plate spacing = 1 cm. C/C_0 represents the degradation efficiency of LEV.

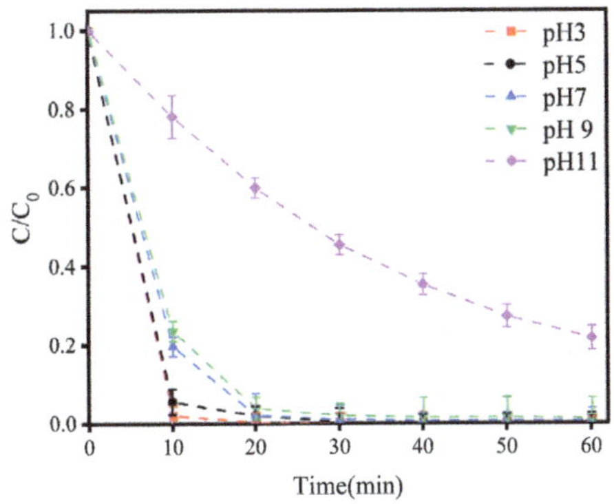

Figure 3. Variations in LEV degradation at different pH values. Experimental conditions: concentration of NaCl = 100 mM; current density = 1.5 mA/cm^2; plate spacing = 1 cm. C/C_0 represents the degradation efficiency of LEV.

3.1.4. Effects of Plate Spacing

Plate spacing is also an important conditioning factor in this type of study, and it exhibits an effect on the degradation of LEV in this system; thus, we investigated it in the present work.

As shown in Figure 4, as the electrode plate spacing increased, the efficiency of LEV degradation significantly decreased. The smaller the electrode plate spacing was, the more efficient the LEV degradation was. The efficiency of LEV degradation within 60 min increased from 80.14% to 96.64% when the electrode plate spacing decreased from 3 cm to 1 cm. Considering that the best degradation results were obtained at a plate spacing of 1 cm, this spacing was chosen as the optimal condition.

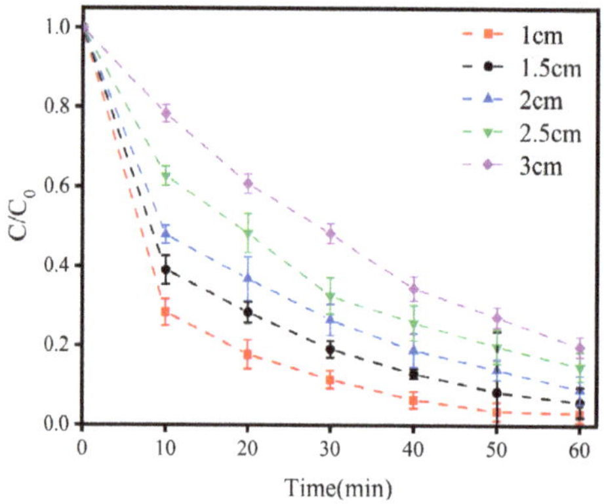

Figure 4. Variations in LEV degradation with different polar plate spacings. Experimental conditions: concentration of NaCl = 100 mM; current density = 1.5 mA/cm^2; solution pH = 7.0. C/C_0 represents the degradation efficiency of LEV.

3.1.5. Effects of Coexisting Substances

In AOPs, the presence of other substances affects the degradation efficiency. These coexisting substances, which are mainly inorganic anions (including SO_4^{2-}, PO_4^{3-}, HCO_3^-, and NO_3^-) and natural organic matter (NOM), may compete with active species and decrease their degradation efficiency. In high-salinity wastewater, there may be several coexisting substances that compete with active chlorine species and decrease the efficiency of LEV degradation. For this reason, the effects that coexisting substances had on the system were investigated. In order to reveal the specific effects of coexisting substances on E-Cl, their effects on the degradation of the pollutant were investigated. Figure 5a shows the effect of SO_4^{2-} on the E-Cl process. When the reaction time reached 60 min, the efficiency of the degradation of LEV from the solution reached 97%, regardless of whether SO_4^{2-} had been added or not, proving that sulfate had little effect on the E-Cl process. The same phenomenon can be observed in Figure 5b; when the reaction time reached 60 min, the efficiency of LEV degradation reached 98%, regardless of whether NO_3^- had been added or not, indicating that NO_3^- had virtually no effect on the process. As shown in Figure 5c, in the first 10 min, the presence of HCO_3^- had an inhibitory effect on the degradation efficiency of LEV. The degradation efficiency decreased from 82.6% to 60.47% when the concentration of HCO_3^- increased from 0 mg/L to 400 mg/L. After 60 min, the efficiency of LEV degradation reached 95% regardless of the HCO_3^- concentration. As shown in Figure 5d, in comparison with the abovementioned coexisting ions, PO_4^{3-} exerted a more significant effect within 20 min, especially in the first 10 min, during which the degradation efficiency was hindered and decreased from 82.6% to 50.74% as the PO_4^{3-} concentration increased from 0 mg/L to 50 mg/L. After 60 min, the degradation efficiency reached 96%. A likely reason for this is that PO_4^{3-} causes a change in the pH of the solution, thus altering the equilibrium of HClO and ClO^- conversion and inhibiting the degradation efficiency in the first 20 min.

In conventional AOPs, HA scavenges the free radicals that compete with organic pollutants and reduces the degradation efficiency. A similar phenomenon exists in the E-Cl process. As shown in Figure 5e, the degradation efficiency was inhibited during the first 10 min, and as the concentration of HA increased from 0 mg/L to 50 mg/L, the efficiency of LEV degradation decreased from 82.6% to 47.7%. The reason for this phenomenon was that HA consumed the RCS and •OH in the system and interacted with the active chlorine species to inhibit the degradation of LEV. However, as the reaction time reached 20 min, regardless of the HA concentration, the LEV degradation efficiency reached 93%, and eventually reached 96% after 60 min.

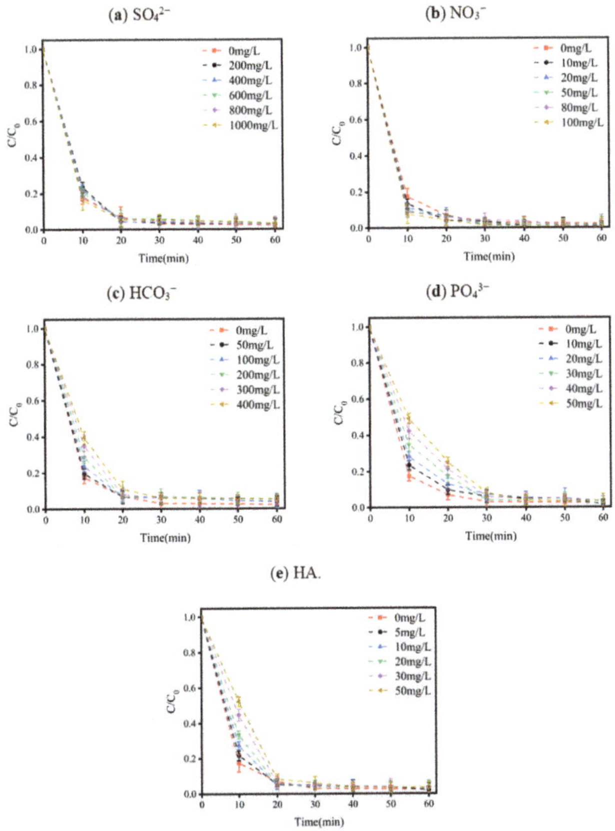

Figure 5. Variations in LEV degradation when background matter of different types and concentrations was added to the system: (**a**) SO_4^{2-}; (**b**) NO_3^-; (**c**) HCO_3^-; (**d**) PO_4^{3-}; (**e**) HA. Experimental conditions: concentration of NaCl = 100 mM; current density = 1.5 mA/cm^2; solution pH = 7.0; plate spacing = 1 cm. C/C_0 represents the degradation efficiency of LEV.

Considering the above results, the degradation of LEV via the E-Cl process was not significantly affected by the presence of coexisting substances, indicating that the E-Cl process is more robust to variations in water properties compared with conventional AOP processes, making it more applicable in actual practice.

3.2. *Identification of Reactive Species*

In EO wastewater treatment, anodic oxidation can produce active species, such as superoxide anions ($O_2^{\bullet-}$), hydroxyl radicals ($^{\bullet}OH$), sulfate ($SO_4^{\bullet-}$), chlorine ($Cl^{\bullet}$), and oxychloride ($ClO^{\bullet}$) [30].

When wastewater contains a high concentration of salt, a large amount of Cl^- is converted into reactive chlorine species. The identification of reactive chlorine species is essential in order to reveal the mechanism of LVE degradation in the E-Cl process [31].

Firstly, $Na_2S_2O_3$ has been reported to be capable of quenching free radicals such as $^{\bullet}OH$, $Cl^{\bullet}$, $Cl_2^{\bullet-}$, $ClO^{\bullet}$, and free chlorine [32]. As shown in Figure 6, the efficiency of LEV degradation was reduced by 75% in comparison with that when $Na_2S_2O_3$ was absent, proving that the degradation of LEV did not occur due to electrochemical oxidation alone, but required free radicals and free chlorine.

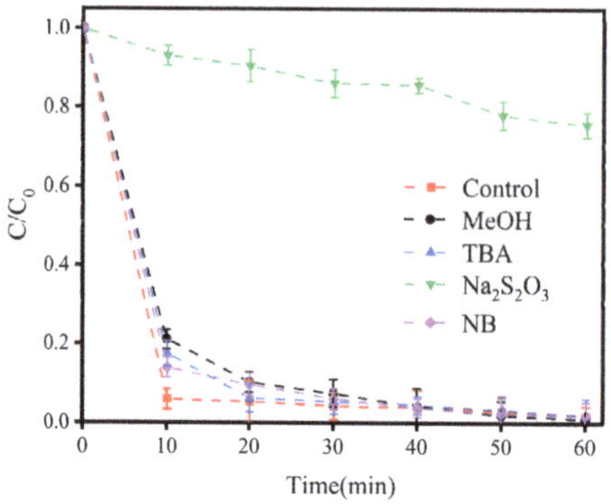

Figure 6. Variations in LEV degradation in the presence of different scavengers ([$Na_2S_2O_3$] = [MeOH] = [TBA] = [NB] = 50 mM). Experimental conditions: concentration of NaCl = 100 mM; current density = 1.5 mA/cm^2; solution pH = 7.0; plate spacing = 1 cm. C/C_0 represents the degradation efficiency of LEV.

MeOH is widely applied to quench $^{\bullet}$OH and $Cl_2^{\bullet-}$ [33]. In the first 10 min following the addition of 50 mM MeOH, the efficiency of LEV degradation decreased by 15.08%. However, after 60 min of the reaction, the efficiency of LEV degradation reached 99%. This result demonstrated that the scavenging of $^{\bullet}$OH and $Cl_2^{\bullet-}$ delayed LEV degradation but did not affect it in the long run. TBA is generally used to quench •OH, ClO$^{\bullet}$, and Cl$^{\bullet}$ [34]. In the first 10 min after the addition of TBA, the LEV degradation efficiency reduced by 11.46%, but when the reaction time reached 60 min, the efficiency of LEV degradation reached 97%. Moreover, NB has been shown to quench $^{\bullet}$OH [35]. Although the addition of NB reduced LEV's degradation efficiency by 7.95% in the first 10 min after its addition, the efficiency of LEV degradation ultimately reached 97%, indicating that $^{\bullet}$OH was not the prominent active substance in the system. The above results suggest that free chlorine may be the RCS contributing to the degradation of LEV in the E-Cl process.

3.3. Possible Pathways of LEV Degradation

Three-dimensional EEM can be used to characterize and analyze organic compound species and observe their changes [36]. To better understand the degradation of LEV, 3D EEM fluorescence spectra of the LEV samples were observed before and after the reaction. As shown in Figure 7a, the initial LEV solution displayed fluorescence peaks at Ex/Em = 200–300/400–550 nm and Ex/Em = 300–350/400–500 nm [37]. As shown in Figure 7b, after 60 min, no fluorescence signals could be detected, indicating that the functional group of LEV was destroyed [38] and that LEV had been degraded via the E-Cl process.

However, the degradation efficiencies of LEV when studied using TOC and HPLC are quite different. As shown in Figure 8, the degradation efficiency of LEV determined using HPLC was 99%, but when detected using TOC, the degradation efficiency of LEV dramatically dropped to 57.17%, Similarly, Wang et al. [32] used the E-Cl process to treat atrazine (ATZ) in pesticide wastewater. Although the degradation efficiency of ATZ reached 98.66% after 60 min, the chemical oxygen demand (COD) removal efficiency was 86.43% and an obvious gap was observed between the results obtained from different analysis methods. It can be deduced that the mineralization of organic pollutant by E-Cl was incomplete and the products of LEV's degradation must be urgently studied.

To reveal the degradation pathway of LEV, LC-MS was also applied to analyze the degradation products. Some possible pathways of LEV degradation in this system are proposed.

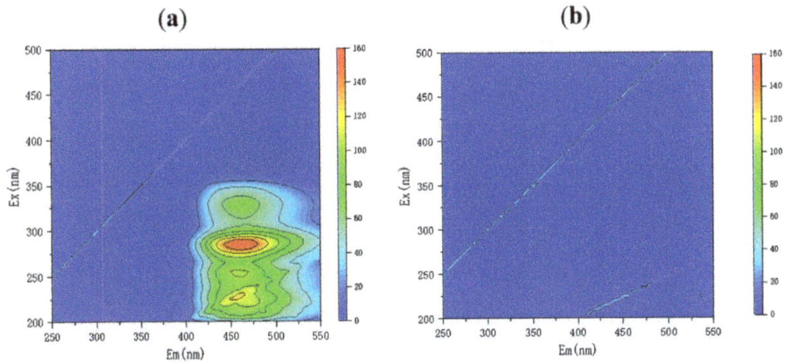

Figure 7. Three-dimensional EEM of the levofloxacin solution after 0 min (**a**) and 60 min (**b**) of degradation via the E-Cl process.

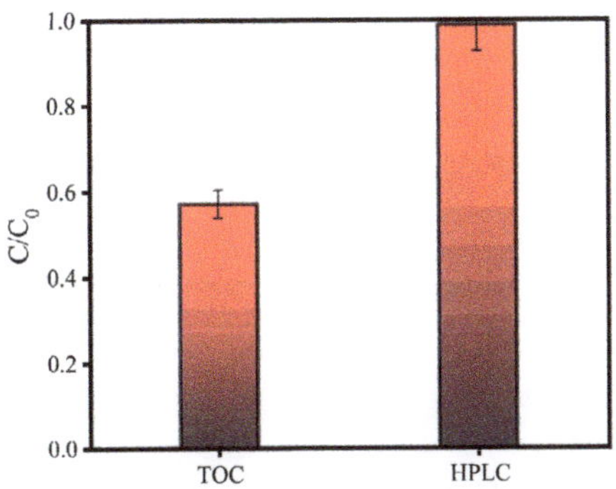

Figure 8. Efficiencies of LEV degradation, as detected using TOC and HPLC.

As shown in Figure 9, in pathway I, the piperazine in LEV is attacked by •OH, opening the piperazine ring to form P1 ($m/z = 336$); then, P1 ($m/z = 336$) is converted into P2 ($m/z = 279$) through deamination and dealkylation reactions. P3 ($m/z = 239$) is generated from P2 ($m/z = 279$) by opening the cycloxazine ring and via dealkylation reactions [39,40]. In pathway II, LEV is converted into the intermediate product P4 ($m/z = 318$) via a decarboxylation reaction, and P4 ($m/z = 318$) is oxidized into P5 ($m/z = 338$); P5 ($m/z = 338$) is converted into P6 ($m/z = 262$) through defluorination, decarboxylation, and demethylation reactions [41,42]. In pathway III, LEV is converted into P8 ($m/z = 348$) via the demethylation of the piperazine ring, and P8 ($m/z = 348$) is degraded into P9 ($m/z = 364$) through dehydroxylation reactions. The byproducts generated in AOP-based water treatment are attracting increasing attention. Wang et al. [32] developed the E-Cl process and treated high-salinity wastewater containing atrazine (ATZ). To compare the variations in the acute toxicity of wastewater before and after treatment, a method was employed using a bio-toxicity assay with bioluminescent bacteria. The results showed that the ATZ showed a 37.53% inhibitory effect on bioluminescence at high concentrations, and the bioluminescence was restored to 77.49% with the decomposition of ATZ by the E-Cl process. He et al. [43] prepared a Zn_nCoO_x anode for electrocatalytic generation of RCS to remove 4-chlorophenol (4-cp) from wastewater containing Cl^-. The potential ecotoxicity of 4-cp and its degradation intermediates was assessed with the ECOSER procedure. The results showed that the acute and chronic toxicity of the treated wastewater containing 4-cp decreased in all four pathways of degradation after treatment. When AOPs were applied to LEV degradation, the byproducts of P3, P6, and P9 could often be detected, and these byproducts were able to further convert into small molecules [44–46]. Although LEV degradation by the E-Cl process generates a number of intermediate products, they are expected to be further degraded with a longer reaction time; therefore, the environmental risks of the byproducts can be limited.

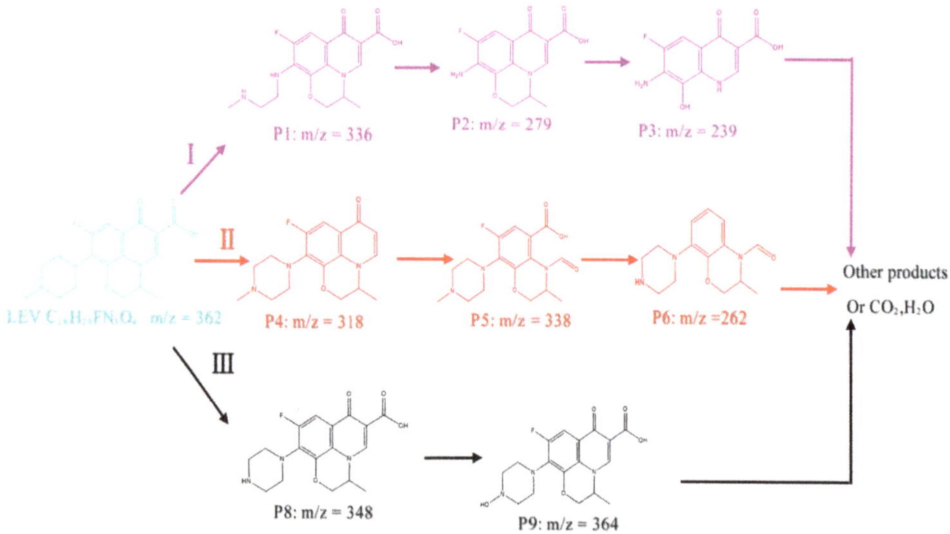

Figure 9. Possible pathways of LEV degradation in the E-Cl system.

4. Conclusions

The E-Cl process was successfully performed for the treatment of LEV, and efficient degradation was achieved. The optimal conditions for E-Cl were determined by selecting a NaCl concentration of 100 mM, a current density of 1.5 mA/cm^2, a pH of 7, and a polar plate spacing of 1 cm. The degradation efficiency of LEV via the E-Cl process is not significantly limited by the presence of other substances in the water matrix, demonstrating that the E-Cl process possesses a potentially higher applicability than conventional AOP processes. When wastewater contains high concentrations of Cl^-, this Cl^- can be converted into reactive chlorine species; among these, free chlorine was identified as the dominant reactive species. Despite a nearly complete degradation of LEV being achieved using the E-Cl process, a number of intermediate products were detected using LC-MS. Therefore, the mineralization efficiency of LEV should be further increased. In conclusion, the E-Cl process provides a new solution to treating organic pollutants in high-salinity wastewater, which is especially advantageous over traditional AOP processes due to its resistance to the presence of coexisting substances in the water matrix.

Author Contributions: Investigation, J.L. and B.J.; Writing—original draft, M.W.; Writing—review & editing, X.L. and G.L. All authors have read and agreed to the published version of the manuscript.

Funding: This research was supported by the Shandong Provincial Natural Science Foundation (ZR2021QE223).

Data Availability Statement: Data are contained within the article.

Conflicts of Interest: The authors declare no conflict of interest.

References

1. Tawabini, B.S.; Plakas, K.V.; Fraim, M.; Safi, E.; Oyehan, T.; Karabelas, A.J. Assessing the efficiency of a pilot-scale GDE/BDD electrochemical system in removing phenol from high salinity waters. *Chemosphere* **2020**, *239*, 124714. [CrossRef]
2. Cao, W.R.; Hu, C.; Zhang, P.; Qiu, T.; Wang, S.G.; Huang, G.H.; Lyu, L. Salinity-mediated water self-purification via bond network distorting of H_2O molecules on DRC-surface. *Proc. Natl. Acad. Sci. USA* **2023**, *120*, e2311920120. [CrossRef]

3. Jiang, Y.; Shi, X.; Ng, H.Y. Aerobic granular sludge systems for treating hypersaline pharmaceutical wastewater: Start-up, long-term performances and metabolic function. *J. Hazard. Mater.* **2021**, *412*, 125229. [CrossRef]
4. Van Doorslaer, X.; Dewulf, J.; Van Langenhove, H.; Demeestere, K. Fluoroquinolone antibiotics: An emerging class of environmental micropollutants. *Sci. Total Environ.* **2014**, *500*, 250–269. [CrossRef]
5. Lv, H.; Han, P.; Li, X.; Mu, Z.; Zuo, Y.; Wang, X.; Tan, Y.; He, G.; Jin, H.; Sun, C.; et al. Electrocatalytic Degradation of Levofloxacin, a Typical Antibiotic in Hospital Wastewater. *Materials* **2021**, *14*, 6814. [CrossRef]
6. John, Y.; David, V.E., Jr.; Mmereki, D. A Comparative Study on Removal of Hazardous Anions from Water by Adsorption: A Review. *Int. J. Chem. Eng.* **2018**, *2018*, 3975948. [CrossRef]
7. Gonzalez-Pleiter, M.; Gonzalo, S.; Rodea-Palomares, I.; Leganes, F.; Rosal, R.; Boltes, K.; Marco, E.; Fernandez-Pinas, F. Toxicity of five antibiotics and their mixtures towards photosynthetic aquatic organisms: Implications for environmental risk assessment. *Water Res.* **2013**, *47*, 2050–2064. [CrossRef]
8. Hanna, N.; Sun, P.; Sun, Q.; Li, X.W.; Yang, X.W.; Ji, X.; Zou, H.Y.; Ottoson, J.; Nilsson, L.E.; Berglund, B.; et al. Presence of antibiotic residues in various environmental compartments of Shandong province in eastern China: Its potential for resistance development and ecological and human risk. *Environ. Int.* **2018**, *114*, 131–142. [CrossRef]
9. Xiong, J.-Q.; Kurade, M.B.; Patil, D.V.; Jang, M.; Paeng, K.-J.; Jeon, B.-H. Biodegradation and metabolic fate of levofloxacin via a freshwater green alga, Scenedesmus obliquus in synthetic saline wastewater. *Algal Res.-Biomass Biofuels Bioprod.* **2017**, *25*, 54–61. [CrossRef]
10. Yang, K.D.; Han, P.W.; Liu, Y.N.; Lv, H.X.; Chen, X.F.; Lei, Y.H.; Yu, L.; Ma, L.; Duan, P.Z. Boosted Electrocatalytic Degradation of Levofloxacin by Chloride Ions: Performances Evaluation and Mechanism Insight with Different Anodes. *Molecules* **2024**, *29*, 662. [CrossRef]
11. Bhat, A.P.; Gogate, P.R. Degradation of nitrogen-containing hazardous compounds using advanced oxidation processes: A review on aliphatic and aromatic amines, dyes, and pesticides. *J. Hazard. Mater.* **2021**, *403*, 123657. [CrossRef]
12. Fonseca Pierangeli, G.M.; Ragio, R.A.; Benassi, R.F.; Gregoracci, G.B.; Subtil, E.L. Pollutant removal, electricity generation and microbial community in an electrochemical membrane bioreactor during co-treatment of sewage and landfill leachate. *J. Environ. Chem. Eng.* **2021**, *9*, 106205. [CrossRef]
13. Shenvi, S.S.; Isloor, A.M.; Ismail, A.F. A review on RO membrane technology: Developments and challenges. *Desalination* **2015**, *368*, 10–26. [CrossRef]
14. Zhou, M.; Chen, J.; Yu, S.; Chen, B.; Chen, C.; Shen, L.; Li, B.; Lin, H. The coupling of persulfate activation and membrane separation for the effective pollutant degradation and membrane fouling alleviation. *Chem. Eng. J.* **2023**, *451*, 139009. [CrossRef]
15. Li, X.; Chen, W.; Ma, L.; Huang, Y.; Wang, H. Characteristics and mechanisms of catalytic ozonation with Fe-shaving-based catalyst in industrial wastewater advanced treatment. *J. Clean. Prod.* **2019**, *222*, 174–181. [CrossRef]
16. Yang, F.; Sheng, B.; Wang, Z.; Xue, Y.; Liu, J.; Ma, T.; Bush, R.; Kusic, H.; Zhou, Y. Performance of UV/acetylacetone process for saline dye wastewater treatment: Kinetics and mechanism. *J. Hazard. Mater.* **2021**, *406*, 124774. [CrossRef]
17. Zhang, J.; Wang, Y.; Hong, M.; Peng, B.; Bao, C.; Xu, X.; Li, D.; Chen, J.; Wang, B.; Zhang, Q. Ferrocene-based resin as heterogeneous fenton-like catalyst for efficient treatment of high salinity wastewater at acidic, neutral, and basic pH. *Chem. Eng. J.* **2023**, *464*, 142450. [CrossRef]
18. Radjenovic, J.; Sedlak, D.L. Challenges and Opportunities for Electrochemical Processes as Next-Generation Technologies for the Treatment of Contaminated Water. *Environ. Sci. Technol.* **2015**, *49*, 11292–11302. [CrossRef]
19. Wang, A.Q.; Chen, Z.; Zheng, Z.K.; Xu, H.; Wang, H.; Hu, K.; Yan, K. Remarkably enhanced sulfate radical-based photo-Fenton-like degradation of levofloxacin using the reduced mesoporous MnO@MnO_x microspheres. *Chem. Eng. J.* **2020**, *379*, 122340. [CrossRef]
20. Li, M.K.; Huang, F.L.; Hu, L.; Sun, W.; Li, E.P.; Xiong, D.L.; Zhong, H.; He, Z.G. Efficient activation of peroxymonosulfate by a novel catalyst prepared directly from electrolytic manganese slag for degradation of recalcitrant organic pollutes. *Chem. Eng. J.* **2020**, *401*, 126085. [CrossRef]
21. Nabgan, W.; Saeed, M.; Jalil, A.A.; Nabgan, B.; Gambo, Y.; Ali, M.W.; Ikram, M.; Fauzi, A.A.; Owgi, A.H.K.; Hussain, I.; et al. A state of the art review on electrochemical technique for the remediation of pharmaceuticals containing wastewater. *Environ. Res.* **2022**, *210*, 112975. [CrossRef]
22. Nejumal, K.K.; Manoj, P.R.; Aravind, U.K.; Aravindakumar, C.T. Sonochemical degradation of a pharmaceutical waste, atenolol, in aqueous medium. *Environ. Sci. Pollut. Res.* **2014**, *21*, 4297–4308. [CrossRef]
23. Hai, H.; Xing, X.; Li, S.; Xia, S.; Xia, J. Electrochemical oxidation of sulfamethoxazole in BDD anode system: Degradation kinetics, mechanisms and toxicity evaluation. *Sci. Total Environ.* **2020**, *738*, 139909. [CrossRef]
24. Sheng, B.; Huang, Y.; Wang, Z.; Yang, F.; Ai, L.; Liu, J. On peroxymonosulfate-based treatment of saline wastewater: When phosphate and chloride co-exist. *RSC Adv.* **2018**, *8*, 13865–13870. [CrossRef]
25. Cho, K.; Hoffmann, M.R. Urea Degradation by Electrochemically Generated Reactive Chlorine Species: Products and Reaction Pathways. *Environ. Sci. Technol.* **2014**, *48*, 11504–11511. [CrossRef]
26. Kuang, W.J.; Yan, Z.; Chen, J.X.; Ling, X.T.; Zheng, W.X.; Huang, W.J.; Feng, C.H. A Bipolar Membrane-Integrated Electrochlorination Process for Highly Efficient Ammonium Removal in Mature Landfill Leachate: The Importance of $ClO^\bullet$ Generation. *Environ. Sci. Technol.* **2022**, *57*, 18538–18549. [CrossRef]

27. Huo, Z.Y.; Winter, L.R.; Wang, X.X.; Du, Y.; Wu, Y.H.; Hubner, U.; Hu, H.Y.; Elimelech, M. Synergistic Nanowire-Enhanced Electroporation and Electrochlorination for Highly Efficient Water Disinfection. *Environ. Sci. Technol.* **2022**, *56*, 10925–10934. [CrossRef]
28. Khelifa, A.; Aoudj, S.; Moulay, S.; Hecini, M.; De Petris-Wery, M. Degradation of EDTA by in-situ electrogenerated active chlorine in an electroflotation cell. *Desalination Water Treat.* **2009**, *7*, 119–123. [CrossRef]
29. Bonola, B.; Aguilar, D.; Fuentes-Camargo, I.; Sosa-Rodríguez, F.S.; Altamirano, R.H.; Luna-Sánchez, R.; Vazquez-Arenas, J. Implications to enhance sulfamethoxazole degradation using statistical optimization of either active chlorine concentration or ORP in an electrochemical reactor. *J. Environ. Chem. Eng.* **2020**, *8*, 104179. [CrossRef]
30. Xie, J.Z.; Zhang, C.Y.; Waite, T.D. Hydroxyl radicals in anodic oxidation systems: Generation, identification and quantification. *Water Res.* **2022**, *217*, 118425. [CrossRef]
31. Wojnarovits, L.; Takacs, E. Rate constants of dichloride radical anion reactions with molecules of environmental interest in aqueous solution: A review. *Environ. Sci. Pollut. Res.* **2021**, *28*, 41552–41575. [CrossRef]
32. Wang, Z.; Li, K.; Guo, J.; Yang, C.; Yan, L.; Feng, W.; Zhang, Y.; Wang, J. Elimination of pesticide from high salinity wastewater by electrochlorination process: Active chlorine species and scale-up performance. *Sep. Purif. Technol.* **2023**, *306*, 122572. [CrossRef]
33. Song, T.H.; Gao, Y.J.; Pang, S.Y.; Hu, R.H.; Li, H.B. Study of Fe_3O_4/PS System in Degrading BPA in Aqueous Solution. *J. Braz. Chem. Soc.* **2021**, *32*, 2264–2271. [CrossRef]
34. Chen, H.; Lin, T.; Wang, P.F.; Wang, Y.C.; Wei, W.; Zhu, S.G. A novel solar-activated chlorine dioxide process for atrazine degradation in drinking water. *Water Res.* **2023**, *239*, 120056. [CrossRef]
35. Xiang, Y.Y.; Fang, J.Y.; Shang, C. Kinetics and pathways of ibuprofen degradation by the UV/chlorine advanced oxidation process. *Water Res.* **2016**, *90*, 301–308. [CrossRef]
36. Lu, X.; Wu, L.; Liang, L.; Liu, D.; Chen, Y.; Zeng, Y.; Zhong, M.; Jia, B. Levofloxacin degradation by porous Co_x/CN activated peroxymonosulfate: Investigation of efficiency, mechanism, and degradation pathways. *J. Water Process Eng.* **2023**, *56*, 104427. [CrossRef]
37. Wen, X.-J.; Niu, C.-G.; Guo, H.; Zhang, L.; Liang, C.; Zeng, G.-M. Photocatalytic degradation of levofloxacin by ternary Ag_2CO_3/CeO_2/AgBr photocatalyst under visible-light irradiation: Degradation pathways, mineralization ability, and an accelerated interfacial charge transfer process study. *J. Catal.* **2018**, *358*, 211–223. [CrossRef]
38. Zhang, M.X.; Xing, M.Y.; Dong, B.; Sun, X.J.; Zhang, H.X.; Wang, C.L.; Zhu, H.X. Preparation of $BiVO_4/CO_3^{2-}$-$Bi_2O_2CO_3$ heterojunctions for enhanced photocatalytic activity in the degradation of levofloxacin under visible light. *J. Alloys Compd.* **2023**, *965*, 171471. [CrossRef]
39. Liu, X.H.; Liu, Y.; Lu, S.Y.; Wang, Z.; Wang, Y.Q.; Zhang, G.D.; Guo, X.C.; Guo, W.; Zhang, T.T.; Xi, B.D. Degradation difference of ofloxacin and levofloxacin by UV/H_2O_2 and UV/PS (persulfate): Efficiency, factors and mechanism. *Chem. Eng. J.* **2020**, *385*, 123987. [CrossRef]
40. Wu, J.; Li, J.B.; Owens, G.; Chen, Z.L. Toward green nano adsorbents and catalysts: Highly active Fe/Mn nanoparticles for enhanced oxidation of oxytetracycline and levofloxacin. *J. Colloid Interface Sci.* **2023**, *632*, 299–310. [CrossRef]
41. Fang, C.; Shao, C.; Wang, S.; Wu, Y.; Liu, C.; Huang, Q. Simultaneous removal of levofloxacin and sulfadiazine in water by dielectric barrier discharge (DBD) plasma: Enhanced performance and degradation mechanism. *Process Saf. Environ. Prot.* **2023**, *171*, 459–469. [CrossRef]
42. Peng, J.; He, Y.Y.; Zhang, Z.Y.; Chen, X.Z.; Jiang, Y.L.; Guo, H.; Yuan, J.P.; Wang, J.H. Removal of levofloxacin by an oleaginous microalgae Chromochloris zofingiensis in the heterotrophic mode of cultivation: Removal performance and mechanism. *J. Hazard. Mater.* **2022**, *425*, 128036. [CrossRef]
43. He, L.; Ma, Y.Y.; Lei, X.; Zhou, H.J.; Yuan, Y.; Du, W.T.; Liu, Z.L.; Miao, R.R.; Guan, Q.Q. Degradation of 4-chlorophenol through in situ generation of reactive chlorine species in a Zn_2CoO_x electrocatalytic system. *J. Environ. Chem. Eng.* **2024**, *12*, 111911. [CrossRef]
44. An, T.; Yang, H.; Song, W.; Li, G.; Luo, H.; Cooper, W.J. Mechanistic Considerations for the Advanced Oxidation Treatment of Fluoroquinolone Pharmaceutical Compounds using TiO_2 Heterogeneous Catalysis. *J. Phys. Chem. A* **2010**, *114*, 2569–2575. [CrossRef]
45. Mahjoore, M.; Honarmand, M.; Aryafar, A. Plant-based green fabrication of CuO-CdO-bentonite S-scheme heterojunction with enhanced photocatalytic performance for the degradation of levofloxacin. *Environ. Sci. Pollut. Res.* **2023**, *30*, 44439–44456. [CrossRef]
46. Kiki, C.; Rashid, A.; Zhang, Y.Q.; Li, X.; Chen, T.Y.; Adéoye, A.B.E.; Peter, P.O.; Sun, Q. Microalgal mediated antibiotic co-metabolism: Kinetics, transformation products and pathways. *Chemosphere* **2022**, *292*, 133438. [CrossRef]

processes

MDPI

Article

Adsorption of Rhodamine B in Aqueous Solution by Mg/CS Composite Modified Attapulgite

Mei Zhang, Hanjie Chen, Ziying Bu and Ying Fang *

College of Materials Science and Engineering, Nanjing Tech University, Nanjing 211816, China;
202261203270@njtech.edu.cn (M.Z.); 202261203281@njtech.edu.cn (H.C.); 202361203168@njtech.edu.cn (Z.B.)
* Correspondence: powder8817@njtech.edu.cn

Abstract: The safety of people's everyday water consumption has been gravely challenged by wastewater from printing and dyeing, however, research on effective contaminants removal from wastewater is encouraging. In this experiment, attapulgite modified with Mg^{2+} was calcined, and chitosan was added to create attapulgite composites. By refining the experimental parameters of the preparation, the adsorption performance of rhodamine B in wastewater was enhanced. A fully automated specific surface area and porosity investigation, a method known as X-ray diffraction, and nitrogen adsorption desorption equilibrium temperatures at 77 K were all performed. The original and composite attapulgite samples were evaluated using BET, Fourier transform infrared spectroscopy, and scanning electron microscopy. Composite materials were prepared using n (Mg^{2+})/m (AP) = 30 mmol/g, m (CS)/m (AP) = 1/3, a calcination temperature of 300 °C, and 1 h. After a series of adsorption experiments, manifesting that adding 0.2 g of attapulgite, at an initial concentration of 800 mg/L, pH of 4, temperature of 25 °C, and stirring for half an hour resulted in a maximum adsorption capacity of 325.73 mg/g. According to thermodynamic and kinetic equations, the pseudo-second-order kinetic theory is more consistent with the Langmuir adsorption isotherm (R^2 = 0.999). Upon examining the thermodynamic properties of ΔS, ΔH, and ΔG, it was found that the reaction is a spontaneous endothermic process that could potentially be utilized to eliminate Rh B.

Keywords: attapulgite; adsorption; rhodamine B; chitosan

1. Introduction

The dye manufacturing industry's fast expansion has led to an explosion in the discharge of dye effluent, which has gradually increased environmental pollution and posed a major threat to human health. Due to its complex chemical structure, high stability, and poor biodegradability, dyes are difficult to degrade through light, chemical, and biological treatment [1–3]. Adsorption is a beneficial remediation process that is used not only for bacterial inactivation but also for removing pollutants from water [4,5]. Rhodamine B (Rh B) is commonly used as a colorant in a variety of industrial processes, including papermaking, textile printing and dyeing, and lacquer pigments. Discharging waste into the water environment not only affects the aesthetics, but also reduces water clarity and endangers the survival of aquatic flora and fauna [6–8]. Meanwhile, a small number of azo dyes release carcinogens after interacting with human metabolites, which may pose a serious threat to human safety [9–12]. Thus, it is crucial to eliminate or minimize the presence of organic dyes in industrial wastewater before releasing it into the water.

Some of the commonly used treatment procedures for dyeing effluent to date are physical, chemical, and biological methods [13]. However, these techniques have some limitations, including high production costs, low operational efficiency, and synthetic design problems [14,15]. Among these methods, adsorption has become one of the primary approaches for treating organic dye wastewater because of its low processing cost, simplicity, and ability to reuse the adsorbent. It is considered the safest and most reliable method for wastewater treatment [16].

Attapulgite (AP) is a natural clay mineral rich in magnesium aluminum silicate, and it is non-toxic and inexpensive [17,18]. AP has multiple internal channels, a large surface area, and numerous active sites. Consequently, it is frequently employed as a cost-effective adsorbent for the removal of Rh B from wastewater. The structural characteristics of the material result in a limited removal efficiency of Rh B by AP. Thus, the removal rate of Rh B was improved through ion exchange and increasing interlayer spacing. The high ion exchange properties of AP permit the replacement of large-radius cations by metal ions with different valence states. Ions with a small radius, such as Mg^{2+}, are more readily able to enter the interlayer and exchange with Si^{4+}. Most academics have refined its adsorption performance through acid modification, salt modification, and organic modification. Wang et al. [19–21] used hydrochloric acid, nitric acid, silicotungstic acid, etc. to remove excess carbonate impurities in attapulgite and increase the specific surface area. Y. et al. [22–24] modified AP with metal salts to improve their ion exchange capacity. Song et al. [25–27] modified attapulgite by HDTMA and CTMAB to effectively remove phenol from wastewater. Chitosan (CS) is a product of the N-deacetylation of chitin. It has the advantages of abundant raw materials, non-toxicity, and biodegradability. Under acidic conditions, many amino and hydroxyl groups protonate positively in the CS molecular skeleton. Chitosan can be effectively loaded onto negatively charged AP surfaces, thereby enhancing their adsorption capacity [28–30]. Sun, WK et al. [31] prepared CS and polyvinyl alcohol (PVA) by co-mixing and crosslinking GO-ATP@CS-PVA adsorbent. The results showed that the maximum adsorption capacity for crystal violet was 136.06 mg/g. Zhou et al. [32,33] combined ternary grafted modified chitosan with attapulgite and achieved a paint removal rate of 89.3%. The single modification method can indeed improve the adsorption capacity of attapulgite, yet the efficiency is low. To overcome the disadvantages of low adsorption capacity and high cost of traditional adsorbents, it is necessary to develop new materials with improved properties. A low-cost, environmentally friendly, and recyclable high-efficiency adsorption material (CS-T-Mg-AP) was prepared using an inorganic organic modification method. The combination of $MgCl_2$, CS, and AP meets the requirements for industrial production. The impact of modification conditions on adsorption performance and mechanisms was investigated through characterization and isothermal adsorption experiments.

2. Materials and Methods

2.1. Materials

Attapulgite ($Si_8Mg_5O_{20}(OH)_2(OH_2)_4{\cdot}4H_2O$, 200 mesh) is from Dingbang Mineral Products Technology Co., Ltd. in Changzhou, Jiangsu Province, China; $MgCl_2{\cdot}6H_2O$, CS (medium viscosity, 200–400 MPs), and Rh B ($C_{28}H_{31}ClN_2O_3$, 479) were obtained from Myriad in Shanghai, China. CH_3COOH (Suzhou Taicang Hushi Reagent Co., Ltd., Taicang, China, >99.5% purity) and HCl (Yonghua Chemical Co., Ltd., Suzhou, China, high purity) were used. All chemical reagents, except for HCl, were of analytical grade.

2.2. Synthesis of Adsorbents

To start with, AP was acidified and pretreated with 3 M HCL. $MgCl_2$ was mixed with acidified attapulgite in a ratio of n (Mg^{2+})/m (AP) = 30 mmol/g. The mixture was then stirred magnetically at 25 °C for 2 h, washed, and dried (Mg-AP). The sample was then calcined at 300 °C for an hour (T-Mg-AP) using a double-temperature tubular furnace (TL1700-1400, Nanjing Boyuntong Instrument Technology Co., Ltd., Nanjing, China). A certain amount of CS was dissolved in 5% acetic acid solution under magnetic stirring at 60 °C. The calcined T-Mg-AP was then added to the CS solution while continuously stirring. This was then stirred for 3 h and centrifuged (4000 r/min) for 3 min. The precipitate was washed three times with water, dried at 105 °C for 2 h, ground, and sieved (200 mesh) to obtain a composite sample (CS-T-Mg-AP).

2.3. Adsorption Experiments

The adsorption capacity of CS-T-Mg-AP for dye Rh B under different conditions was tested. A specific amount of modified AP was added to a 100 mL diluted Rh B solution. The mixture was agitated with a magnetic bar for 30 min at 25 °C, allowed to stand still, and then centrifuged at 4000 rpm. The supernatant was collected, and its absorbance was determined using a UV spectrophotometer at 554 nm. This was calculated as the difference between the initial concentration of the solution of Rh B and the measurements to determine the optimum adsorption capacity q_e (mg/g). The most appropriate adsorption factors were discovered through investigation of the effects of pH (2–12), adsorbent dosage (0.5–5 g/L), adsorption the period (0–90 min), temperature (25–35 °C), and launching dye concentration (50–90 mg/L) on the method of adsorption in the ensuing batch experiments. When repeating the experiment, 0.2 g of AP with an initial concentration of 300 mg/L Rh B at pH of 4 was added, 25 °C, and stirred for half an hour. An amount of 0.1 M HCl was used as the eluent for Rh B dye, and the process was repeated five times. Each was repeated three times, and the mean was taken. The formula used is expressed as:

$$R = \left(\frac{C_0 - C_e}{C_O}\right) \quad (1)$$

$$q_e = \frac{(C_O - C_e)V}{M} \quad (2)$$

where C_0 (mg/L) is the original pigment concentration, C_e (mg/L) is the steady-state dye configuration, M (g) is the sorbent mass, and V (L) is the volume of the pigment solution.

2.4. Characterization

An X-ray diffractometer (XRD, D2 Phaser, Bruker, Bremen, Germany) was used to characterize the structure of the sample, which was scanned under the following conditions: Cu Kα radiation, voltage 40 kV, current 40 mA, scanning range 5~70°, rated power of 9 kW, scanning speed of 10°/min. Fourier transform infrared spectroscopy (FTIR, Nicolet iS20, Thermos Scientific, Waltham, MA, USA) has been implemented to detect variations in functional groups on the exterior of the adsorbent in the 400–4000 cm^{-1} wavenumber range. A fully automated analyzers for specific surface area and porosity, the BET (Micromeritics ASAP 2460, Malvern, PA, USA) was used to measure particle size and specific surface area. The weight change of the sample was measured using a thermogravimetric analyzer (TG). A field emission scanning electron microscope (FESEM, Thermos Fisher Scientific, Waltham, MA, USA) operating at an accelerating voltage of 3 kV was incorporated to investigate the surface morphology. XPS was tested using Thermo Scientific ESCALAB 250Xi from the United States.

3. Results and Discussion

3.1. Modified Attapulgite

Figure 1 illustrates how the removal rate and adsorption capacity of Rh B are trending upward with the addition of modifiers. Magnesium modification upped the removal rate of Rh B, which varied from 65% to 83%. Calcination stripped the water on the surface of Mg-AP, and the recovery rate was up by 6%. When CS was combined with T-Mg-AP, the adsorption site was extended, and the removal rate was elevated by 5%. As such, CS-T-Mg-AP was selected as the final modifier.

3.2. Specific Surface Area Analysis

The specific surface area of AP under different modification conditions is shown in Table 1. Acidification dissolves impure cement between attapulgite particles and carbonate impurities, dredging the pores. The specific surface area rises when a large-radius metal cation takes the place of H^+ in the structure. After calcination at 300 °C, some of the adsorbed water and crystal water were removed. The surface water molecular film was destroyed, leading to the breaking of chemical bonds inside the lattice and in the zeolite pores. The micropores broadened. CS is a type of polymer polysaccharide, with the majority being deposited on the surface of the AP, while a small portion enters the interlayer. This is beneficial for dye adsorption as it increases the average pore diameter and specific surface area.

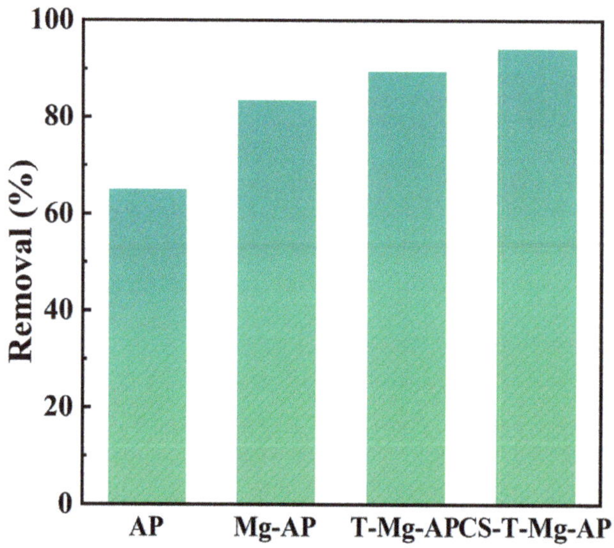

Figure 1. Comparison of adsorption performance of AP.

Table 1. BET surface area analysis of AP, T-Mg-AP, and CS-T-Mg-AP.

Adsorbate	BET Surface Area (m^2/g)	External Surface Area (m^2/g)	Micropore Volume (cm^3/g)	Average Pore Diameter (nm)
AP	114.709	78.667	0.016	5.5
T-Mg-AP	120.818	86.725	0.015	5.6
CS-T-Mg-AP	125.702	100.734	0.012	6.0

3.3. *TGA Analysis*

Both AP and CS-T-Mg-AP indicate a three-stage thermal decomposition process, as illustrated in Figure 2a. As the temperature raised to 520 °C, there was a 4.65% mass loss of AP, which mainly resulted by surface adsorbed water and water of crystallization detachment. Between 520 and 720 °C, AP showed a significant drop in weight of around 14%. This temperature range was thought to be linked to the structural collapse and breakdown of oxygen-containing functional groups in the pores. At 320 °C, the CS-T-Mg-AP composites and AP undergo a similar weight loss stage. The stability of the CS-T-Mg-AP structure was significantly better than that of AP. At 800 °C, the residuals of CS-T-Mg-AP and AP were 86.35% and 81.03%. So, this composite is very favorable for improving the stability of AP.

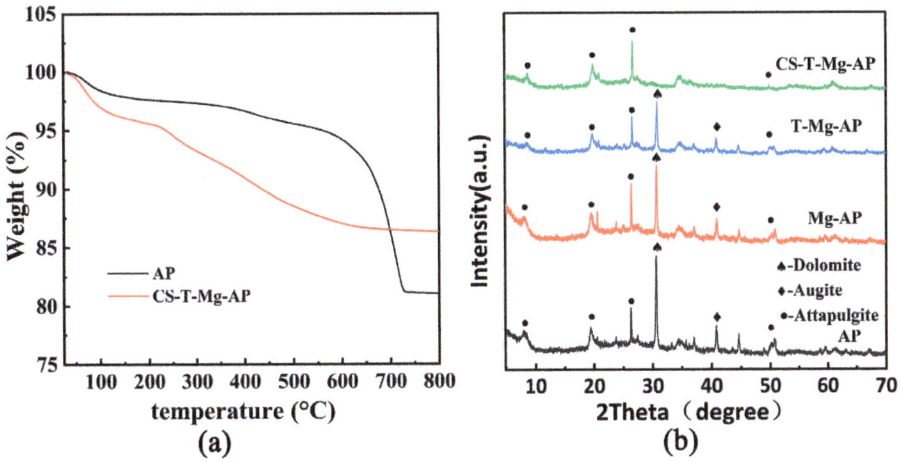

Figure 2. (**a**) TGA curves of AP and CS-T-Mg-AP; (**b**) XRD diffraction pattern of AP before and after modification.

3.4. XRD Analysis

XRD diffraction patterns of the AP both before and upon modification are reported in Figure 2b. It is evident that the AP typical diffraction peaks are situated at 2θ = 8.3°, 19.5°, 20.5°, 26.3°, and 34.6°, which are in line with the standard card (JCPDS No. 21-0958), corresponding to (110), (040), (231), and (440) crystal planes, respectively. The characteristic peaks of dolomite and augite appeared at 2θ = 30.6° and 2θ = 40.7°. The diffraction peaks of dolomite were stronger, and the AP was found to be impure. The characteristic peaks were weakened after modification, indicating successful loading of chitosan. The acid treatment and high temperature treatment dissolved and eliminated some impurities in AP, but the overall shape of the AP diffraction peaks did not change significantly, suggesting that the modification would not significantly affect its main structure. Magnesium binds to AP and enters the interlayer, enlarging the interlayer gap. The grain size of CS-T-Mg-AP is 19 nm. Using the Debye Scherrer equation to calculate [34–36]:

$$D = \frac{k\lambda}{\beta \cos\theta} \tag{3}$$

In the above formula, D represents the particle diameter, β indicates the peak width at half height, θ signifies the diffraction angle at the peak location, and λ is the wavelength of the X-ray (λ: 0.1540 nm).

3.5. SEM Analysis

SEM tests were carried out to assess the size and shape of the AP composites. From Figure 3a, it can be observed that the natural attapulgite exhibits a lamellar structure with distinct contour lines, a high density of particles, rough and porous surfaces, significant spaces between layers, pore channels, and uneven surfaces, resulting in a large surface area [37]. After modification (Figure 3b), CS was loaded onto the AP surface. This resulted in an irregular shape with a reduced degree of laminar flakiness and some collapsed holes, but the main structure remained unchanged.

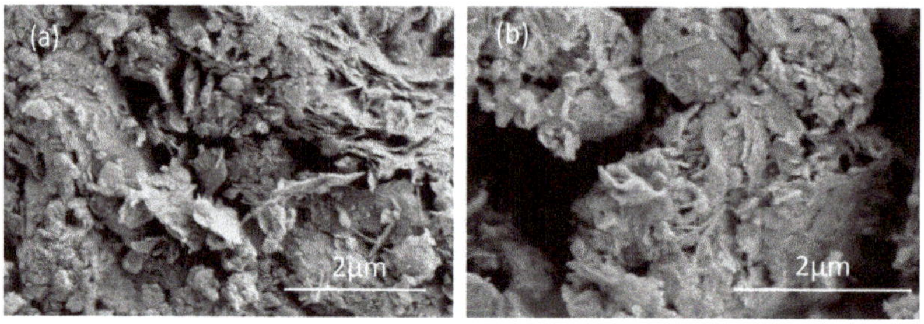

Figure 3. SEM photographs of the adsorbents: (**a**) AP; (**b**) CS-T-Mg-AP.

3.6. FTIR Analysis

FTIR spectra of AP, CS, and CS-T-Mg-AP adsorbents are pictured in Figure 4. The hydroxyl expansion and bending vibration peaks of water adsorbed on the AP surface were separated at 3410 cm^{-1} and 1630 cm^{-1}. The adduction peak at 1030 cm^{-1} represents the unbalanced Si-O bond stretching vibration peak in the Si-O tetrahedron, the Si-Fe bending vibration peak at 470 cm^{-1} and the Si-O-Mg bending vibration peak at 520 cm^{-1}. The signals at 2930 cm^{-1} and 2860 cm^{-1} are equivalent to the C-H stretching peaks in -CH_2 and -CH_3. After loading CS, the characteristic peaks of AP remained. What is more, the C-H telescopic vibration peak was increased at 1390 cm^{-1}, and the characteristic absorption peak of C=O was augmented at 1640 cm^{-1}. This means that the CS has been successfully loaded [38,39].

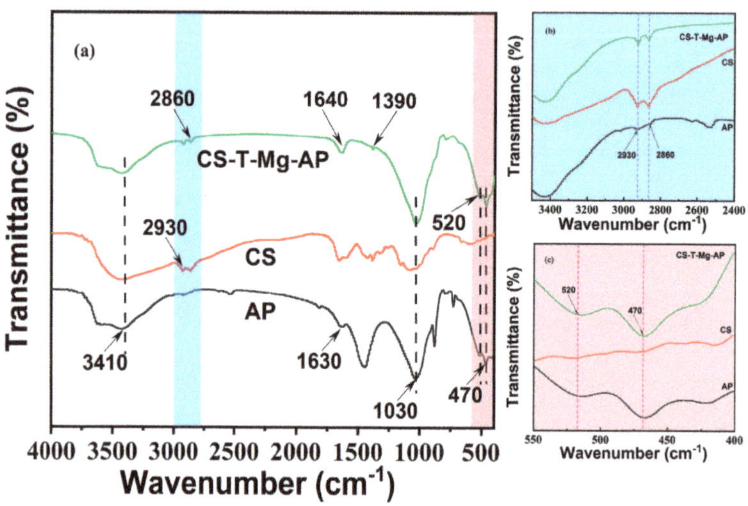

Figure 4. (**a**) FTIR plots of AP, CS, and CS-T-Mg-AP. (**b**,**c**) enlarged image of FTIR spectrum section.

3.7. XPS Analysis

The chemical composition and elemental valence states were characterized using X-ray photoelectron spectroscopy (XPS). Figure 5 depicts the deconvolution spectra of C1s, O1s, N1s, and Mg1s before and after CS-T-Mg-AP adsorption. Three peaks were observed in the C1s spectrum (Figure 5a) at 284.8 eV, 286.4 eV, and 287.3 eV, which are attributed to the C-C, C-O-H, and C-O-C bonds in CS [40,41]. Figure 5b illustrates the presence of two peaks at 399.6 eV and 402 eV, respectively, which can be attributed to the C-N and N-H bonds in the CS-T-Mg-AP system [42]. The presence of nitrogen and carbon elements confirms the successful combination of materials. This is consistent with the results of FTIR. In the CS-T-Mg-AP system, the Si^{4+} ion is replaced by the low-cost Mg^{2+} ion, leading to excellent adsorption performance of the material. After adsorption, the Mg 1s spectrum exhibits a significant decrease in the Mg-O peak after 5 h, suggesting that Mg^{2+} is involved in material adsorption and is positively associated with adsorption performance [43].

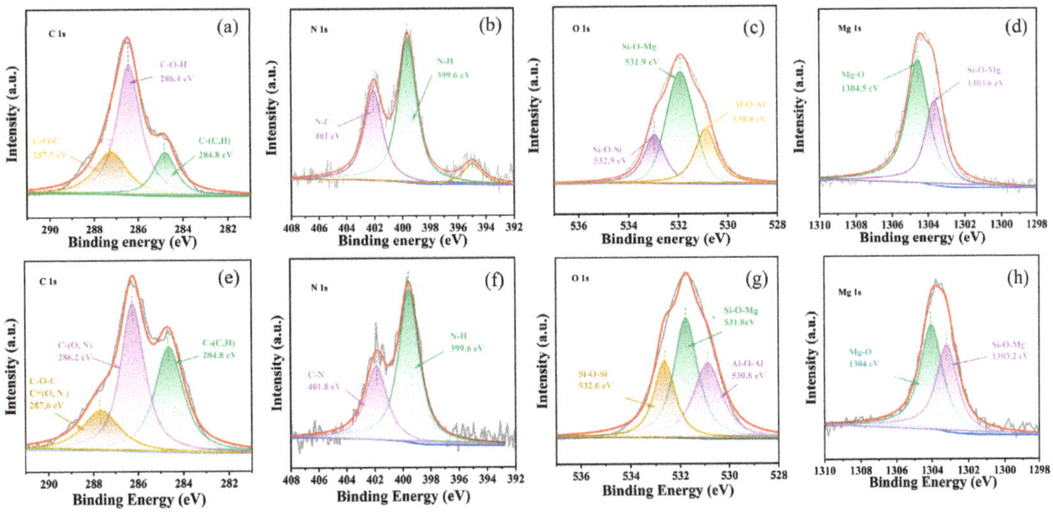

Figure 5. XPS analysis spectra of CS-T-Mg-AP before (**a–d**) and after (**e–h**) adsorption.

3.8. Analysis of Adsorption Results

3.8.1. Effect of Solution pH

The pH of the solution is a key element of the adsorption process and behavior. It affects adsorption results by influencing the surface-active site of the sorbent and the chemical properties of Rh B. At 25 °C and when the concentration of the Rh B solution was 300 mg/L, the removal rate and adsorption capacity of the dye declined as the pH of the solution changed from 2 to 12 (Figure 6a). Considering the cost and other factors, the optimal condition was at pH 4, and the removal rate of Rh B still exceeded 91%. The structure of Rh B is known to vary with pH. Under acidic conditions, Rh B is prone to generate tertiary amine and quaternary amine ions with positive charges, which neutralize with the negative charges on the AP surface and promote adsorption. The carboxyl group on the Rh B molecule ionizes to form rhodamine B in zwitterionic forms as the pH ascends. This produces competitive adsorption and a decrease in adsorption capacity. Under alkaline conditions, many OH^- and AP in solution create electrostatic repulsion. Rh B is also prone to neutralize with anions or OH^- in aqueous solutions to form uncharged neutral molecules and precipitate. This degrades the adsorption capacity. Similarly, the polymerization of Rh B molecules increases the difficulty of entering the AP, which also affects the adsorption efficiency [44].

3.8.2. Effect of Adsorbent Dosage

To investigate the interaction of adsorbates with adsorbent binding sites, it is essential to optimize the sorbent dosage (Figure 6b). Moreover, the ideal sorbent needs to be cost effective and capable of adsorbing large amounts of dye with a small amount of agent. Hence, the effect of adsorbent dosage on the percentage of Rh B absorption was assessed, all other conditions being held constant. The scores revealed that the removal of Rh B promoted from 60.49% to 91.14% with a reduced dose of modified AP. Furthermore, with a higher dosage of adsorbent, the removal of Rh B strengthened slowly until the saturation of surface-active sites reached equilibrium. The change in removal rate is due to the growth of the specific surface area and the number of surface active sites as the adsorbent dose is incremented [45].

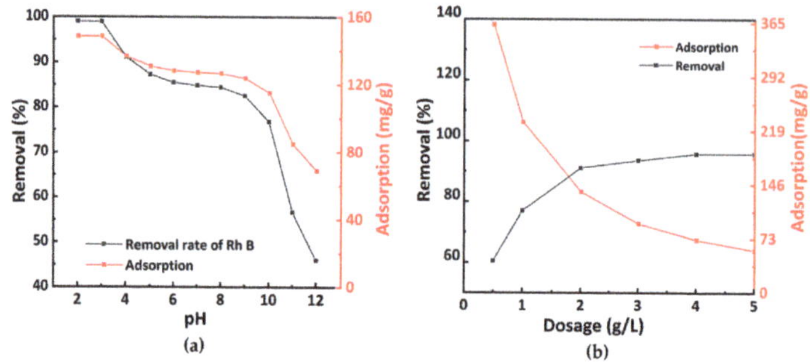

Figure 6. (**a**) Effect of pH on the adsorption effect of Rh B; (**b**) effect of adsorbent dosage on adsorption performance of Rh B.

3.8.3. Adsorption Isotherms

At different temperatures, the adsorption isotherm portrays the relationship between the amount of dye adsorbed per unit weight of adsorbent (q_e) and the equilibrium concentration of dye remaining in solution (C_e). Figure 7 illustrates the process of adsorption of isotherms of Rh B by CS-T-Mg-AP under the pre-optimized experiment at different temperatures (the start-up concentration of Rh B is anywhere from 50 mg/L to 800 mg/L). The amount of dye adsorbed was measured to depend on the ambient temperature and was found to be significantly higher at 25 °C than at 45 °C, demonstrating that the adsorption process is heat-absorbing [46]. The description of the adsorption isotherms was investigated using the Langmuir and Freundlich models.

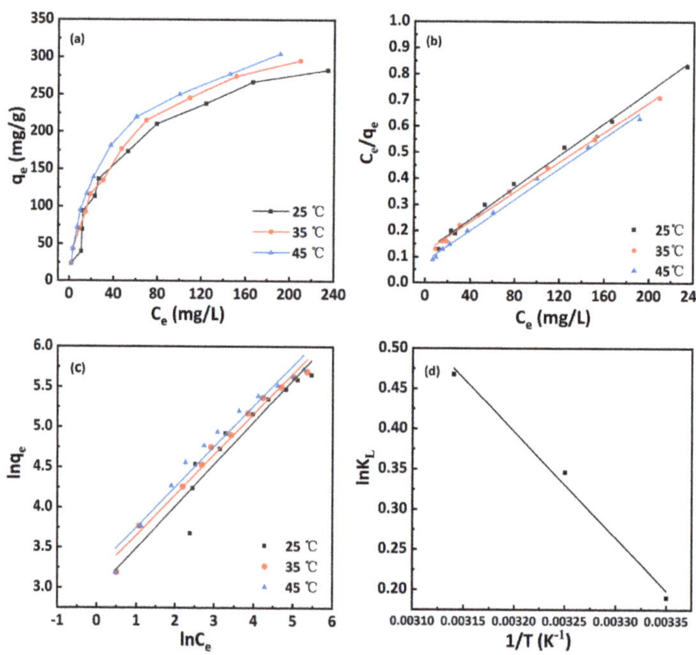

Figure 7. Thermodynamic diagram of adsorption of Rh B by AP composites; (**a**) isothermal adsorption line; (**b**) Langmuir isotherm; (**c**) Freundlich isotherm; and (**d**) plots of thermodynamic parameters.

The Langmuir model is an ideal monolayer admixing model where the adsorbent surface is homogeneous [47]. The Langmuir model has the following formula:

$$\frac{C_e}{q_e} = \frac{1}{q_m b} + \frac{C_e}{q_m} \tag{4}$$

The Langmuir coefficients q_m (mg/g) and b (L/mg) can be calculated from the intersection and slope of the linear plots of C_e/q_e and C_e, respectively, from which the average adsorption capacity can be derived.

The applicability of the Langmuir adsorption isotherm is determined by the separation factor or equilibrium parameter, R_L. It is derived from the following relationship:

$$R_L = \frac{1}{1 + bC_0} \tag{5}$$

where C_0 is the initial concentration of Rh B in solution (mg/L). Adsorption is optimized when the R_L equation is between 0 and 1.

The adsorption isotherms Freundlich model is derived from empirical equations, which assume that the surface of the sorbent is non-homogeneous. The adsorption capacity depends on the density of the adsorbate. The polarized mathematical equation is expressed as:

$$lnq_e = lnk_f + \frac{1}{n} \ln C_e \tag{6}$$

The status of k_f and $1/n$ are Freundlich coefficients (pertaining to the degree of adsorption and force) computed from the midpoint of the lnq_e and lnC_e curves. The Freundlich constant ($1/n$) likewise forecasts the compatibility of the adsorbent and the favorability of the technique for adsorption over a broad spectrum of adsorbent concentrations. If $1/n$ exceeds 1. 0, this adsorbent will only work for higher adsorbate concentrations and not for lower concentrations. Conversely, if $1/n$ is less than 1.0, the adsorbent is effective throughout the concentration range tested [48].

Figure 7b,c plots the Langmuir and Freundlich isotherms at each temperature. The angle of incidence and slope of the quadratic equations served to compute numerous Freundlich and Langmuir isotherm parameters, which are documented in Table 2. At all three temperatures, the association coefficients for the model with Langmuir fits were greater than 0.99, which was considerably greater than the values captured for the Freundlich model fit. Therefore, the Langmuir model fits with the isotherm files better. The maximum adsorption capacity varied from 325.73 mg/g to 350.87 mg/g at 25 and 45 °C.

Table 2. Langmuir and Freundlich constants for adsorption of Rh B on AP composites.

T (°C)	Langmuir Isotherm Constants				Freundlich Isotherm Constants		
	b (L/mg)	q_e (mg/g)	R^2	R_L	K_f ((mg/g) (L/mg))	1/n	R^2
25	0.0270	325.7329	0.9948	0.0467–0.4397	19.3699	0.5246	0.9261
35	0.0270	349.6053	0.9968	0.0500–0.4572	23.4001	0.5041	0.9827
45	0.0339	350.8772	0.9972	0.0356–0.3711	25.8430	0.5046	0.9637

Moreover, for all three temperatures examined, the separation factor (R_L) values spanned between 0 and 1. This verifies that the adsorption of Rh B on CS-T-Mg-AP is favorable. Additionally, the 1/n values for the three different temps were less than one, demonstrating how this composite can adsorb Rh B over the whole span of concentrations examined. Compared with other literature, the adsorption capacity of CS-T-Mg-AP in this paper is more excellent and the research prospect is broader, which is summarized in Table 3.

Table 3. Comparison between the present adsorption results and those already reported.

Adsorbent	Contaminant	Maximum Adsorption Capacity ($mg{\cdot}g^{-1}$)	Equilibrium Time (min)	Ref.
Cd-MOF/GO-2	Rh B	100.09	60	[49]
Mn/Ti/BC	Rh B	19.0607	90	[50]
α-Ag_2WO_4/SBA-15	Rh B	150	15	[51]
ZIF-8 derived porous carbon	Rh B	66	-	[52]
CS-T-Mg-AP	Rh B	325.73	30	This work

3.8.4. Thermodynamic Parameters

The van't Hoff equation was implemented to figure out the thermodynamic criteria:

$$lnk_f = \frac{\Delta S^0}{R} + \frac{-\Delta H^0}{RT} \quad (7)$$

$$\Delta G^0 = -RTlnk_f \quad (8)$$

$$K_L = \frac{q_e}{C_e} \quad (9)$$

where K_L denotes the divide of the coefficient. ΔS^0 is the level of entropy change. ΔH^0 is the enthalpy change. ΔG^0 is the free energy change. q_e remains the volume of Rh B retained per unit mass of adsorbent (mg/g). C_e is the equilibrium concentration (mg/L), T is the absolute temperature in terms of K, and R can be the gas constant (8.314 J/ (K mol)).

The slopes and intercepts of the ln K_L and $1/T$ linear plots (Figure 7d) can be leveraged to derive the values of ΔH^0 and ΔS^0. Table 4 reveals the computed values of the variation in enthalpy (ΔH^0), fluctuation in entropy (ΔS^0), and free alterations in energy (ΔG^0). The kinetic parameters of Rh B adsorption on the surface of CS-T-Mg-AP, k_1 and k_2, are positive, while ΔG^0 is negative. The values of ΔH^0 and ΔS^0 were 10.9765 KJ/mol and 0.0384 $kJ{\cdot}mol^{-1}{\cdot}K^{-1}$. The upward trending value of ΔH^0 clarifies that the adsorption is heat absorption. A positive value of ΔS^0 denotes an amplification of the disorder at the solid–liquid interface [53,54]. The ΔG^0 tumbled from −0.4699 at 25 °C to −1.2377 at 45 °C favoring the spontaneous reaction. The ΔG^0 value drops as the temperature goes up, which means that the higher the temperature, the better the adsorption.

Table 4. Thermodynamic parameters of Rh B adsorption on AP composites.

ΔG^0 (kJ/mol)			ΔH^0 (kJ/mol)	ΔS^0 ($kJ{\cdot}mol^{-1}{\cdot}K^{-1}$)
298.15 K	308.15 K	318.15 K		
−0.4699	−0.8868	−1.2377	10.9765	0.0384

3.8.5. Adsorption Kinetics

The effects of initial concentration and adsorption time on Rh B absorption are plotted in Figure 8. The dynamics of adsorption has been investigated at 50, 70, and 90 mg/L dye concentrations. Obviously, the pigment was absorbed relatively quickly, reaching equilibrium in about 20 min. In comparison to previous reported adsorbent systems, the time taken to establish adsorption equilibrium in this system was found to be extremely short [8,55]. The rapid equilibrium of adsorption may be due to the low concentration of the dye, which is effectively absorbed by the CS-T-Mg-AP. On the other hand, by making the surface larger and the diffusion path shorter, the distance between the adsorbent and the active center is lessened. Furthermore, many active sites in the composites are located on the outer surface, making them easily accessible to the adsorbed species.

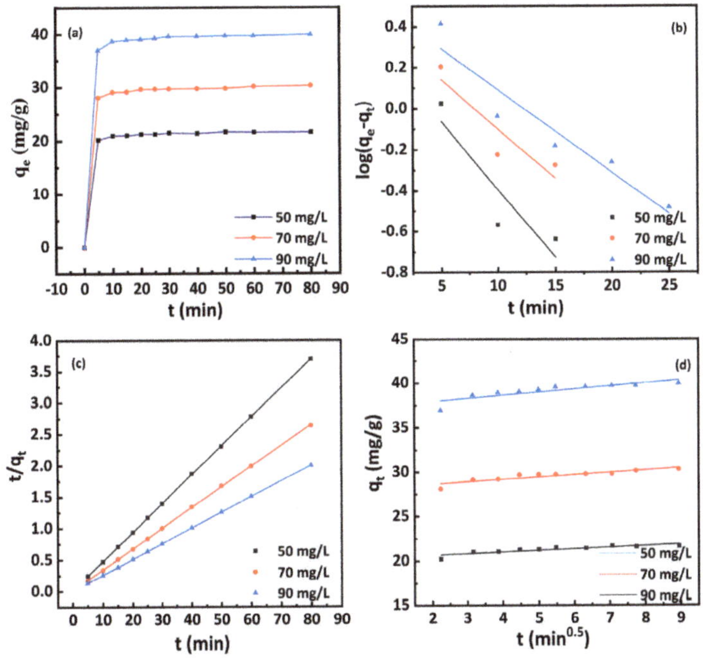

Figure 8. Plots of (**a**) kinetic data, (**b**) pseudo-primary model, (**c**) pseudo-secondary model, and (**d**) intraparticle diffusion model for adsorption of Rh B dye by AP composites.

The strong correlation between the experimental and kinetic data elucidates the kinetic mechanism. A kinetic analysis of the Rh B adsorption process on CS-T-Mg-AP was conducted implementing the pseudo first order, pseudo second order, and the Intraparticle diffusion [48,56]. The kinetic pseudo-first-order equation is denoted by Equation (10):

$$log(q_e - q_t) = logq_e - \frac{tk_1}{2.303} \tag{10}$$

where q_t (mg/g) is the dye digestion at time t, q_e is the state of equilibrium capacity for absorption (mg/g), and k_1 (1/min) is the pseudo-primary velocity constant. The intercept and slope of the curve for k_1 and q_e were incorporated for figuring out the operates log ($q_e - q_t$) as an indicator of t (Figure 8b). The adsorption capacity, rate constant, and correlation coefficient are calculated and are given in Table 5. Additionally, it was noted that there was a discrepancy within the laboratory measurements of q_e and the q_e values derived from the application of the pseudo-first-order velocity model. The pseudo-first-order rate model is inadequate to adequately explain the removal of Rh B by CS-T-Mg-AP, confirmed by the lower values of R^2 and q_e.

Table 5. Table of different kinetic parameters obtained from kinetic model fitting.

C_0 (mg/L)	q_e Experimental (mg/g)	Pseudo First Order			Pseudo Second Order			Intraparticle Diffusion		
		k_1 (1/min)	q_e Calculated (mg/g)	R^2	k_1 (1/min)	q_e Calculated (mg/g)	R^2	k_i (mg/g·min$^{0.5}$)	C	R^2
50	21.27	0.1527	1.8578	0.8271	0.1027	21.7628	0.9999	0.1793	20.2860	0.7517
70	29.71	0.1103	2.4025	0.8288	0.0566	30.3048	0.9991	0.2638	28.1232	0.7999
90	39.58	0.0929	3.1362	0.9110	0.0529	40.0565	0.9998	0.3459	37.2564	0.7225

The remaining equations symbolize the pseudo-second-order dynamics model's integral form:

$$\frac{t}{q_t} = \frac{1}{k_2 q_e^2} + \frac{t}{q_e} \tag{11}$$

where k_2 (g/ (mg min)) is the pseudo-second-order velocity variable. The points of contact and charts of t/q_t and t are employed for calculating k_2 and q_e. The fit of the measured information at multiple dye levels to the pseudo-second-order hypothesis is seen in Figure 8c and shows excellent linear results. For all three temperatures studied, the R^2 value is 0.999. The pseudo-second-order mechanism most accurately sums up the trajectory of adsorption of the adsorbed Rh B on the external layer of the CS-T-Mg-AP, and the estimated q_e values are in good alignment with the empirical findings.

There are several stages of adsorption, such as the diffusion of the dye from the aqueous phase to the surface of the adsorbent and from the interface to the interior of the pore size. The intraparticle propagation approach was implemented to pinpoint adsorption's restricting stage. The most popular model for the dispersion is the intraparticle model, which was put forth by Weber and Morris [57]. It can be summarized essentially as follows:

$$q_t = k_i t^{0.5} + C \tag{12}$$

where k_i (mg/ (g min$^{0.5}$)) is the metabolic constant. If the relationship curve between q_t and $t^{0.5}$ is a straight line and passes through the origin, the adsorption process is controlled by intraparticle diffusion. Three distinct dye concentrations' q_t against $t^{0.5}$ graphs are depicted in Figure 8d. The correlation coefficient values obtained from the plot of q_e vs. t are poor. The line does not pass through the zero point, which means that intraparticle diffusion is not the only rate-limiting factor in the kinetic process. Table 5 lists several distinct parameters related to kinetics that were discovered while fitting the momentum framework.

3.9. Recycling and Stability of CS-T-Mg-AP

Excellent adsorption capacity and high stability are particularly important for adsorbents. Figure 9 presents the adsorption desorption cycle experiment of CS-T-Mg-AP. An amount of 0.1 M HCL was used as the eluent of Rh B dye. After five cycles, the removal rate of CS-T-Mg-AP still reached more than 80%, which revealed that it had good stability and recyclability. With the loss of clay mass and the decrease of adsorption sites in the experiment, the adsorption capacity of CS-T-Mg-AP gradually tends to be saturated. On the whole, it still has a high reusability.

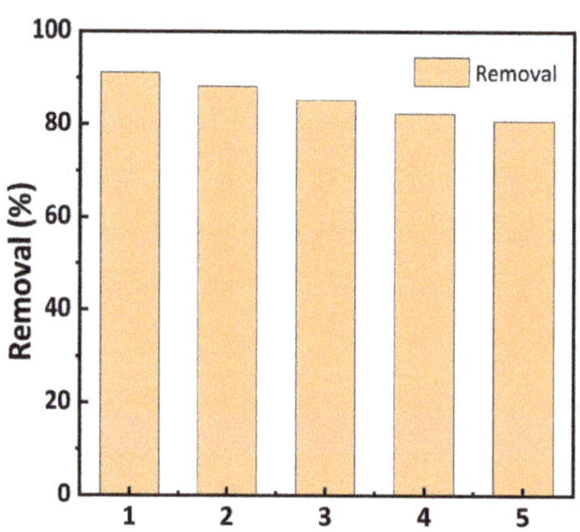

Figure 9. CS-T-Mg-AP removes rhodamine B in five adsorption desorption cycles.

3.10. Adsorption Mechanism

As is well known, the adsorption capacity depends on factors such as specific surface area, adsorption sites, and the interaction between the adsorbent and Rh B. Thermodynamic equation parameters indicate that the process is a spontaneous endothermic reaction. The adsorption effect is enhanced as the temperature increases. The Langmuir equation fits well with the kinetic model and confirms that it belongs to single-layer adsorption. Acidification treatment increased the specific surface area of AP. The addition of Mg^{2+} increases the interlayer distance, and chitosan molecules adsorb onto the surface of attapulgite to enhance the adsorption performance. Compared with attapulgite, CS-T-Mg-AP offers a significant number of adsorption sites, which enhance the adsorption process (Figure 10).

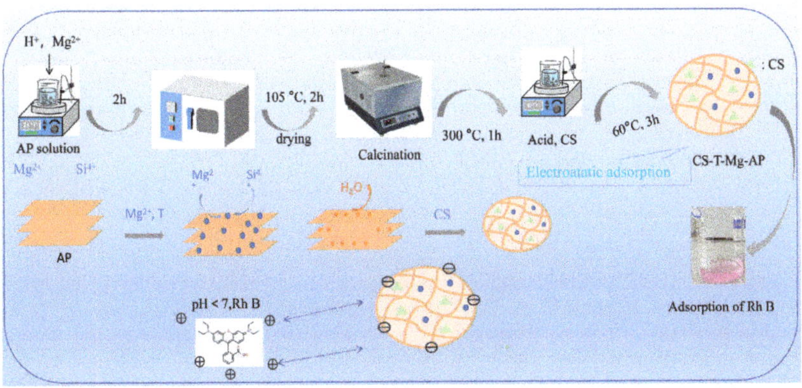

Figure 10. Diagram of the synthesis process of CS-T-Mg-AP and its adsorption mechanism for Rh B.

4. Conclusions

In this study, the successful preparation of CS-T-Mg-AP was demonstrated. At 25 °C, 0.2 g of AP, with an initial concentration of 800 mg/L and a pH of 4, was stirred for half an hour. The maximum adsorption capacity for Rh B reached 325.73 mg/g, which is a significant improvement over the adsorption capacity of most AP-based adsorbents. By fitting kinetic, thermodynamic linear, and nonlinear equations, it was ultimately determined that the adsorption process conforms to the Langmuir isotherm model and pseudo second-order kinetic model. Furthermore, the material exhibits high stability and reusability in repeated experiments. Overall, CS-T-Mg-AP is a multifunctional adsorbent integrating green synthesis, strong adsorption capacity, and environmental friendliness.

Author Contributions: Conceptualization, M.Z. and Y.F.; methodology, M.Z.; software, M.Z.; validation, M.Z.; formal analysis, M.Z. and Z.B.; investigation, M.Z. and H.C.; resources, Y.F.; data curation, M.Z.; writing—original draft preparation, M.Z.; writing—review and editing, M.Z. and Y.F.; visualization, M.Z.; supervision, H.C.; project administration, M.Z.; funding acquisition, Y.F. All authors have read and agreed to the published version of the manuscript.

Funding: This work was supported by the Priority Academic Program Development of Jiangsu Higher Education Institutions (PAPD).

Data Availability Statement: The original contributions presented in the study are included in the article, further inquiries can be directed to the corresponding author.

Conflicts of Interest: The authors declare no conflicts of interest.

References

1. Cui, H.; Xu, Y.; Sun, H. Application of MXene in dye wastewater treatment. *New Chem. Mater.* **2022**, *50*, 270–273.
2. Qin, B.; Gu, J.; Yin, P.; Zhang, Y. Research progresses on dye wastewater treatment technology. *Environ. Prot. Chem. Ind.* **2021**, *41*, 9–18.
3. Chen, S.; Zhang, M.; Chen, H.; Fang, Y. Removal of Methylene Blue from Aqueous Solutions by Surface Modified Talc. *Materials* **2023**, *16*, 3597. [CrossRef]
4. Barreca, S.; Oliveri, I.P.; Lo Presti, F.; Oliveri, V.; Giannakis, S.; Tuccitto, N.; Spampinato, V.; Auditore, A. An innovative "Up-and-Down" adsorption processes for pyrene removal from acid wastewater as new approach in water remediation. *Sep. Purif. Technol.* **2024**, *348*, 127516. [CrossRef]
5. Barreca, S.; Velez Colmenares, J.J.; Pace, A.; Orecchio, S.; Pulgarin, C. Escherichia coli inactivation by neutral solar heterogeneous photo-Fenton (HPF) over hybrid iron/montmorillonite/alginate beads. *J. Environ. Chem. Eng.* **2015**, *3*, 317–324. [CrossRef]
6. Zhang, L.; Zhou, G.; Liu, Y.; Fu, Y. Degradation of Rhodamine B in Wastewater by Zero-valent Cobalt Activated Peroxymonosulfate. *Environ. Sci. Technol.* **2022**, *45*, 101–107.
7. He, L.; Yang, Y.; Xu, C.; Ye, Z.; Liu, S. Catalytic Degradation of Rhodamine B in Wastewater by Goethite. *Environ. Prot. Chem. Ind.* **2008**, *28*, 396–400.
8. He, M.; Lü, R.; Song, M.; Wang, Y.; Wang, K.; Yao, S. Synthesis and Catalytic Performance of Zeolite Using Attapulgite from Gansu. *Non-Met. Mines* **2022**, *45*, 1–5.
9. Qi, P.; Wen, Z.; Wang, G. Progress in Semiconductor Titanium Dioxide Photocatalyzed Degradation of Azo Dyes. *Spec. Petrochem.* **2007**, *24*, 75–79.
10. Dong, Z.; Liu, L.; Hao, Y.; Jiang, J.; Liu, W. Research Progress on the Treatment of Azo Dye Containing Wastewater. *Technol. Water Treat.* **2017**, *43*, 6–10.
11. Cheng, A.; Zhang, J. Research on treatment of azo dye wastewater by advanced reduction process. *Mod. Chem. Ind.* **2019**, *39*, 112–115.
12. Wu, Y.; Chu, J.; Wu, C.; Li, L.; Liu, Z. Comparison between treatment of azo dyeing wastewater by two biological reactors. *Technol. Water Treat.* **2007**, *33*, 41–44.
13. Shabir, M.; Yasin, M.; Hussain, M.; Shafiq, I.; Akhter, P.; Nizami, A.-S.; Jeon, B.-H.; Park, Y.-K. A review on recent advances in the treatment of dye-polluted wastewater. *J. Ind. Eng. Chem.* **2022**, *112*, 1–19. [CrossRef]
14. Ren, N.; Zhou, X.; Guo, W.; Yang, S. A review on treatment methods of dye wastewater. *J. Chem. Ind. Eng.* **2013**, *64*, 84–94.
15. Cui, Y.; Yin, R. The progress of treatment methods of dye wastewater. *Sci. Technol. Rev.* **2021**, *39*, 79–87.
16. Lu, Z.; Shen, L.; Zhang, Q. Research development of technics and mechanism of dye wastewater treatment by adsorption. *Ind. Water Treat.* **2004**, *24*, 13–16.
17. Yu, Z.; Lu, Z.; Zhao, X.; Li, C.; Pan, J.; Yan, Y. Attapulgite Modification and Its Application in Environmental Water Treatment. *Bull. Chin. Ceram. Soc.* **2010**, *29*, 1367–1372.
18. Wang, C.; Cao, J.; Xie, K.; Shang, K. Research progress of the modification of attapulgite and its application in water treatment. *Appl. Chem. Ind.* **2016**, *45*, 1575–1578.
19. Shi, D.; Qiao, R.; Xu, Q. Preparation of Acid Modified Attapulgite and Its Performance of Mercury Removal. *Chin. J. Synth. Chem.* **2015**, *23*, 720–724.
20. Huang, J.; Song, Z.; Jin, Q.; Liu, Y.; Wang, X. Study on the adsorption of tannic acid on palygorskite. *Sci. Technol. Food Ind.* **2010**, *31*, 255–257, 260.
21. Wang, S.; Chen, J.; Chen, T.; Wei, C. Treatment technology of phenol-containing wastewater with modified attapulgite. *Ind. Water Treat.* **2012**, *32*, 55–58.
22. Wang, H.; Zhang, D.; Zhao, Y.; Xie, M. Cationic surfactant modified attapulgite for removal of phenol from wastewater. *Colloids Surf. A-Physicochem. Eng. Asp.* **2022**, *641*, 128479. [CrossRef]
23. Wang, B.; Li, M.; Xu, H.; Wang, J.; He, L. Preparation of attapulgite loaded with ferrihydrite adsorbent and its adsorption propreties of As(V). *Environ. Chem.* **2014**, *33*, 656–662.
24. Li, Y.; Li, W.; Yao, C.; Chen, Z.; Zhang, Y.; Wang, W. Attapulgite Modified by Quaternary Ammonium Salt and Its Application to Oily Waste Water. *Non-Met. Mines* **2010**, *33*, 58–60.
25. Song, Y.; Wang, H.; Zhu, Z.; Sun, J. Preparation and characterization of modified attapulgite and its adsorption to 4-chlorophenol. *Chin. J. Environ. Eng.* **2012**, *6*, 4481–4486.
26. Zhang, P.; Shi, Y.; Zhang, G.; Lv, C.; Qin, Y.; Zhao, Y.; Chen, L. Preparation and adsorption of Cd^{2+} of attapulgite modified by AMPS. *J. Funct. Mater.* **2016**, *47*, 9085–9090.
27. Lei, C.; Li, Y.; Yao, Q.; Tang, L.; Zhang, F. Dynamic adsorption study of micro-polluted phenol wastewater with microwave modified attapulgite. *Technol. Water Treat.* **2012**, *38*, 90–93.
28. Xu, H.; Zhang, Y.; Cheng, Y.; Tian, W.; Zhao, Z.; Tang, J. Polyaniline/attapulgite-supported nanoscale zero-valent iron for the rival removal of azo dyes in aqueous solution. *Adsorpt. Sci. Technol.* **2019**, *37*, 217–235. [CrossRef]
29. Peng, F.; Guo, H.; Zhang, H.; Wang, C.; Yang, H.; Chen, X. Study on the decoloriZation of corn straW dilute acid hydrolysate with thermal modified attapulGite. *Appl. Chem. Ind.* **2014**, *43*, 2149.

30. Teng, Y.; Jiang, Z.; Yu, A.; Yu, H.; Huang, Z.; Zou, L. Optimization of preparation parameters for environmentally friendly attapulgite functionalized by chitosan and its adsorption properties for Cd(2). *Environ. Sci. Pollut. Res. Int.* **2021**, *28*, 44064–44078. [CrossRef]
31. Sun, W.; Ou, H.; Chen, Z. Study on Preparation of Chitosan/Polyvinyl Alcohol Aerogel with Graphene-Intercalated Attapulgite(GO-ATP@CS-PVA) and Adsorption Properties of Crystal Violet Dye. *Nanomaterials* **2022**, *12*, 3931. [CrossRef]
32. Hu, S.; Su, Y.; Tian, D.; Zhang, S. Preparation and characterisation of chitosan/attapulgite nanocomposites. *New Chem. Mater.* **2013**, *41*, 149–151.
33. Zhou, J.; Dong, R.; Dong, Y.; Fu, P. Flocculation of paint wastewater by composite of modified chitosan and attapulgite. *Chin. J. Environ. Eng.* **2014**, *8*, 4674–4680.
34. Moslehi, M.H.; Zadeh, M.S.; Nateq, K.; Shahamat, Y.D.; Khan, N.A.; Nasseh, N. Statistical computational optimization approach for photocatalytic-ozonation decontamination of metronidazole in aqueous media using $CuFe_2O_4/SiO_2/ZnO$ nanocomposite. *Environ. Res.* **2024**, *242*, 117747. [CrossRef]
35. Moslehi, M.H.; Eslami, M.; Ghadirian, M.; Nateq, K.; Ramavandi, B.; Nasseh, N. Photocatalytic decomposition of metronidazole by zinc hexaferrite coated with bismuth oxyiodide magnetic nanocomposite: Advanced modelling and optimization with artificial neural network. *Chemosphere* **2024**, *356*, 141770. [CrossRef]
36. Azqandi, M.; Nateq, K.; Amarzadeh, M.; Yoosefian, M.; Yaghoot-Nezhad, A.; Ahmad, A.; Ramavandi, B.; Nasseh, N. Intensified photo-decontamination of tetracycline antibiotic by S-scheme spinel manganese ferrite-grafted ZIF-8 heterojunction photocatalyst: Mechanism conception, degradation pathway and DFT studies. *J. Environ. Chem. Eng.* **2024**, *12*, 112875. [CrossRef]
37. Li, M.; Wu, Z.; Kao, H. Study on preparation, structure and thermal energy storage property of capric–palmitic acid/attapulgite composite phase change materials. *Appl. Energy* **2011**, *88*, 3125–3132. [CrossRef]
38. Guo, X.; Qiu, F.; Dong, K.; Rong, X.; He, K.; Xu, J.; Yang, D. Preparation and application of copolymer modified with the palygorskite as inhibitor for calcium carbonate scale. *Appl. Clay Sci.* **2014**, *99*, 187–193. [CrossRef]
39. Barreca, S.; Colmenares, J.J.V.; Pace, A.; Orecchio, S.; Pulgarin, C. Neutral solar photo-Fenton degradation of 4-nitrophenol on iron-enriched hybrid montmorillonite-alginate beads (Fe-MABs). *J. Photochem. Photobiol. A Chem.* **2014**, *282*, 33–40. [CrossRef]
40. Zhang, Y.; Zhu, L.; Chen, L.; Liu, L.; Ye, G. Influence of Magnesia on Demoulding Strength of Colloidal Silica-Bonded Castables. *Rev. Adv. Mater. Sci.* **2019**, *58*, 32–37. [CrossRef]
41. Dabrowska, S.; Migdalski, J.; Lewenstam, A. A Breakthrough Application of a Cross-Linked Polystyrene Anion-Exchange Membrane for a Hydrogencarbonate Ion-Selective Electrode. *Sensors* **2019**, *19*, 1268. [CrossRef]
42. Zhang, G.; Li, S.; Shuang, C.; Mu, Y.; Li, A.; Tan, L. The effect of incorporating inorganic materials into quaternized polyacrylic polymer on its mechanical strength and adsorption behaviour for ibuprofen removal. *Sci. Rep.* **2020**, *10*, 5188. [CrossRef] [PubMed]
43. Aboutalebianaraki, N.; Neal, C.J.; Seal, S.; Razavi, M. Biodegradable Mg-Sc-Sr Alloy Improves Osteogenesis and Angiogenesis to Accelerate Bone Defect Restoration. *J. Funct. Biomater.* **2022**, *13*, 261. [CrossRef] [PubMed]
44. Shi, C.; Feng, J.; Peng, S.; Hu, Z. Adsorption propreties of Rhodamine B by activater carbon fiber in aqueous solutions. *Environ. Chem.* **2013**, *32*, 394–401.
45. Sahar, J.; Naeem, A.; Farooq, M.; Zareen, S.; Farida; Sherazi, S. Kinetic studies of graphene oxide towards the removal of rhodamine B and congo red. *Int. J. Environ. Anal. Chem.* **2019**, *101*, 1258–1272. [CrossRef]
46. Mohan, D.; Singh, K.P.; Singh, V.K. Removal of hexavalent chromium from aqueous solution using low-cost activated carbons derived from agricultural waste materials and activated carbon fabric cloth. *Ind. Eng. Chem. Res.* **2005**, *44*, 1027–1042. [CrossRef]
47. Hameed, B.H. Grass waste: A novel sorbent for the removal of basic dye from aqueous solution. *J. Hazard. Mater.* **2009**, *166*, 233–238. [CrossRef]
48. Chiou, M.S.; Li, H.Y. Equilibrium and kinetic modeling of adsorption of reactive dye on cross-linked chitosan beads. *J. Hazard. Mater.* **2002**, *93*, 233–248. [CrossRef]
49. Han, X.; Wang, L.-Y.; Geng, F.-J.; Xi, G.-Q. Adsorption of rhodamine B by benzimidazole-based metal-organic framework/graphene oxide composites. *Chin. J. Inorg. Chem.* **2023**, *39*, 1159–1168. [CrossRef]
50. Yang, J.; Wei, Q.; Li, D.; Yu, J.; Cai, Z. Study on adsorption performance of manganese titanium oxide composite biochar for removal of Rhodamine B. *J. Indian Chem. Soc.* **2023**, *100*, 100958. [CrossRef]
51. Silva, F.C.M.; Silva, L.K.R.; Santos, A.G.D.; Caldeira, V.P.S.; Cruz-Filho, J.F.; Cavalcante, L.S.; Longo, E.; Luz, G.E., Jr. Structural Refinement, Morphological Features, Optical Properties, and Adsorption Capacity of α-Ag_2WO_4 Nanocrystals/SBA-15 Mesoporous on Rhodamine B Dye. *J. Inorg. Organomet. Polym. Mater.* **2020**, *30*, 3626–3645. [CrossRef]
52. Li, K.; Yu, L.; Cai, J.; Zhang, L. C@TiO_2 core-shell adsorbents for efficient rhodamine B adsorption from aqueous solution. *Microporous Mesoporous Mater.* **2021**, *320*, 111110. [CrossRef]
53. Zhang, X.; Li, F.; Zhao, X.; Cao, J.; Liu, S.; Zhang, Y.; Yuan, Z.; Huang, X.; De Hoop, C.F.; Peng, X.; et al. Bamboo Nanocellulose/Montmorillonite Nanosheets/Polyethyleneimine Gel Adsorbent for Methylene Blue and Cu(II) Removal from Aqueous Solutions. *Gels* **2023**, *9*, 40. [CrossRef] [PubMed]
54. Ma, Y.L.; Xu, Z.R.; Guo, T.; You, P. Adsorption of methylene blue on Cu(II)-exchanged montmorillonite. *J. Colloid Interface Sci.* **2004**, *280*, 283–288. [CrossRef] [PubMed]
55. Zhou, P.; Dai, Z.; Lu, T.; Ru, X.; Ofori, M.A.; Yang, W.; Hou, J.; Jin, H. Degradation of Rhodamine B in Wastewater by Iron-Loaded Attapulgite Particle Heterogeneous Fenton Catalyst. *Catalysts* **2022**, *12*, 669. [CrossRef]

56. Khamizov, R.K. A Pseudo-Second Order Kinetic Equation for Sorption Processes. *Russ. J. Phys. Chem. A* **2020**, *94*, 171–176. [CrossRef]
57. Weber Walter, J.; Morris, J.C. Kinetics of Adsorption on Carbon from Solution. *J. Sanit. Eng. Div.* **1963**, *89*, 31–59. [CrossRef]

water

MDPI

Article

CTAB Surfactant Promotes Rapid, Efficient, and Simultaneous Removal of Cationic and Anionic Dyes through Adsorption on Glycerol/Citrate Polyester

Andrés F. Chamorro [1,*], Tulio A. Lerma [2,3] and Manuel Palencia [3,*]

1 Research Group of Electrochemistry and Environment (GIEMA), Faculty of Basic Sciences, Universidad Santiago de Cali, Cali 760035, Colombia
2 Mindtech Research Group (Mindtech-RG), Mindtech SAS, Cali 760026, Colombia; t.lerma@mindtech.com.co
3 Research Group in Science with Technological Applications (GICAT), Department of Chemistry, Faculty of Natural and Exact Science, Universidad del Valle, Cali 760042, Colombia
* Correspondence: andres.chamorro03@usc.edu.co (A.F.C.); manuel.palencia@correounivalle.edu.co (M.P.); Tel.: +57-2-5183000 (A.F.C.); +57-3205271934 (M.P.)

Abstract: The discharge of dye waste by industries has caused environmental impacts on water properties, aquatic species, and human life. The production of eco-friendly materials for dye removal from wastewater has gained increasing interest, particularly through adsorption, as it is an efficient method for removing pollutants. However, the selectivity and limited adsorption capacity of materials for simultaneously adsorbing cationic and anionic dyes make it challenging to address the prevailing dye effluent issue. In this work, a poly(glycerol citrate) polyester was prepared without a solvent, using microwave heating, and was combined with cetyltrimethylammonium bromide (CTAB) to modulate the simultaneous adsorption of cationic (Rhodamine B, RB) and anionic dyes (acid blue 113, AB and acid red 27, AR). The formation of the polyester was successfully confirmed by ATR-FTIR and the thermal properties were evaluated by TGA and DCS. In the presence of CTAB 5 mM, the material removes almost 100% of the dyes, reaching the kinetics equilibrium in 30 min following the Pseudo-second order model. Additionally, dye adsorption on the polyester in the presence of CTAB was described by the Freundlich isotherm model, indicating a heterogeneous polyester surface that promotes a multi-layer adsorption driven by electrostatic interaction and hydrogen bonds. The material showed an adsorption capacity of 73.7 ± 3.2, 82.1 ± 4.4, and 21.2 ± 0.6 mg/g for RB, AB, and AR, respectively. Our results provide evidence that the poly(glycerol citrate)/CTAB has a higher potential for application in wastewater treatment to remove both anionic and cationic dyes.

Keywords: adsorption; cationic and anionic dyes; citric acid; glycerol; biodegradable polyester

1. Introduction

Organic and synthetic dyes play a crucial role in several industries, including textiles, food, cosmetics, and pharmaceutical industries, among others. Currently, there are over 1000 different dyes and pigments used in countless industrial formulations, with the market projected to reach approximately $15 billion by 2026 [1]. However, the extensive use of dyes and pigments contributes to environmental contamination, particularly in water sources. The textile and dyeing industries are the major contributors to dye contamination in effluents, accounting for approximately 54% and 21% of the presence of dyes in wastewater, respectively [2–4]. For instance, in the textile industry, about 10–15% of the synthetic dyes used are lost and discharged into water sources, such as rivers and lakes, without adequate treatment [5,6]. Even low concentrations of dyes in water (1 mg/L) can have a significant impact on the aquatic environment, as these pigments can absorb and reflect sunlight, affecting the development of aquatic species and plants. Additionally, dyes are resistant to thermal degradation and light in natural conditions, making them persistent pollutants in wastewater [2,7]. Typically, dyes contain aromatic, azo, carboxyl, and sulphonated groups, which are highly toxic to human health and can cause skin issues, cardiovascular diseases, and mutations, among other adverse effects [8,9]. Therefore, efficient methods for the removal of dyes from contaminated water are necessary before their release into the environment.

Several methods have been explored for the removal and/or reduction of dye concentrations in contaminated water, including photocatalysis [10], electrochemical treatments [11], bacteria treatment [12], oxidation [13], and solvent extraction [14]. Among these methods, the adsorption process is considered an efficient strategy for removing contaminants from water sources. It is often preferred due to its simplicity, cost-effectiveness, and energy-saving properties, compared to other removal methods [15,16]. However, the use of hazardous and toxic solvents, such as aromatic solvents, chlorinated solvents, and organic solvents, in the production of adsorbent materials is considered problematic due to the increased environmental pollution. In the adsorption process, the functional groups ($-NH_2$, $-SO_3H$, $-COOH$, $-OH$) of the dyes chemically and/or physically interact with the surface of the adsorbents through hydrogen bonding, electrostatic and hydrophobic interactions, and π–π stacking [17]. Therefore, various materials, such as carbon nanotubes, metallic nanoparticles, hydrogels, and metallic composites with functional groups, have been explored for the removal of dye pollutants from water sources [18,19]. However, many of these materials have high production costs and require multiple preparation steps, which limit their application in industrial processes. For example, carbon-based materials have demonstrated the ability to retain larger quantities of dyes, but their preparation methods involve pyrolysis steps and long retention times [20]. Polyesters are inexpensive alternative materials for adsorbing cationic dyes, with higher removal efficiency, and can be formed by the reaction of polyfunctional acids and alcohols, such as citric acid (CA) and glycerol [21]. CA is an affordable material found in several natural sources, such as vegetables and citrus fruits like lemons, oranges, and limes [22]. When combined with glycerol under heat, it forms a three-dimensional poly(glycerol citrate) material, which is considered a benign and biodegradable material [21,23]. Poly(glycerol citrate) has a higher content of unreacted carboxylate and hydroxyl groups, providing the material with anionic sites to interact with cationic dyes in wastewater. However, to our knowledge, there are no reported applications of this material in the removal of cationic and/or anionic dyes. Similar materials, such as hydrogels formed with glycerol, CA, and acrylic acid, have been explored for the adsorption of methylene blue (a cationic dye). The results showed that the hydrogel had an adsorption capacity ranging from 202 to 245 mg/g, which could be attributed to electrostatic interactions and electron exchange between the dye and the material [24]. However, due to the anionic surface, the material is limited to adsorbing anionic dyes.

Cationic surfactants, such as cetyltrimethylammonium bromide (CTAB), have been demonstrated to interact and adsorb on surfaces with an opposite charge to the surfactant's cationic head, forming monolayers and hemimicelles [25]. The saturation of the surface by the surfactant promotes hydrophobic interactions between the surfactant coils, forming bilayers where the outer surface is composed of the cationic head of the surfactant [25,26]. Therefore, the surfactant bilayer can interact with anionic contaminates such as dyes to remove them from wastewater. The use of CTAB bilayers as a strategy to remove cationic dyes has been explored with TiO_2 nanoparticles for the removal of alizarin red and indigo carmine, but the adsorption capacity of the materials is limited [27]. Based on the fact that CTAB can adsorb and interact with the anionic dyes, this work aimed to obtain solvent-free poly(glycerol citrate) material through microwave heating, to form an environmentally friendly and cost-effective material and evaluate its dye removal capability, with and without CTAB. Rhodamine B (RB), acid blue 113 (AB), and acid red 27 (AR) were used as model dyes to assess the removal efficiency of the material (Figure 1). Furthermore, to understand the adsorption mechanism, the sorbent–sorbate equilibrium and kinetic adsorption were evaluated using isotherm and kinetics models, respectively.

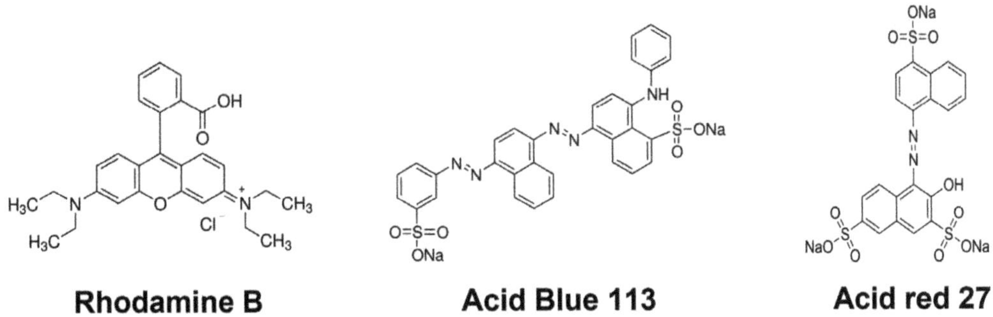

Figure 1. Chemical structure of model dyes tested in this study: rhodamine B, acid blue 113, and Acid red 27.

2. Materials and Methods

2.1. Materials

Citric acid (CA, 98%), cetyltrimethylammonium bromide (CTAB, 99%), rhodamine B (RB, 99%), acid blue 113 (AB, 99%), and acid red 27 (AR, 99%) were purchased from Sigma-Aldrich, Burlington, MA, USA. Industrial grade glycerol was obtained from a local market. All the chemicals were used without further purification. Distilled water was used throughout all the experiments.

2.2. Preparation of Poly(Glycerol Citrate) Material

Biodegradable poly(glycerol citrate) was obtained from CA and glycerol without solvent, using microwave heating. To form the material, 96.8 g of glycerol and 240.0 g of CA were mixed, resulting in a molar ratio of 1.0:1.0 of CA to glycerol. The mixture was stirred manually until a homogeneous material was obtained. Next, the mixture was placed in a microwave oven and heated for 15 min at a power of 360 W. Finally, the obtained polyester was milled and sieved through a 2.0 mm diameter.

2.3. Physiochemical Characterization of Poly(Glycerol Citrate) Material

The poly(glycerol citrate) material was characterized using various techniques. Fourier-transformed infrared spectroscopy with attenuated total reflectance (ATR-FTIR) was performed using an IR-Affinity-1 instrument (Shimadzu, Kyoto, Japan) to verify the esterification reaction. The data were acquired from 4000 to 700 cm^{-1} at a scanning rate of 2 cm^{-1} per point, and each sample underwent 20 scans. Thermogravimetric analysis (TGA) was carried out using TA Instrument SDT-Q600 (TA instrument, New Castle, DE, USA). The poly(glycerol citrate) weighed between 5 and 10 mg and was heated from room temperature to 800 °C at a constant heating rate of 10 °C/min, under a nitrogen flow of 20 mL/min. The results were evaluated by analyzing the first-order derivative of the weight loss thermogram. Differential scanning calorimetry (DSC) was performed using a TA Instrument DSC25 (TA instrument, Delaware, USA). Prior to analysis, any thermal memory of the material was eliminated by heating from room temperature to 170 °C, at a ramp rate of 15 °C/min, followed by isothermal heating at 170 °C for 3 min. Subsequently, the material was cooled from 180 °C to −30 °C at a cooling ramp rate of 25 °C/min. Finally, the DSC was carried out by heating the material from −30 °C to 170 °C at a ramp rate of 15 °C/min. In addition to spectroscopy and thermal analysis, digital photographs were taken of the materials before and after immersion in water distilled water. These photographs were captured using a flexible digital electronic microscope (Suzhou Jingtong Instrument Co., Ltd., Suzhou, China) to visually assess any changes in the samples upon water exposure.

2.4. Dye Adsorption Studies

2.4.1. Dye Removal Efficiency

To evaluate the adsorption capacity of poly(glycerol citrate), AB, RB, and AR were used. Approximately 0.3 g of adsorbed poly(glycerol citrate) was added to 5.0 mL of a dye solution (500 mg/L), then shaken at 100 rpm in a water bath for 24 h until equilibrium was achieved. Afterward, the samples were centrifuged at 5000 rpm to separate the poly(glycerol citrate)/dye material. The concentration of the dye solutions in the supernatant was determined using an ultraviolet visible spectrophotometer (Drawell DU-8800DS, Chongqing Drawell Instrument Co., Ltd, Chongqing, China), with a linear calibration curve specific to each dye used. The linear calibration curve ranged from 0.5 to 4.0 mg/L for AB, from 3.0 to 20.0 mg/L for RB, and from 2.0 to 50.0 mg/L for AR (Figure S1, Supplementary Materials). The concentrations of AB, RB, and AR were measured at equilibrium at wavelengths of 566 nm, 554 nm, and 520 nm, respectively. These values correspond to the absorbance maxima according to literature data [28–30]. The percentage removal efficiency (%RE) was calculated using Equation (1):

$$\%RE = \frac{C_1 - C_2}{C_1} \times 100 \quad (1)$$

where C_1 and C_2 are the concentrations of the dye solution in mg/L before and after the adsorption process, respectively [31]. The %RE was also determined in the presence of different concentrations of CTAB (1–40 mM) to evaluate the effectiveness of the surfactant in promoting dye removal. Additionally, the materials were recovered after dye adsorption (with and without CTAB 5 mM) and dispersed in water (10 mg/mL) to conduct zeta potential measurements (DLS, Malvern Panalytical, Zetasizer LAB, Malvern, UK), performed in triplicate.

2.4.2. Adsorption Capacity

To study the adsorption process in the absence and presence of CTAB, the same methodology described in Section 2.4.1 was employed at 298 K. However, the adsorption capacity at equilibrium (q_e, mg/g) for different initial dye concentrations (0–5000 mg/L) was calculated using Equation (2):

$$q_e = \frac{(C_o - C_e)V}{m} \quad (2)$$

where C_o (mg/L) and C_e (mg/L) are the initial and equilibrium concentrations of the dye, respectively. V (L) is the volume of dye solution used and m (g) is the weight of the dried poly(glycerol citrate) material. The Langmuir, Freundlich, and Temkin adsorption isotherms, described by Equations (3)–(5), were evaluated to understand the interaction mechanism between the dye and the polyester [6,31].

$$\frac{C_e}{q_e} = \frac{1}{(q_m K_L)} + \frac{C_e}{q_m} \quad (3)$$

$$\ln q_e = \ln K_F + \frac{1}{n} \ln C_e \quad (4)$$

$$q_e = B_T \ln K_T + B_T \ln C_e \quad (5)$$

where q_m and q_e (mg/g) are the maximum adsorption capacity and adsorption capacity at equilibrium, respectively. C_e (mg/L) is the equilibrium dye solution concentration. In the Langmuir model, K_L (L/mg) is the Langmuir constant, which relates to both the free energy and adsorption affinity. The Freundlich constants K_F (L/ mg) and n pertain to adsorption capacity and intensity, respectively. Lastly, K_T is the binding constant of the

Temkin isotherm, and BT is the Temkin constant, which is influenced by the adjusted uptake temperature (J mol^{-1}).

2.4.3. Adsorption kinetics

The effect of the CTAB surfactant on the dye adsorption was also evaluated at various times (30–1440 min), allowing the calculation of the adsorption capacity at each time interval (q_t), using Equation (6).

$$q_t = \frac{(C_o - C_t)V}{m} \tag{6}$$

where the dye concentrations at time t and the initial dye concentration are represented by C_t and C_0 (mg/L), respectively. V (mL) is the volume of the dye solution and m (g) is the mass of the adsorbent. To evaluate the adsorption kinetics mechanism, the experimental data were fitted to the pseudo-first-order model, pseudo-second-order model, and Boyd liquid-film diffusion model, which follow Equations (7)–(9).

$$\log(q_e - q_t) = \log q_e - \frac{k_1}{2.303}t \tag{7}$$

$$\frac{t}{q_t} = \frac{1}{k_2 q_e^2} + \frac{t}{q_e} \tag{8}$$

$$-\ln\left(1 - \frac{q_t}{q_e}\right) = k_{FD}t \tag{9}$$

where q_e (mg/g) and q_t (mg/g) are the adsorption capacities at equilibrium time and time t (min), respectively. k_1 (/min) represents the pseudo-first-order rate constant, k_2 (g (mg/min)) the pseudo-second-order rate constant, and k_{FD} (1/min) represents the adsorption constant corresponding to the Boyd liquid-film diffusion model [17,31].

2.4.4. Selectivity and Remotion Mix of Dyes

The adsorption selectivity of the poly(glycerol citrate) for the dyes was investigated using the methodology described in Section 2.4.1. The %RE was evaluated for three combinations of dyes: RB and AR, RB and AB, AB and AR. The %RE was calculated using Equation (1) for each system. Additionally, digital photographs were taken of the systems and the solid residue after centrifugation, using a flexible digital electronic microscope. The mass ratio was evaluated for adsorption with and without CTAB 5 mM.

3. Results and Discussion

3.1. *Synthesis and Characterization of Poly(Glycerol Citrate)*

Figure 2 shows the ATR-FTIR spectra of glycerol, CA, and the poly(glycerol citrate) material. Figure 2A shows the principal bands of the glycerol, highlighting the bands in the fingerprint at 845 cm^{-1}, attributed to the stretching of the C–C bonds. Additionally, bands at 993 cm^{-1}, 1036 cm^{-1}, and 1109 cm^{-1} correspond to the C–O stretching of the three C–O bonds in the glycerol structure. The O–H deformation and stretching bands are observed in the range of 1400 cm^{-1} to 1100 cm^{-1} and at 3320 cm^{-1}, respectively [32,33]. However, at 1640 cm^{-1}, there is a band due to the presence of water, as the glycerol used is of industrial grade and not reactive grade, making it an economical material with easy application in the industry. On the other hand, the ATR-FTIR spectra of CA (Figure 2B) exhibit three bands between 1700 cm^{-1} and 1600 cm^{-1}, which are attributed to the stretching of the carboxylic (C=O) groups in CA [21,33].

The ATR-FTIR spectra of the poly(glycerol citrate) material (Figure 2C) show a band at 1726 cm^{-1}, indicating the presence of the ester group and confirming the successful reaction between glycerol and CA [21,33]. However, there is also a band at 1626 cm^{-1}, suggesting the presence of carboxylate groups from CA and indicating the incomplete reaction of acid groups. Similar bands were observed by [33] in the polycondensation reaction between glycerol and CA. This characteristic of the material is interesting, because it suggests that the carboxylate groups with a negative charge can interact with cationic dyes through electrostatic interactions, providing active sites on the polyester for the chemisorption of the dyes. Additionally, a band at 1159 cm^{-1} is observed, which is attributed to the O–C–O stretching in the polyester. These results confirm the polycondensation reaction achieved by microwave heating, demonstrating that the produced polyester is a low-cost material that follows the principles of green chemistry, as it utilizes economical reactants, does not require organic solvents, and has low energy demand.

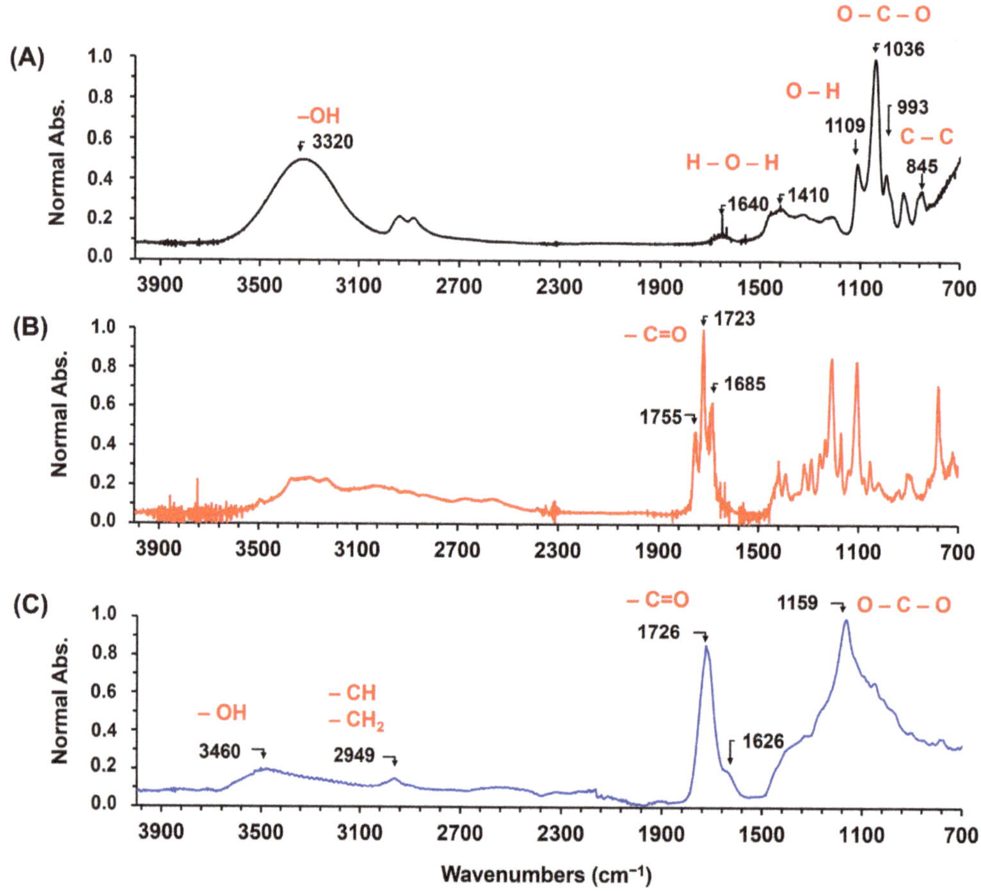

Figure 2. ATR-FTIR spectra of (**A**) glycerin, (**B**) citric acid, and (**C**) poly(glycerol citrate).

The TGA analysis of the poly(glycerol citrate) material revealed a weight loss as the temperature increased. The first significant weight loss occurred at 123.1 °C, which can be attributed to the loss of water through a condensation reaction within the material (Figure 3A). DTA analysis indicated two degradation events at higher temperatures (346.1 and 400.5 °C), suggesting that the poly(glycerol citrate) can withstand higher temperature degradation, compared to the boiling points of CA (175 °C) and glycerol (290 °C) compounds [21]. Additionally, the glass transition of the material was found at 88.9 °C (Figure 3B). Furthermore, the material displayed a rigid structure (Figure 4) and even after immersion in distilled water for 24 h, the structure remained intact, without dispersion. However, a gelatinization appearance, resembling a hydrogel, was observed, likely due to the absorption and/or entrapment of water within the material. The insolubility of the material is an important characteristic that enables its application in the adsorption of pollutants.

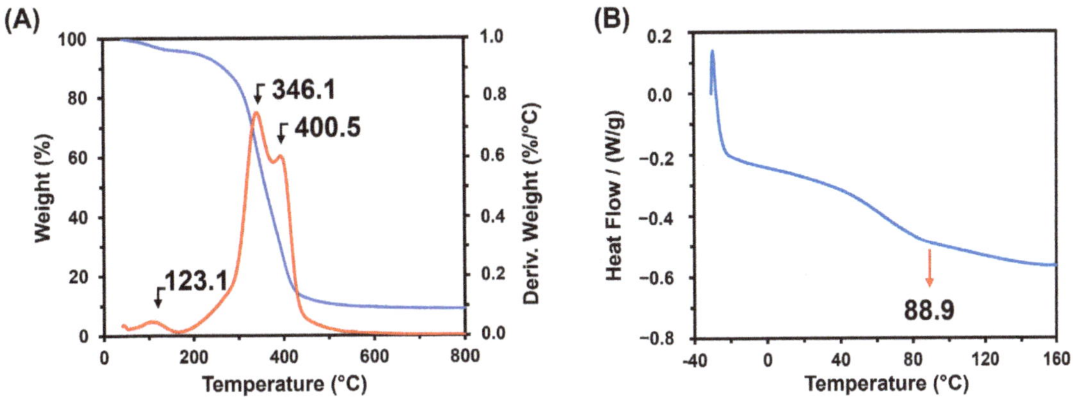

Figure 3. (**A**) TGA (-) and DTA (-), (**B**) DSC of poly(glycerol citrate).

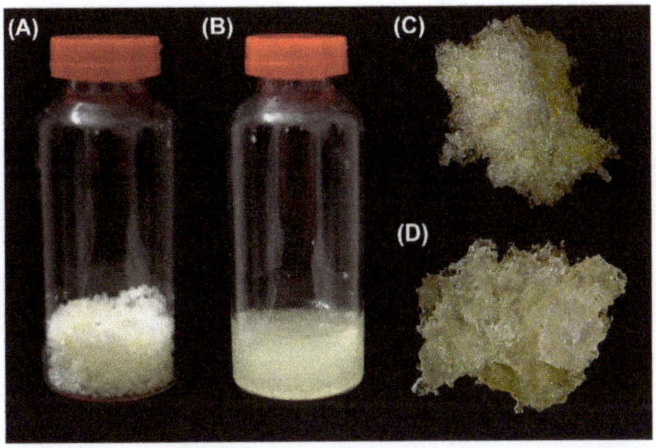

Figure 4. Digital photographs of the poly(glycerol citrate) synthetized: (**A**,**C**) dried, (**B**) immersed in water, and (**D**) after immersion in water.

3.2. Adsorption Studies

The poly(glycerol citrate) material exhibits a high density of unreacted carboxyl groups, which enables strong interactions with cationic dyes. This is evident in Figure 5A, where the cationic dye RB dye (Figure 1) is effectively removed by the poly(glycerol citrate), with an efficiency of almost 100% (98.5 ± 0.1%). This demonstrates the efficient capacity of the polyester to interact with RB and adsorb it at its interface. However, for anionic dyes such as AB and AR (Figure 5B,C), the removal efficiency is lower, with values of 67.9 ± 2.5% and 11.6 ± 6.4%, respectively. This can be attributed to the chemical repulsion between the dyes and the carboxylate groups in the poly(glycerol citrate). AB contains two sulfate groups and AR contains three sulfate groups in their chemical structures, leading to electrostatic repulsion with the polyester, reducing the adsorption capacity.

To explore the potential application of the polyester, the dye removal efficiency was evaluated at different concentrations of CTAB (1–40 mM). In the case of RB, the removal efficiency decreases as the CTAB concentration increases, with a lower value of 50.1 ± 2.6% observed at CTAB 40 mM. However, for the anionic dyes AB and AR, the results exhibit the opposite behavior. As the CTAB concentration increases (Figure 5B,C), the removal efficiency also increases, reaching values of up to 99.1%. This indicates that the CTAB surfactant promotes efficient dye removal for these colorants, within the range of 5–10 mM. Beyond that concentration, the removal efficiency decreases with increasing CTAB concentration. These findings are further supported by the digital photographs of dye removal at 0, 5, and 40 mM CTAB concentrations (Figure 5D), highlighting the efficient removal of RB and AR in 5 mM of CTAB.

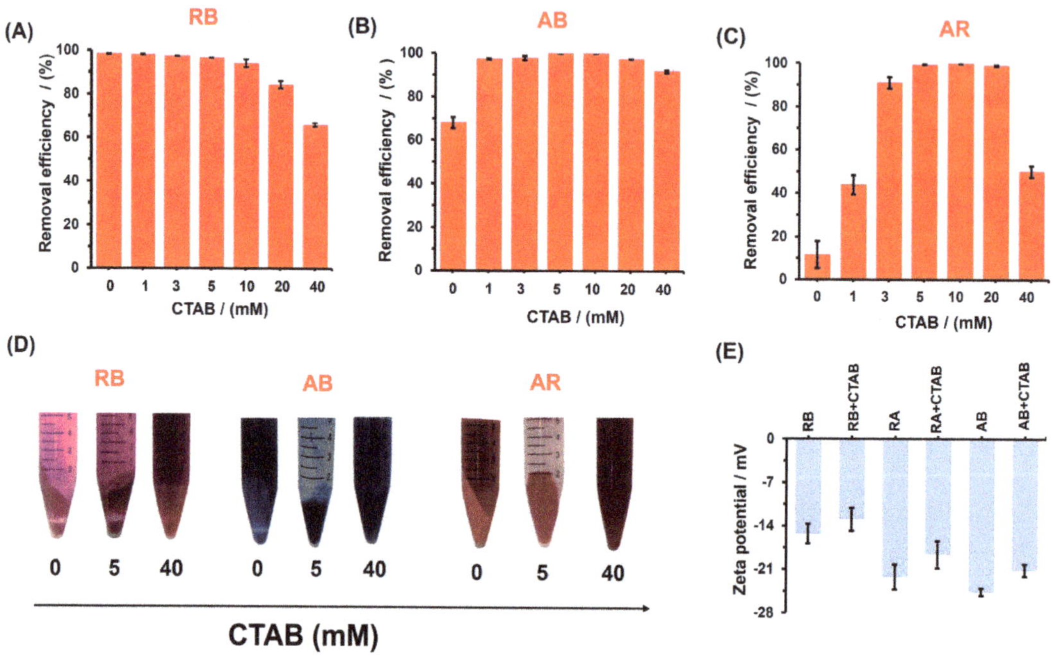

Figure 5. Effect of CTAB in the dye %RE of the poly(glycerol citrate) of (**A**) RB, (**B**) AB, and (**C**) AR. A dye concentration of 500 mg/L was used in each assay. (**D**) Digital photographs of the effect of CTAB (0, 5, and 40 mM) on dyes (RB, AB, and AR) remotion by poly(glycerol citrate) synthetized. (**E**) Zeta potential of poly(glycerol citrate) with dyes (500 mg/L) tested without and without CTAB 5 mM.

CTAB, as a cationic surfactant, exhibits a critical micelle concentration (CMC) of around 1 mM in pure water [34]. In the presence of CTAB at 5 mM, the formation of CTAB micelles and aggregates significantly affects the adsorption behavior of the dyes. It is likely that at low CTAB concentrations, interaction occurs between the CTAB aggregates (with positive charge) and the poly(glycerol citrate) material. This was corroborated by zeta potential measurement, where the CTAB 5 mM adsorbed on the material surface, increasing the zeta potential value from -16.8 ± 2.7 mV to -13.2 ± 1.1 mV, facilitating the removal of the dyes. The negative charge of the polyester particles indicates that CTAB does not interact with all of the particle surface, leaving carboxylate and/or acid carboxylic functional groups disponible to interact with the cationic dye, thus explaining the higher adsorption of RB with CTAB at 5 mM. Additionally, the adsorption of RB on the material surface with and without CTAB 5 mM increases the zeta potential to -15.3 ± 1.6 mV and -12.9 ± 1.8 mV, indicating the adsorption of the RB dye and CTAB on the polyester surface (Figure 5E).

In the case of AB and AR, the removal of the dyes may involve favorable electrostatic interactions between the negative sulfate groups on the dyes and positive ammonium groups on the surfactant head, as well as hydrophobic interactions between the organic chain of CTAB and hydrophobic functional groups of the dyes. When calculating the molar proportion between the AB and AR dyes with CTAB at 5 mM, it was found that the removal was more effective at the proportions of 3.6:25 and 4.9:25, respectively. This indicates that the same amount of CTAB could remove a larger amount of AR dye than AB. This is likely because the AR dye has three negative charges in its chemical structure, whereas AB contains only two negative charges, thus favoring the electrostatic interaction with the CTAB adsorber in the material. Consequently, it can be hypothesized that CTAB acts as a vehicle or mediator to trap the dyes and interact with the surface of the poly(glycerol citrate) material for their removal. This hypothesis is supported by the zeta potential values of the poly(glycerol citrate) after dye adsorption with and without CTAB at 5 mM (Figure 5E). In the cases of RA and RB (anionic dyes), the zeta potential was −22.2 ± 1.9 mV and −24.6 ± 0.6 mV, respectively, and both systems increased the zeta potential in CTAB 5 M, indicating that the cationic surfactant can enhance the retention of the dyes by increasing the particle charge. However, at higher CTAB concentrations, efficient dye removal is not observed, even for RB dyes. This may be attributed to two reasons: (i) the higher density of surfactant micelles, leading to the favorable encapsulation of the dyes, inhibiting their adsorption on the polyester; (ii) the saturation of the polyester surface by CTAB, resulting in repulsive interactions between the cationic RB dye and the cationic surfactant head. To validate these hypotheses, further investigations were conducted, including adsorption kinetics and isotherm studies.

3.3. Adsorption Kinetics

To investigate the adsorption kinetics of the dyes on the polyester, the adsorption capacity of the dyes was evaluated at various time intervals, ranging from 30 to 1440 min at 500 ppm at 298 K (Figure 6). For the RB dye, the adsorption capacity reaches approximately 8 mg/g after 30 min of contact, and remains nearly constant as the contact time increases, regardless of the presence of CTAB. In the case of the AB and AR dyes, in the absence of CTAB, the adsorption capacity reaches values of 6.8 ± 0.2 mg/g and 1.4 ± 0.1 mg/g, respectively, after 720 min. However, the addition of CTAB at 5 mM significantly enhances the adsorption capacity of both dyes, reaching approximately 8 mg/g. Furthermore, the presence of CTAB reduces the time required to achieve the adsorption equilibrium to 30 min for the systems.

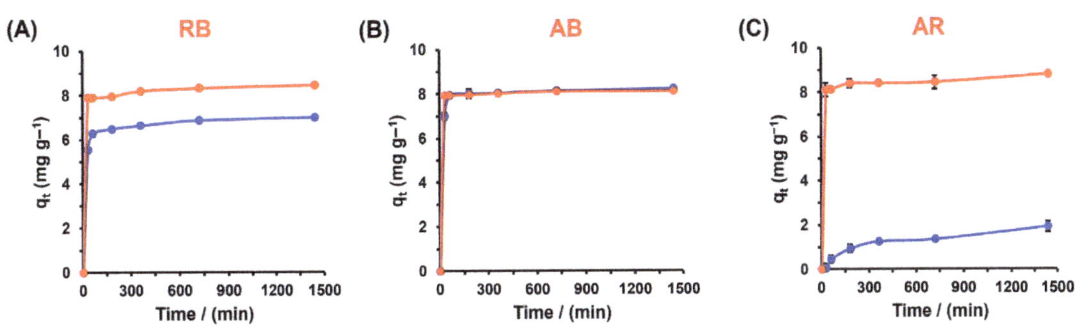

Figure 6. Effect of time on the dye adsorption of RB (**A**), AB (**B**), and AR (**C**) in distilled water (•) and CTAB 5 mM (•) at 298 K.

The kinetics data were analyzed using the pseudo-first-order, pseudo-second-order, and Boyd liquid-film diffusion models, to gain insight into the adsorption mechanisms on the polyester surface. Table 1 presents the kinetics parameters for each dye, revealing that the pseudo-second order model exhibits the highest R^2 values. The adsorption process involves several steps, including the transportation of the dye from the aqueous solution to the surface of the polyester, diffusion of the dye from the outer surface to active adsorption sites on the material, and the physical and/or chemical interaction between the dye and the active functional groups present [35,36].

Table 1. Kinetic model parameters for dyes on poly(glycerol citrate) material at 298 K.

Model	Parameter	Units	RB	RB-CTAB	AB	AB-CTAB	AR	AR-CTAB
Pseudo-first-order	q_e	mg/g	1.82	0.20	1.09	0.62	2.13	0.62
	K_1	$\times10^{-3}$ (/min)	27.60	2.07	3.22	2.30	0.92	0.92
	R^2		0.6570	0.9891	0.9245	0.9756	0.6443	0.6750
Pseudo-second-order	q_e	mg/g	8.24	8.12	7.04	8.47	3.01	8.80
	K_2	$\times10^{-3}$ (/min)	2.09	5.48	1.09	1.52	0.07	1.23
	R^2		1.0000	1.0000	0.9999	0.9999	0.7746	0.9996
Boyd liquid-film diffusion model	K_{Fd}	$\times10^{-3}$ (/min)	2.70	6.40	3.10	2.30	0.90	0.90
	R^2		0.6570	0.8933	0.9245	0.9756	0.6443	0.6750

The pseudo-second-order model suggests that the adsorption rate is primarily driven by the chemical adsorption mechanisms involving chemical bounding, electrostatic interaction, hydrogen bonds, and other interactions [31,37]. Therefore, the kinetics model confirms the hypothesis that the adsorption of the dyes occurs through electrostatic interactions between the negative carboxylate functional groups on the polyester surface and the tested dyes. In the case of RB dye, the cationic NH_4^+ groups can interact with the carboxylate groups through electrostatic interactions, facilitating the adsorption of the RB dye (Figure 7A). The presence of CTAB alters this behavior, as CTAB can exist in both liquid solid phases. In the aqueous solution, CTAB can exist as monomers and micelles aggregates, which are in equilibrium with the monomers and aggregates (hemimicelle, bilayers, and micelles) formed on the polyester surface through Coulombic attractive forces (Figure 7B,C) [38]. This suggests the hypothesis that the CTAB aggregates adsorbed on the polyester surface may induce repulsive interactions with the RB dye, reducing the adsorption capacity. However, no significant difference was observed in the presence of CTAB, indicating that the polyester possesses sufficient active sites to interact with the dye at the tested CTAB concentration.

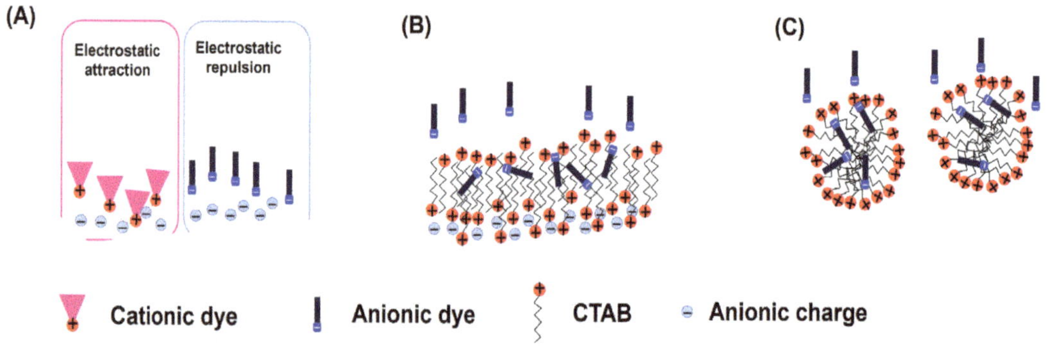

Figure 7. Representation of possible interactions between cationic and anionic dyes with the polyester surface without (**A**) and with CTAB surfactant (**B**,**C**) at 298 K.

On the other hand, for the anionic dyes (AB and AR), without CTAB, the presence of negative charges on the sulfate groups prolongs the time required to reach the kinetic equilibrium. This effect is more pronounced in the case of AR dye, due to its higher negative charge, which increases the electrostatic repulsion with the negative charge on the polyester surface. In contrast, in the presence of CTAB, the anionic dyes likely strongly interact with the positive charge on the external surface of the CTAB aggregates on the polyester surface.

This interaction reduces the time required to reach kinetics equilibrium and increases the adsorption capacity of the material for the anionic dyes under the tested conditions. Additionally, the anionic dyes may be hosted inside the CTAB aggregates through favorable hydrophobic interactions between the CTAB coils and the dyes, providing protection for the dyes. Similar behavior was observed by [27], in the adsorption of indigo carmine (anionic dye) on CTAB-modified TiO_2 nanoparticles, where CTAB acted as a bridge through bilayer aggregation between the negatively charged surface of the TiO_2 nanoparticles and the dye, reaching adsorption equilibrium within 100 min.

3.4. Adsorption Isotherm Study

The constructed adsorption isotherms in water and CTAB 5 mM at different initial dye concentrations from 50 to 5000 mg/L provide further insights into the characteristics of the adsorption processes (Figure 8). All the systems tested showed an increase in the q_e as the C_e concentration increased (Tables S1–S3, Supplementary Materials). However, a saturation profile of the polyester surface with the tested dye concentrations was not observed. Comparing the adsorption results at the maximum initial concentration point, it was found that CTAB decreased the q_e capacity for the RB dye from 75.7 ± 3.9 mg/g to 73.7 ± 3.2 mg/g. In contrast, for the AB and AR dyes, the q_e increased significantly from 72.9 ± 3.3 to 82.1 ± 4.4 mg/g for the AB, and from 18.3 ± 1.6 to 21.2 ± 0.6 mg/g for the AR. These are particularly interesting results, as the presence of the CTAB surfactant enhanced the removal efficiency of the AR dye. This demonstrates that the poly(glycerol citrate) material is efficient in removing both anionic and cationic dyes from water sources, especially with CTAB, which has a notable affinity for dyes with a higher negative charge density in their molecular structure. The experimental results obtained in this study were compared with various adsorbent materials reported in the literature for RB, AB, and AR adsorption (Table 2). The adsorption capacity values obtained in this work for RB were found to be higher than those reported for RB retention in other materials, except for the graphene oxide/Poly(NIPAM-co-AA) hydrogel [39] and chitosan/electrospun sodium alginate nanofiber [40], which showed three and ten times higher adsorption capacity for RB, respectively. In the cases of AB and AR, the materials reported in the literature also exhibited higher values compared to those reported in this work, especially for the materials in the nanometric scale. However, the high cost of production for the nanomaterial is a disadvantage when considering their application. Therefore, considering the successful dye removal results, the low cost, and ease of production of poly(glycerol citrate), this material has great potential for application in the wastewater treatment industry.

Table 2. Comparing poly(glycerol citrate) with various adsorbents to remove RB in water, and AB and AR dyes in CTAB 5 mM.

Adsorbent	RB (mg/g)	AB (mg/g)	AR (mg/g)	Ref.
Poly(glycerol citrate)	73.7 ± 3.2	82.1 ± 4.4	21.2 ± 0.6	This work
Kaolinite	46.1	-	-	[41]
Coconut fibers	22.0	-	-	[42]
Graphene oxide/Poly(NIPAM-co-AA) Hydrogel	193.1	-	-	[39]
Chitosan/electrospun sodium alginate nanofiber	695.4 ± 17.0	926.2 ± 25.7	-	[40]
Mesoporous activated carbon	-	9.20	-	[43]
Chitosan magnetized by Fe_2O_3 nanoparticles	-	124.2	-	[44]
Hydroxyapatite	-	153.8	-	[45]
Nano-Fe_2O_3-based magnetic cationic hydrogel	-	-	833	[46]
Polyethyleneimine grafted chitosan beads	-	-	48.3	[47]
Polypyrrole/$SrFe_{12}O_{19}$/graphene oxide nanocomposite	-	-	294.1	[48]

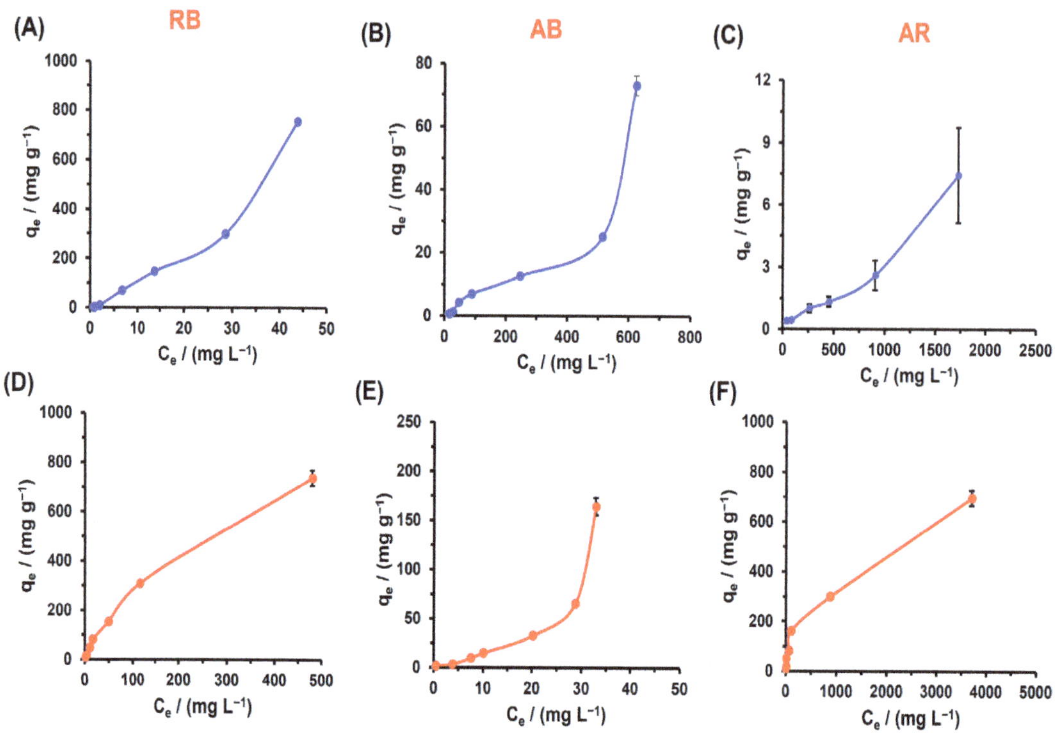

Figure 8. Dye adsorption isotherm of RB (**A**,**D**), AB (**B**,**E**), and AR (**C**,**F**) in deionized water (•) and CTAB 5 mM (•) at 298 K.

On the other hand, to better understand the adsorption process of cationic and anionic dyes tested with and without CTAB, adsorption isotherm models such as Langmuir, Freundlich, and Temkin were fitted to the dye adsorption isotherm data, allowing the determination of the adsorption isotherm parameters for each model tested (Table 3). Based on the R^2 values, the Freundlich isotherm showed the best correlation compared to the other models. The Freundlich isotherm has been applied to dye adsorption on the materials, as it is not limited by a number of specific adsorption sites and does not saturate the material surface [49]. This is in agreement with the adsorption behavior observed in Figure 8, where the material surface it is not saturated, describing an IUPAC isotherm classification type II with and without CTAB, except for the system AR with CTAB which shows type I behavior, indicating surface saturation. In general, the K_F parameters, which are related to the adsorption capacity, were found to be higher in the presence of CTAB 5 mM than in water for the tested AB and AR dyes, indicating that the surfactant increases the adsorption capacity.

The Freundlich adsorption model is an empirical equation that assumes a heterogeneous adsorbent surface, promoting multilayer coverage over the surface with different energy values [50,51]. It suggests that the adsorbate interacts with the active adsorption center with maximum energy first, followed by adsorption on centers with lower energy. Thus, the poly(glycerol citrate) material has a heterogeneous surface, with different active centers, to interact with the tested dyes, indicating incomplete reaction of the acid groups. This is in accordance with the FTIR-ATR results (Figure 2C), as previously described. Additionally, the Freundlich adsorption model has been used to describe multilayer adsorption, suggesting the deposition of the dye on the material surface at higher dye concentrations (3000 mg/L of dye), producing a multilayer adsorption profile. This behavior was also observed in the adsorption of acid blue 80 (anionic dye) on white mud when higher dye concentrations were used [52]. On the other hand, the parameter n in this model provides information about the intensity of interaction between the adsorbent and adsorbate. When n < 1, the adsorption process occurs through chemical interaction, involving the sharing of electrons to form covalent or ionic bonds. When n > 1, the process occurs through physical adsorption. In general, the n values obtained show that the adsorption of the dye-CTAB systems occurs through physical adsorption on the polyester surface [31,39]. Additionally, the 1/n values found in the range of 0–1 reflect the favorable adsorption of the dyes in the presence of CTAB, compared to the AB and AR dyes, although this favorability decreases in the absence of the surfactant. These results are consistent with the kinetics results, where CTAB influences the adsorption process and reduces the time required to reach equilibrium, due to the highly favorable electrostatic interactions between the CTAB micelle, the dye, and the polyester surface with carboxylate groups. Despite the low R^2 values obtained for the Langmuir and Temkin isotherms, the K_L and K_T parameters show a similar trend to the Freundlich, as they also exhibit higher values in the presence of CTAB for all the dyes tested.

Table 3. Adsorption isotherm parameters for dyes on poly(glycerol citrate) material at 298 K.

Model	Parameter	Units	RB	RB-CTAB	AB	AB-CTAB	AR	AR-CTAB
Langmuir isotherm	q_m	mg/g	98.0	104.2	119.0	38.8	20.2	21.5
	K_L ($\times 10^{-3}$)	L/mg	13.56	4.58	0.45	16.50	0.09	20.34
	R^2		0.9359	0.9476	0.1124	0.2455	0.1124	0.9981
Freundlich isotherm	n		1.36	1.29	0.90	1.00	1.07	4.10
	$1/n$		0.74	0.78	1.11	1.00	0.94	0.24
	K_F ($\times 10^3$)	L/mg	173.2	73.03	3.73	99.39	0.36	28.87
	R^2		0.9548	0.9904	0.9494	0.8530	0.9003	0.7951
Temkin isotherm	B_T		11.30	11.40	13.64	12.40	2.84	1.71
	K_T	L/mg	0.72	0.26	0.04	0.66	0.008	27.51
	R^2		0.8241	0.7909	0.5969	0.4065	0.4969	0.8686

3.5. Selectivity, Separation, and Dye Remotion Studies

According to adsorption isotherms, the adsorption capacity increases with the dye concentration, however, in order to know the amount of dye removed by the polyester materials, the %RE versus the dye concentration was plotted to evaluate the applicability of the material (Figure 9). The RB showed a higher affinity for the polyester material in the absence of the cationic surfactant, resulting in nearly 100% removal for all RB concentrations. However, in the presence of CTAB, the %RE slightly decreases as the dye concentration increases, likely due to competitive interactions between the cationic dye and the CTAB on the anionic surface, and/or the entrapment of the dye in the aqueous phase by the CTAB micelles. A different behavior was observed for the AB and AR dyes. Without CTAB, for the AR dye, the %RE did not exceed 30%. In the presence of CTAB, there were two distinct trends: for concentrations ranging from 0 to 1000 mg/L, the polyester material removed almost 100% of the dye, but after 1000 mg/L, the %RE decreased as the dye concentration increased. This suggests that the CTAB aggregate phase adsorbed on the polyester material starts to saturate with the dye at 1000 mg/L, leading to a decrease in the %RE.

Since the adoption capacity of the RB was higher than that of AR in the absence of CTAB, it can be inferred that RB and AR can be selectively absorbed and separated by the polyester material. To evaluate the selectivity of the material, adsorption experiments were conducted using dye mixtures with and without CTAB, as shown in Figure 10. In the mixture of RB and AR dyes without CTAB, the %RE for AR increased compared to the results in Figure 10A, and the dye removal increased as the mass ratio of RB increased. RB dye has a positive charge, making it favorably adsorb onto the anionic surface. Additionally, according to the zeta potential results reported in Section 3.2, the adsorption of RB increases the zeta potential of the material surface, facilitating the adsorption of AR dye, which has three negative charges in its structure. Thus, the AR adsorption possibly occurs through electrostatic interaction between the AR and RB dyes, allowing AR to approach the polyester surface. However, the material exhibited greater adsorption capacity for the cationic dye, as corroborated by the pink color of the polyester after adsorption (Figure 10D). When analyzing the structure of RB and AR, it can be inferred that favorable electrostatic interactions occur between the cationic charge of RB and the negatively charged sulfate groups. Additionally, both dyes contain aromatic rings that promote favorable π–π stacking interactions [37]. In the presence of CTAB 5 mM, both dyes exhibited a higher adsorption capacity, with %RE values ranging between 89 and 90% for both dyes. The color of the aqueous phase confirmed that under these conditions, complete removal of the dyes did not occur where dye mixtures were used (Figure 10D).

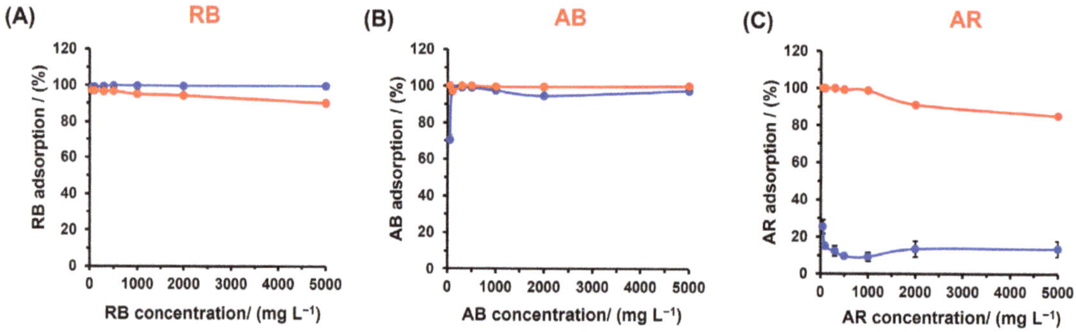

Figure 9. Dye remotion by the poly(glycerol citrate) material of (**A**) RB, (**B**) AB, and (**C**) AR in distilled water (•) and CTAB 5 mM (•).

In the RB + AB system (Figure 10B), higher %RE values were obtained for both dyes, with and without CTAB, at different mass ratios. Without CTAB, RB promoted the removal of AB because AB dye has two negative charges in its structure, decreasing the repulsive interaction with the polyester surface. This indicates the more selective retention of AB than AR, when mixed with RB dye. However, complete removal of the dyes was not achieved, indicating that interactions between RB and AB negatively affected the adsorption process. Similar behavior was observed for the adsorption of AB + AR dyes without CTAB (Figure 10C), where the %RE for AR increased to approximately 80% as the AB mass ratio increased. These results can be explained by the repulsive electrostatic interaction of the negatively charged surface of the material with anionic dyes, where the repulsion interaction is higher for AR (three negative charges) than AB (two negative charges), indicating a more selective interaction with AB than with AR (supported by the visual observation of the solid phase). In the presence of CTAB, both dyes showed almost 100% removal, resulting in a purple coloration of the solid phase after adsorption. Additionally, the material showed an uncolored region, indicating that the material was not completely saturated. These results demonstrate that the polyester material exhibits a more selective adsorption capability for RB and AB dyes, compared to AR dye. However, in the presence of CTAB, all dyes showed a higher %RE, indicating that CTAB is an effective solute for promoting the simultaneous adsorption of cationic and anionic dyes. Therefore, poly(glycerol citrate) has the potential to be used as an adsorbent for wastewater treatment.

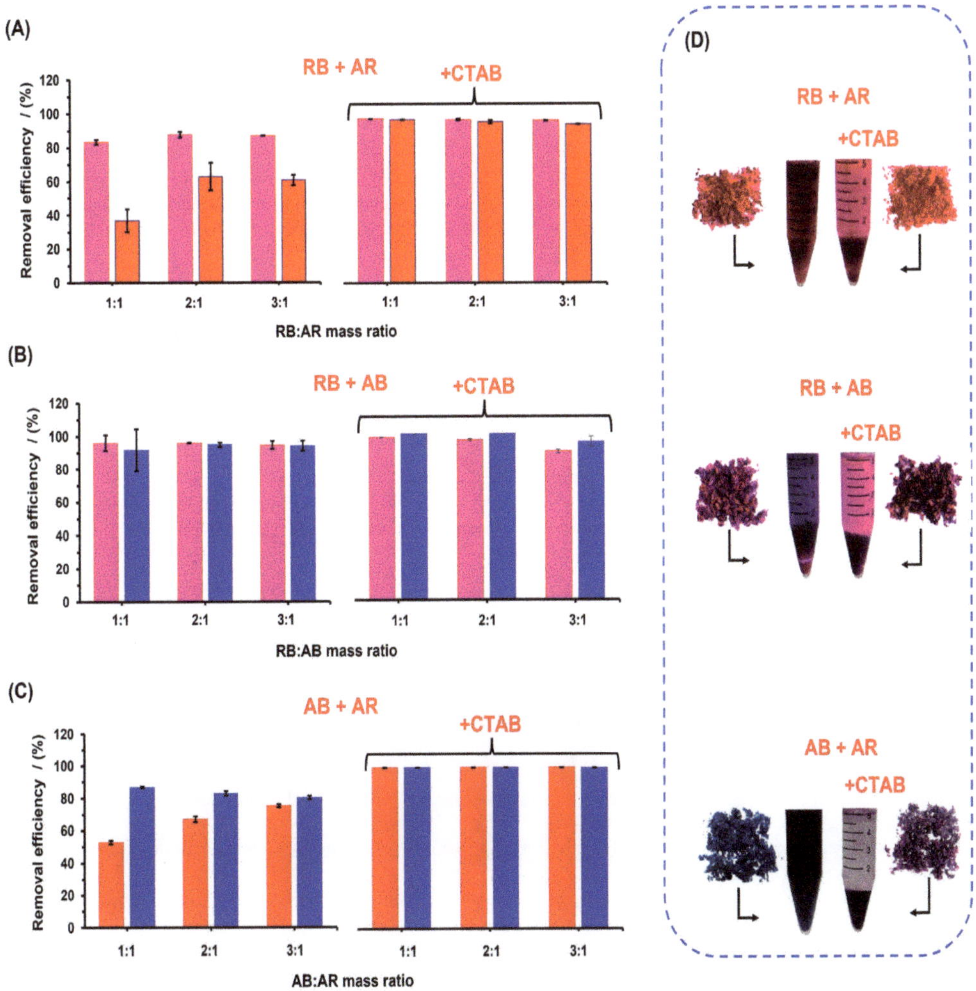

Figure 10. %RE on poly(glycerol citrate) material, with and without CTAB, of mixture dyes at different mass ratios of: (**A**) RB:AR, (**B**) RB:AB, and (**C**) AB:AR. (**D**) Digital photography of adsorption systems and polyester material after adsorption at 3:1 mass ratio, with and without CTAB. The pink, red, and blue bars represent the RB, AR, and AB dyes, respectively. Mass ratio indicates the dye mass (mg) in the system: 1:1 (1.2 mg:1.2 mg), 2:1 (2.4 mg:1.2 mg), and 3:1 (3.6 mg:1.2 mg).

4. Conclusions

A poly(glycerol citrate) material was formed by microwave heating of glycerol and citric acid, without the use of solvents. ATR-FTIR and TGA results confirmed the successful formation of the ester with high temperature resistance, making the material stable at temperatures below 300 °C. The presence of anionic groups on the material facilitated a selective interaction with the cationic dye RB, resulting in the removal of nearly 100% of the tested dye concentrations. In the case of anionic dyes, the adsorption capacity increased significantly in the presence of CTAB. This can be attributed to the ability of the surfactant to adsorb on the polyester surface in the form of hemimicelles and bilayers, imparting a positive charge to the polyester surface, which promotes favorable interactions with the anionic dyes. This was supported by the zeta potential results, indicating that

CTAB increases the surface charge of the polyester. The kinetics results indicated that the adsorption of both cationic and anionic dyes followed a Pseudo-second-order model, suggesting that electrostatic interactions played a crucial role in the adsorption process. Furthermore, the Freundlich isotherm model showed the best fit with the adsorption data, except in the adsorption of RA with CTAB. The Freundlich isotherm model indicates a heterogeneous adsorption process, involving the formation of multilayers on the polyester surface. In the presence of CTAB, the poly(glycerol citrate) material demonstrated an efficient capacity (>95%) to remove both cationic and anionic dyes. Overall, this low-cost and environmentally-friendly poly(glycerol citrate)/CTAB system holds promise for the development of robust methods of wastewater treatment.

Supplementary Materials: The following supporting information can be downloaded at: https://www.mdpi.com/article/10.3390/w16131860/s1, Figure S1: Calibration curve of: (A) rhodamine B (RB), (B) acid blue 113 (AB), and (C) acid red 27 (AR); Table S1: Equilibrium data of the adsorption of rhodamine B on poly(glycerol citrate) with and without CTAB; Table S2: Equilibrium data of the adsorption of acid blue 113 on poly(glycerol citrate) with and without CTAB; Table S3: Equilibrium data of the adsorption of acid red 27 on poly(glycerol citrate) with and without CTAB.

Author Contributions: T.A.L. Investigation, Formal analysis, writing—review and editing: A.F.C. Investigation, Formal analysis, writing—review and editing. M.P. Conceptualization, Investigation, Formal analysis, writing—review and editing. All authors have read and agreed to the published version of the manuscript.

Funding: This research has been funded by Dirección General de Investigaciones of Universidad Santiago de Cali under call No. DGI-01-2024 and grant number No. 939-621122-030.

Data Availability Statement: Data is contained within the article (and Supplementary Materials).

Conflicts of Interest: The authors declare that they have no known competing financial interests or personal relationships that could have appeared to influence the work reported in this paper.

References

1. Benkhaya, S.; M'rabet, S.; El Harfi, A. Classifications, properties, recent synthesis, and applications of azo dyes. *Heliyon* **2020**, *6*, e03271. [CrossRef] [PubMed]
2. Rápó, E.; Tonk, S. Factors affecting synthetic dye adsorption—Desorption studies: A review of results from the last five years (2017–2021). *Molecules* **2021**, *26*, 5419. [CrossRef] [PubMed]
3. De Gisi, S.; Lofrano, G.; Grassi, M.; Notarnicola, M. Characteristics and Adsorption Capacities of Low-Cost Sorbents for Wastewater Treatment: A Review. *Sustain. Mater. Technol.* **2016**, *9*, 10–40. [CrossRef]
4. Bouabidi, Z.B.; El-Naas, M.H.; Cortes, D.; McKay, G. Steel-Making Dust as a Potential Adsorbent for the Removal of Lead (II) from an Aqueous Solution. *Chem. Eng. J.* **2018**, *334*, 837–844. [CrossRef]
5. Pei, Y.C.; Wang, J.J.; Xuan, X.P.; Fan, J.; Fan, M. Factors affecting ionic liquids-based removal of anionic dyes from water. *Environ. Sci. Technol.* **2007**, *41*, 5090–5509. [CrossRef] [PubMed]
6. Qin, P.; Yang, Y.; Zhang, X.; Niu, J.; Yang, H.; Tian, S.; Lu, M. Highly efficient, rapid, and simultaneous removal of cationic dyes from aqueous solution using monodispersed mesoporous silica nanoparticles as the adsorbent. *Nanomaterials* **2017**, *8*, 4. [CrossRef] [PubMed]
7. Li, C.; Xiong, Z.; Zhang, J.; Wu, C. The Strengthening Role of the Amino Group in Metal-Organic Framework MIL-53 (Al) for Methylene Blue and Malachite Green Dye Adsorption. *J. Chem. Eng. Data* **2015**, *60*, 3414–3422. [CrossRef]
8. Farhadi, S.; Mahmoudi, F.; Amini, M.M.; Dusek, M.; Jarosova, M. Synthesis and characterization of a series of novel perov-skite-type $LaMnO_3$/Keggin-type polyoxometalate hybrid nanomaterials for fast and selective removal of cationic dyes from aqueous solutions. *Dalton. Trans.* **2017**, *46*, 3252–3264. [CrossRef] [PubMed]
9. Ardila-Leal, L.D.; Poutou-Piñales, R.A.; Pedroza-Rodríguez, A.M.; Quevedo-Hidalgo, B.E. A Brief History of Colour, the Environmental Impact of Synthetic Dyes and Removal by Using Laccases. *Molecules* **2021**, *26*, 3813. [CrossRef]
10. Chahar, D.; Kumar, D.; Thakur, P.; Thakur, A. Visible light induced photocatalytic degradation of methylene blue dye by using Mg doped Co-Zn nanoferrites. *Mater. Res. Bull.* **2023**, *162*, 112205. [CrossRef]
11. Belal, R.M.; Zayed, M.A.; El-Sherif, R.M.; Ghany, N.A. Electrochemical Degradation and Degree of Mineralization of the BY28 Dye in a Supporting Electrolyte Mixture Using an Expanded Dimensionally Stable Anode. *Electrocatalysis* **2022**, *13*, 26–36. [CrossRef]
12. Jain, D.; Navariya, J.K.; Bhojiya, A.A.; Singh, A.; Mohanty, S.R.; Upadhyay, S.K. Bioprospecting of novel ligninolytic bacteria for effective bioremediation of agricultural by-product and synthetic pollutant dyes. *Microbiol. Res.* **2023**, *270*, 127330. [CrossRef] [PubMed]

13. Shamrock, W.; Abdelhamid, H.N. Fenton-like Cerium Metal–Organic Frameworks (Ce-MOFs) for Catalytic Oxidation of Olefins, Alcohol, and Dyes Degradation. *J. Clust. Sci.* **2023**, *34*, 2509–2519.
14. Karma, V.; Gande, V.V.; Pushpavanam, S. Simultaneous extraction and enrichment of sunset yellow dye in an aqueous two-phase system. *Dyes. Pigm.* **2023**, *212*, 111100. [CrossRef]
15. Mu, B.; Wang, A. Adsorption of dyes onto palygorskite and its composites: A review. *J. Environ. Chem. Eng.* **2016**, *4*, 1274–1294. [CrossRef]
16. Tony, M.A. Low-Cost Adsorbents for Environmental Pollution Control: A Concise Systematic Review from the Prospective of Principles, Mechanism and Their Applications. *J. Dispers. Sci. Technol.* **2022**, *43*, 1612–1633. [CrossRef]
17. Salunkhe, B.; Schuman, T.P. Super-Adsorbent Hydrogels for Removal of Methylene Blue from Aqueous Solution: Dye Ad-sorption Isotherms, Kinetics, and Thermodynamic Properties. *Macromoleclues* **2021**, *1*, 256–275. [CrossRef]
18. Azari, A.; Nabizadeh, R.; Mahvi, A.H.; Nasseri, S. Magnetic multi-walled carbon nanotubes-loaded alginate for treatment of industrial dye manufacturing effluent: Adsorption modelling and process optimisation by central composite face-central design. *Int. J. Environ. Anal. Chem.* **2023**, *103*, 1509–1529. [CrossRef]
19. Lourens, A.; Falch, A.; Malgas-Enus, R. Magnetite immobilized metal nanoparticles in the treatment and removal of pollutants from wastewater: A review. *J. Mater. Sci.* **2023**, *58*, 2951–2970. [CrossRef]
20. Liang, C.; Shi, Q.; Feng, J.; Yao, J.; Huang, H.; Xie, X. Adsorption Behaviors of Cationic Methylene Blue and Anionic Reactive Blue 19 Dyes onto Nano-Carbon Adsorbent Carbonized from Small Precursors. *Nanomaterials* **2022**, *12*, 1814. [CrossRef]
21. Tisserat, B.; O'kuru, R.H.; Hwang, H.; Mohamed, A.A.; Holser, R. Glycerol citrate polyesters produced through heating without catalysis. *J. Appl. Polym. Sci.* **2012**, *125*, 3429–3437. [CrossRef]
22. Lee, S.H.; Md Tahir, P.; Lum, W.C.; Tan, L.P.; Bawon, P.; Park, B.-D.; Osman Al Edrus, S.S.; Abdullah, U.H. A Review on Citric Acid as Green Modifying Agent and Binder for Wood. *Polymers* **2020**, *12*, 1692. [CrossRef] [PubMed]
23. Berube, M.A.; Schorr, D.; Ball, R.J.; Landry, V.; Blanchet, P. Determination of in situ esterification parameters of citric acid-glycerol based polymers for wood impregnation. *J. Polym. Environ.* **2018**, *26*, 970–979. [CrossRef]
24. Franklin, D.S.; Guhanathan, S. Investigation of citric acid–glycerol-based pH-sensitive biopolymeric hydrogels for dye removal applications: A green approach. *Ecotoxicol. Environ. Saf.* **2015**, *121*, 80–86. [CrossRef] [PubMed]
25. Zhu, L.; Pan, D.; Ding, L.; Tang, F.; Zhang, Q.; Liu, Q.; Yao, S. Mixed hemimicelles SPE based on CTAB-coated Fe_3O_4/SiO_2 NPs for the determination of herbal bioactive constituents from biological samples. *Talanta* **2010**, *80*, 1873–1880. [CrossRef] [PubMed]
26. Bejaoui, B.; Nefzi, K.; Bouchmila, I.; Koumba, S.; Joly, N.; M'Hamdi, N.; Martin, P. Solid phase extraction of polycyclic aromatic hydrocarbons from water samples using CTAB-TIO_2 modified nanotubes. *Microchemistry J.* **2023**, *193*, 109027. [CrossRef]
27. Zolgharnein, J.; Rajabalipour, F.; Dermanaki Farahani, S. Indigo Carmine Dye Adsorptive Removal by Polyethylene Glycol-Modified Hydroxyapatite Nanoparticles as an Efficient Adsorbent. *Water. Air. Soil. Pollut.* **2023**, *234*, 210. [CrossRef]
28. Dhaif-Allah, M.A.; Taqui, S.N.; Syed, U.T.; Syed, A.A. Kinetic and isotherm modeling for acid blue 113 dye adsorption onto low-cost nutraceutical industrial fenugreek seed spent. *Appl. Water Sci.* **2020**, *10*, 1–16. [CrossRef]
29. Xiao, W.; Garba, Z.N.; Sun, S.; Lawan, I.; Wang, L.; Lin, M.; Yuan, Z. Preparation and evaluation of an effective activated carbon from white sugar for the adsorption of rhodamine B dye. *J. Clean. Prod.* **2020**, *253*, 119989. [CrossRef]
30. de Santana, J.E.; de Andrade, F.G.S.; Ferreira, A.F.; Ghislandi, M.G.; da Motta Sobrinho, M.A. Isotherms, kinetics and thermodynamics of industrial dye acid red 27 adsorption on Sugarcane Bagasse Ash. *Environ. Sci. Pollut. Res. Int.* **2024**, 1–15. [CrossRef]
31. Alharbi, R.A.; Alminderej, F.M.; Al-Harby, N.F.; Elmehbad, N.Y.; Mohamed, N.A. Preparation and Characterization of a New Bis-Uracil Chitosan-Based Hydrogel as Efficient Adsorbent for Removal of Anionic Congo Red Dye. *Polymers* **2023**, *15*, 1529. [CrossRef] [PubMed]
32. Razali, N.A.; Conte, M.; McGregor, J. The role of impurities in the La_2O_3 catalysed carboxylation of crude glycerol. *Catal. Lett.* **2019**, *149*, 1403–1414. [CrossRef]
33. Zahlan, H.; Saeed, W.S.; Alrasheed, R.; Alandes, N.M.; Aouak, T. Synthesis of poly (citric acid-co-glycerol) and its application as an inhibitor of $CaCO_3$ deposition. *Materials* **2019**, *12*, 3800. [CrossRef]
34. Gahrooee, T.R.; Moud, A.A.; Danesh, M.; Hatzikiriakos, S.G. Rheological characterization of CNC-CTAB network below and above critical micelle concentration (CMC). *Carbohydr. Polym.* **2021**, *257*, 117552. [CrossRef] [PubMed]
35. Shi, Y.; Chang, Q.; Zhang, T.; Song, G.; Sun, Y.; Ding, G. A review on selective dye adsorption by different mechanisms. *J. Environ. Chem. Eng.* **2022**, *10*, 108639. [CrossRef]
36. Tan, K.B.; Vakili, M.; Horri, B.A.; Poh, P.E.; Abdullah, A.Z.; Salamatinia, B. Adsorption of dyes by nanomaterials: Recent developments and adsorption mechanisms. *Sep. Purif. Technol.* **2015**, *150*, 229–242. [CrossRef]
37. Zhao, D.; Shen, X. Preparation of Chitosan-Diatomite/Calcium Alginate Composite Hydrogel Beads for the Adsorption of Congo Red Dye. *Water* **2023**, *15*, 2254. [CrossRef]
38. Purnomo, A.S.; Prasetyoko, D.; Bahruji, H. Single-step synthesis and modification of CTAB-hectorite for efficient adsorption of methyl orange dye. *Mater. Chem. Phys.* **2022**, *291*, 126749.
39. Yao, G.; Li, S.; Xu, J.; Liu, H. Dual-responsive graphene oxide/poly (NIPAM-co-AA) hydrogel as an adsorbent for rhodamine B and imidacloprid. *J. Chem. Amp. Eng. Data* **2019**, *64*, 4054–4065. [CrossRef]
40. Zhao, X.; Wang, X.; Lou, T. Simultaneous adsorption for cationic and anionic dyes using chitosan/electrospun sodium alginate nanofiber composite sponges. *Carbohydr. Polym.* **2022**, *276*, 118728. [CrossRef]

41. Khan, T.A.; Dahiya, S.; Ali, I. Use of kaolinite as adsorbent: Equilibrium, dynamics and thermodynamic studies on the adsorption of Rhodamine B from aqueous solution. *Appl. Clay Sci.* **2012**, *69*, 58–66. [CrossRef]
42. Nascimento, R.J.; Pereira, K.R.A.; Avelino, F. Parametric and modeling studies of Rhodamine-B adsorption using coconut coir-based materials as eco-friendly adsorbents. *J. Environ. Chem. Eng.* **2021**, *9*, 105943. [CrossRef]
43. Gupta, V.K.; Gupta, B.; Rastogi, A.; Agarwal, S.; Nayak, A. A comparative investigation on adsorption performances of mesoporous activated carbon prepared from waste rubber tire and activated carbon for a hazardous azo dye—Acid Blue 113. *J. Hazard. Mater.* **2011**, *186*, 891–901. [CrossRef]
44. Al-Musawi, T.J.; Mengelizadeh, N.; Al Rawi, O.; Balarak, D. Capacity and modeling of acid blue 113 dye adsorption onto chitosan magnetized by Fe_2O_3 nanoparticles. *J. Polym. Environ.* **2022**, *30*, 344–359. [CrossRef]
45. Sodhani, H.; Hedaoo, S.; Murugesan, G.; Pai, S.; Vinayagam, R.; Varadavenkatesan, T.; Selvaraj, R. Adsorptive removal of Acid Blue 113 using hydroxyapatite nanoadsorbents synthesized using *Peltophorum pterocarpum pod* extract. *Chemosphere* **2022**, *299*, 134752. [CrossRef]
46. Khan, M.; Lo, I.M. Removal of ionizable aromatic pollutants from contaminated water using nano γ-Fe_2O_3 based magnetic cationic hydrogel: Sorptive performance, magnetic separation, and reusability. *J. Hazard. Mater.* **2017**, *322*, 195–204. [CrossRef]
47. Yusof, N.H.; Foo, K.Y.; Wilson, L.D.; Hameed, B.H.; Hussin, M.H.; Sabar, S. Microwave-assisted synthesis of polyeth-yleneimine grafted chitosan beads for the adsorption of acid red 27. *J. Polym. Environ.* **2020**, *28*, 542–552. [CrossRef]
48. Ebrahimpoor, S.; Kiarostami, V.; Khosravi, M.; Davallo, M.; Ghaedi, A. Bees metaheuristic algorithm with the aid of artificial neural networks for optimization of acid red 27 dye adsorption onto novel polypyrrole/$SrFe_{12}O_{19}$/graphene oxide nano-composite. *Polym. Bull.* **2019**, *76*, 6529–6553. [CrossRef]
49. Tseng, R.L.; Wu, F.C. Inferring the favorable adsorption level and the concurrent multi-stage process with the Freundlich constant. *J. Hazard. Mater.* **2008**, *155*, 277–287. [CrossRef]
50. Debnath, S.; Das, R. Strong adsorption of CV dye by Ni ferrite nanoparticles for wastewater purification: Fits well the pseudo second order kinetic and Freundlich isotherm model. *Ceram. Int.* **2023**, *49*, 16199–16215. [CrossRef]
51. Ferreira, D.C.M.; Dos Santos, T.C.; dos Reis Coimbra, J.S.; de Oliveira, E.B. Chitosan/carboxymethylcellulose polyelectrolyte complexes (PECs) are an effective material for dye and heavy metal adsorption from water. *Carbohydr. Polym.* **2023**, *315*, 120977. [CrossRef] [PubMed]
52. Zhu, M.X.; Lee, L.; Wang, H.H.; Wang, Z. Removal of an anionic dye by adsorption/precipitation processes using alkaline white mud. *J. Hazard. Mater.* **2007**, *149*, 735–741. [CrossRef] [PubMed]

water

MDPI

Article

Anaerobic Digestion of Dye Wastewater and Agricultural Waste with Bio-Energy and Biochar Recovery: A Techno-Economic and Sustainable Approach

Albert Tumanyisibwe [1], Mahmoud Nasr [1,2], Manabu Fujii [3,*] and Mona G. Ibrahim [1,4]

[1] Environmental Engineering Department, Egypt-Japan University of Science and Technology (E-JUST), Alexandria 21934, Egypt; tumanyisibwe.albert@ejust.edu.eg (A.T.); mahmoud.nasr@ejust.edu.eg (M.N.); mona.gamal@ejust.edu.eg (M.G.I.)

[2] Sanitary Engineering Department, Faculty of Engineering, Alexandria University, Alexandria 21544, Egypt

[3] Department of Civil and Environmental Engineering, Tokyo Institute of Technology, Meguro-Ku, Tokyo 152-8552, Japan

[4] Environmental Health Department, High Institute of Public Health, Alexandria University, Alexandria 21544, Egypt

* Correspondence: fujii.m.ah@m.titech.ac.jp

Abstract: While several researchers have investigated the anaerobic digestion (AD) of textile wastewater for dye degradation, their studies suffer from lower biogas productivity due to substrate inhibition and the occurrence of secondary pollution from digestate disposal. Hence, this study focuses on using the extract of wheat straw (WS) as a co-substrate to facilitate the dye AD process, followed by recycling the digestate sludge for biochar production. In the first study, the batch digesters were operated at different dye wastewater (DW)/WS ratios (0–50% *v*/*v*), substrate-to-inoculum ratio of 0.28–0.50 g/g, pH 7.0 ± 0.2, and 37 °C. The digester operated at a DW/WS fraction of 65/35% (*v*/*v*) showed the best chemical oxygen demand (COD) removal efficiency of 68.52 ± 3.40% with bio-CH_4 of 270.52 ± 19.14 mL/g $COD_{removed}$. About 52.96 ± 3.61% of the initial COD mass was converted to CH_4, avoiding inhibition caused by volatile fatty acid (VFA) accumulation. In the second experiment, the dry digestate was thermally treated at 550 °C for 2 h under an oxygen-deprived condition, yielding 0.613 ± 0.031 g biochar/g. This biochar exhibited multiple functional groups, mineral contents, and high stability (O/C = 0.193). The combined digestion/pyrolysis scenario treating 35 m^3/d (106.75 kg COD/d) could maintain profits from pollution reduction, biogas, biochar, and carbon trading, obtaining a 6.5-year payback period.

Keywords: biogas recovery; co-substrate; methylene blue; profits; pyrolysis; sludge pyrolysis

1. Introduction

Textile industries generate large volumes of wastewater containing organic and inorganic pollutants, mainly dyes, surfactants, detergents, solvents, and recalcitrant compounds [1]. The informal disposal of dye wastewater (DW) without proper treatment imposes adverse effects on ecology, agriculture, and public health [2]. For example, a dye layer can float on the water surface and hamper natural processes of sunlight penetration into water bodies [3]. This condition negatively influences the photosynthetic activity of most aquatic plants, further disrupting the availability of dissolved oxygen and nutrients in aquacultures [4]. Furthermore, exposure to toxic pollutants in DW is associated with various health concerns, including vomiting, stomach pain, and damage to the eyes, liver, skin, and kidneys [5]. Moreover, contaminating the water bodies with textile dyes tends to deteriorate the water resource systems available for landscape irrigation; hence, textile wastewater should be adequately treated before reaching the aquatic environment to alleviate the associated risks.

While various physical [6], chemical [7], and electrochemical [1] methods have been employed in DW treatment, these systems entail some drawbacks, especially in low-income countries. For instance, DW filtration using membrane technologies requires high amounts of energy to overcome the osmotic pressure accompanied by the high concentrations of dissolved ions. Moreover, the electrochemical techniques used to capture cationic dyes are infeasible in developing countries with limited access to energy services, and they might generate toxic degradation byproducts [8]. The chemical processes require the addition of reagents that might be expensive and unavailable in some countries, and they could generate sludge containing unfavorable dye byproducts [9]. The biological treatment of DW under anaerobic conditions is considered environmentally friendly because of less chemical consumption and resource (e.g., energy) utilization [10]. Moreover, it could earn profits from bioenergy recovery (based on renewable feedstock resources) owing to dye degradation under the action of some methanogenic activities [11].

The anaerobic digestion (AD) process undergoes multiple biological reactions using microorganisms (e.g., archaea, bacteria) to decompose the biodegradable material of DW under an oxygen-deprived condition [12]. This process can be employed for DW because it involves a consortium of microbes that perform a sequence of biochemical and metabolic pathways [13], including enzymatic hydrolysis, acidogenesis/fermentation, acetogenesis, and methanogenesis [14]. The main products of this AD process include biogas that can be used for electricity production, while the released biogenic CO_2 is considered "carbon neutral" [12]. Digestate, including nutrients and undigested biomass, can also be obtained from the biogas systems, and it can be managed to generate fertilizers [15]. Another study showed that the thermal treatment of solid digestate could generate biochar used for carbon storage and the enhancement of soil texture [16]. The marketing of these outputs could gain revenues that could be used to overcome the initial investment of the AD project. More research is required to estimate the time consumed to recoup the construction and equipment costs of the digesters and determine the amount of money that can be earned per year per investment.

Due to the recalcitrant nature of DW, adding a co-substrate to the fermentation medium has been used to enhance the overall biodegradation process [17]. This co-substrate (e.g., lactate and glucose) serves as an electron donor to facilitate the breakdown of complex dye compounds, avoiding volatile fatty acid (VFA) accumulation in the reactors. Some examples of co-substrates/electron donors used in the AD of textile dyes include henna plant [18], honey processing wastewater [15], pineapple wastewater [19], and ethanol [17]. Supplementing the digester with low-cost co-substrates, e.g., extracts of agricultural wastes, would stimulate microbial activity to stabilize sludge and convert more organic dyes into bio-methane (bio-CH_4) [20].

Different lignocellulosic waste materials, e.g., wheat straw (WS), paddy straw, sugarcane bagasse, wood shavings, and rice husk, have been reported to enhance the biodegradation of DW under anaerobic conditions [2]. Moreover, these crop residues are rich in multiple macro-nutrients, such as potassium (K), phosphorous (P), nitrogen (N), and aluminum (Al), along with carbohydrates and polysaccharides, which could serve as food for microorganisms' growth [20]. Because the AD process involves a diversity of microorganisms, optimizing the substrate-to-inoculum ratio (SIR) is a crucial factor for facilitating the hydrolytic acidification of poor biodegradable compounds [21]. The adaptation of SIR not only promotes process efficiency but also avoids the accumulation of inhibitory compounds in the AD systems [10].

The resultant digestate from the anaerobic co-digestion of DW could further cause environmental pollution, such as soil/groundwater contamination via nutrient leaching [22]. Therefore, the thermal treatment of this secondary waste could eliminate the remaining pollutants (e.g., inorganic elements, pathogens, and aromatic compounds) [3]. In addition, this thermal decomposition has been recognized as a waste-to-resource technique, obtaining a final product that has successful environmental applications. For instance, pyrolysis of digestate from the AD of petrochemical wastewater at 500 °C generates biochar with multiple functional groups, improved surface morphology, and a large surface area suitable for adsorbing dyes and heavy metals from aqueous solutions [12]. Biochar has also been applied to improve soil fertility and physical properties, further maximizing the crop yield pattern [23]. Because of its high carbon content, biochar can be a suitable candidate for mitigating greenhouse gas (GHG) emissions via carbon sequestration. As such, the thermal treatment of solid digestate for generating biochar can add profits to the AD process because the market value of biochar would reach up to 1.6 USD/kg [7].

Although previous studies have demonstrated the anaerobic treatment of DW by co-digestion, there is still a research gap in the economic feasibility of the treatment system, as well as the recyclability of the sludge digestate. Therefore, this study focused on combining the co-digestion/pyrolysis approach for the dual benefit of reducing pollution from textile wastewater and maintaining a sustainable strategy from digestate recycling. Methylene blue (MB) was selected as a cationic and primary thiazine dye because it is extensively used in various industries (e.g., as a colorant) and medical applications [9], and direct exposure to MB results in serious health complexities [4]. The study objectives were threefold: (1) use the extract of WS as a co-substrate to enhance biogas productivity from the AD of MB-laden solutions, (2) determine the characteristics, surface morphology, and surface functional groups of biochar derived from the pyrolysis of solid digestate after DW treatment, and (3) estimate the techno-economic feasibility from implementing the digestion/pyrolysis integrated process, regarding biogas recovery, biochar selling, chemical oxygen demand (COD) shadow price, and carbon credit trading.

2. Materials and Methods

2.1. Substrate and Inoculum Preparation

Synthetic DW was prepared to simulate the composition of real textile wastewater [6]. The components of this DW feed included (in mg/L) methylene blue (50), NH_4Cl (203), KH_2PO_4 (22), K_2HPO_4 (28), $CaCl_2$ (100), NaCl (204), and sodium acetate (205). $NaHCO_3$ solution was used to buffer the fermentation medium and maintain the desirable pH condition, whereas urea was added as a nitrogen source [24]. Trace elements (in mg/L) of $CoCl_2$ (0.1), $MgCl_2$ (4), $NiSO_4$ (0.48), $FeCl_3 \cdot 6H_2O$ (0.24), and KI (10) were used for nutrient supply [25]. The solution recorded an initial total COD of 1810 $\pm$ 131 mg/L. All the chemicals were of analytical grade.

Wheat straw (WS) was collected from an agricultural field located in Borg El Arab City in Alexandria, Egypt. Cleaning was performed carefully with distilled water to eliminate any foreign material adhering to the WS surface. The cleaned samples were reduced into small sizes and dried at 105 °C for 24 h to obtain dry biomass. The dried matter was milled and passed through a 0.1 mm sieve, and the resultant fine powder was stored in plastic bags for further utilization. The biomass was subjected to a pretreatment process to increase the bio-availability of monomeric sugars required to promote the formation of CH_4, following the previously established hypothesis [26]. For this purpose, the samples were dissolved in an alkaline solution containing 1% NaOH and heated at 120–130 °C for 2 h. After pretreatment, the biomass was filtered and washed with deionized water. The solid biomass was air-dried at room temperature and kept for further utilization in subsequent AD experimentations.

Activated sludge as the inoculum was collected from a sewage treatment plant located in Alexandria, Egypt, treating domestic wastewater. The collected sludge was initially settled for 8 h, and the supernatant was withdrawn to acquire well-concentrated microorganisms. Further, the concentrated sludge passed through a 200 μm pore screen for sieving. The inoculum recorded total solids (TS: 21.35 $\pm$ 1.00 g/L) and volatile suspended solids (VSS: 13.88 $\pm$ 0.71 g/L), leading to a VSS/TS ratio of 0.65 (Table 1).

Table 1. Physicochemical characteristics of dye wastewater and sludge inoculum.

Parameter	Unit	Value
	Dye wastewater	
pH	-	11.35 ± 0.53
Total COD	mg/L	1810 ± 131
Soluble COD	mg/L	1180 ± 59
Total organic carbon (TOC)	mg/L	2954 ± 135
Ammonia nitrogen (NH_4-N)	mg/L	34.16 ± 0.56
Volatile fatty acids (VFA)	mg/L	378.26 ± 17.39
	Sludge inoculum	
Total solids	g/L	21.35 ± 1.00
Volatile solids	g/L	16.95 ± 0.85
Volatile suspended solids	g/L	13.88 ± 0.71
Moisture content	(w/w%)	85.65 ± 3.50
Ash content	(w/w%)	17.50 ± 0.72

2.2. Experiemental Set Up

The experimental procedures included the anaerobic co-digestion of DW and WS for biogas recovery, followed by dry digestate pyrolysis for biochar production (Figure 1).

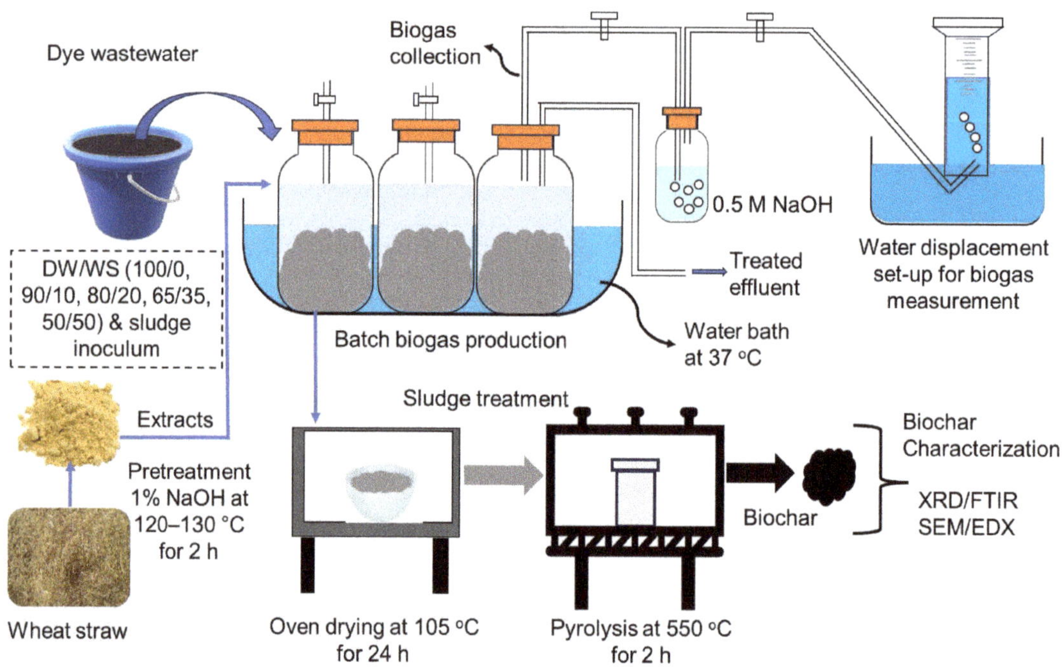

Figure 1. Experimental setup of the anaerobic co-digestion of dye wastewater (DW) and wheat straw (WS) for bio-CH_4 recovery, followed by biomass pyrolysis for biochar production.

The substrates (DW and WS) were mixed according to a previously reported methodology, with a slight modification allowing for up to 50% WS in the mixtures [27]. WS was added to DW in the proportions (%) of 100/0, 90/10, 80/20, 65/35, and 50/50 (DW/WS v/v). The batch experiments were performed using 200 mL working volume in 250 mL serum bottles, allowing for 50 mL headspace. A certain amount of sludge inoculum was added to each reactor, maintaining a SIR range of 0.28–0.50 g COD/g VSS. Both alkaline (0.5 M NaOH) and acidic (0.5 M HCl) solutions were used to adjust the reactors' pH to 7.0 ± 0.2, which is suitable for dominating methane-producing bacteria [24]. The headspaces of batch reactors were initially purged with nitrogen gas for 3–5 min and sealed with rubber stoppers and aluminum crimp caps to create anaerobic conditions. The batch experimentation assays were performed in a temperature-controlled water bath (37 °C) at a 40-rpm mixing speed. The batch tests were conducted in triplicate to obtain mean ± standard deviation values. Methane-rich biogas was obtained by passing the gas through an alkaline solution (0.5 M NaOH) and then quantified with an inverted measuring cylinder using a water displacement technique following a previous study [12].

In the second experiment, anaerobic digestate was collected after terminating the AD process and oven-dried at 105 °C for 24 h. The dry biomass was pyrolyzed at 550 °C for 2 h under an oxygen-limited condition to obtain biochar, as described elsewhere [28]. The obtained carbon material was ground (<0.1 mm particle size) and further studied for elemental compositions, crystallinity, surface morphology, and main functional groups.

2.3. Analytical Methods

Wastewater samples were analyzed for ammoniacal nitrogen (NH_4-N), volatile fatty acids (VFAs), COD, and alkalinity following APHA standard methods [29]. COD was measured using a multiparameter portable calorimeter HACH system (DR900, Loveland, CO, USA). To measure the solution color, the samples were centrifuged at 5000 rpm for 10 min and analyzed according to the procedures described by Bashiri et al. [30]. Decolorization was monitored with UV spectroscopy (Jasco V-630 spectrometer, Jasco Inc., Tokyo, Japan) at the maximum visible absorbance wavelength of 664 nm for MB detection. The sludge inoculum was analyzed for total solids (TSs) and volatile solids (VSs), following the APHA methods [29]. A portable multimeter (Lutron, YK-2001PHA, Taiwan) was used to monitor pH adjustments. The solid phase samples (AD sludge and biochar) were characterized according to previously reported procedures [12]. For example, the variation in surface functional groups was estimated using Fourier transform infrared (FTIR) spectroscopy (Bruker Optics, ALPHA, Ettlingen, Germany) in the 4000–500 cm^{-1} wavenumber range. The surface morphology was determined using a scanning electron microscope (JCM-6000PLUS Neoscope Benchtop SEM, Tokyo, Japan). The sample's main elemental compositions were detected using an energy-dispersive X-ray (EDX) spectroscopy (JOEL JSM-6510LV, Tokyo, Japan). The X-ray diffraction (XRD) instrument (XRD-7000, Shimadzu, Tokyo, Japan) was used to determine the mineral phases present in solid substrates and the associated crystallographic structures. The crystallinity index (CrI) was calculated with Equation (1) [31]:

$$CrI = \frac{I_{200} - I_{am}}{I_{200}} \times 100, \tag{1}$$

where I_{200} is the peak intensity of the 002 lattice plane (i.e., the maximum intensity of the crystalline peak), and I_{am} is the peak intensity of the amorphous phase.

2.4. Determination of the System Performance

The removal efficiency R(%) of each pollutant was calculated with Equation (2) [21]:

$$R(\%) = \frac{C_O - C_t}{C_O}, \tag{2}$$

where C_O and C_t are the initial pollutant concentration and the corresponding value at time t, respectively (mg/L).

The modified Gompertz model (Equation (3)) was used to fit the variation of methane productivity from different reactors following a previous study [24]:

$$P = P(t) \cdot \exp\left\{-\exp\left[\frac{R.e}{P(t)}(\lambda - t) + 1\right]\right\}, \tag{3}$$

where P is the cumulative methane production (mL), P(t) is the CH_4 potential (mL) at an operating time (t in day), R is the maximum CH_4 production rate (mL/d), e is Euler's number (2.71828), and λ is the lag phase to initiate bacterial growth (day) estimated using an optimization algorithm with MATLAB mathematical software (R2019b).

The COD mass balance model was performed to validate the accuracy of data obtained from the biodegradation of organic matter in the bioreactors according to Equation (4), as reported earlier [20]:

$$COD_{balance} = \frac{COD_{soluble} + COD_{biomass} + COD_{CH_4}}{tCOD_{initial}}, \quad (4)$$

where $COD_{soluble}$ is a soluble COD fraction in the final effluent (g), $COD_{biomass}$ is the COD equivalent to biomass growth (in g based on 1.42 g COD/g VSS), COD_{CH_4} is COD-to-CH_4 formation (in g based on 1 g COD/350 mL CH_4), and $tCOD_{initial}$ is the influent total COD.

2.5. Economic Estimation

The economic feasibility of the AD/pyrolysis scheme was computed based on the data derived from the COD mass balance model. Public sources, supplier quotations, and manufacturer specifications were consulted to obtain the prices for materials and equipment as capital input. Capital and operational costs were computed based on a medium-scale textile industry discharging 35 m^3 of wastewater per day (equivalent to 106.75 kg COD/day). The profits related to pollution reduction (environmental benefits) were estimated with Equation (5) [32].

$$ENV_{profit} = COD_{removed} \times P_{COD}, \quad (5)$$

where ENV_{profit} is the profit related to the prevention of environmental damages (USD/y), $COD_{removed}$ is the mass of COD removed (kg/y), and P_{COD} is the shadow price per kg of COD removed (USD/kg).

The revenues from biogas and biochar selling were computed with Equation (6) and Equation (7), respectively [6].

$$Biogas_{profit} = COD_{in} \times Y_{biogas} \times P_{biogas}, \quad (6)$$

$$Biochar_{profit} = M_d \times Y_{char} \times P_{char}, \quad (7)$$

where $Biogas_{profit}$ and $Biochar_{profit}$ are the incomes derived from biogas and biochar selling (USD/y), COD_{in} is the influent COD (kg/y), Y_{biogas} is the biogas yield (m^3/kg COD), M_d is the annual dry sludge for biochar production (kg/y), Y_{char} is the biochar yield (kg/kg), and P_{biogas} (USD/m^3) and P_{char} (USD/kg) are the biogas and biochar selling prices, respectively.

The biogenic carbon footprint was computed for biogas (Equation (8)) and biochar (Equation (9)), as reported elsewhere [20]:

$$CC_{biogas} = CH_4 \times GWP \times \text{methane density} \times CO_{2,trading}, \quad (8)$$

$$CC_{biochar} = Char_{mass} \times Carbon_{content} \times E_c \times CO_{2,trading}, \quad (9)$$

where CC_{biogas} and $CC_{biochar}$ represent the carbon credit related to the biogenic CO_2 of biogas and biochar (USD/y), respectively, CH_4 is the annual methane production (m^3/y), GWP is the global warming potential for CH_4 (i.e., the 100 year GWP for CH_4 is 25), CH_4 density = 0.717 kg/m^3, $CO_{2,trading}$ is the global carbon trading price (USD/kg CO_2), $Char_{mass}$ is the annual biochar production (kg/y), $Carbon_{content}$ is the carbon fraction in biochar (*w*/*w*%), and E_c is the carbon emission coefficient (CO_2/C = 44/12).

The time taken to recover the initial investment was calculated with Equation (10) [33]:

$$PBP = \frac{\text{Capital cost}}{\text{Net Profit}}, \tag{10}$$

where PBP is the payback period (years).

3. Results and Discussion

3.1. Performance of Anaerobic Batch Reactor toward Pollutant Removal and Bio-CH_4 Production from Dye Wastewater (Results of the First Experiment)

3.1.1. Effect of Mixing Ratio on Color and COD Removal Efficiencies

Increasing the co-substrate fraction in the feed wastewater from 10% to 35% v/v positively impacted color R(%) from 65.30 ± 3.13% to 83.70 ± 4.12%, respectively. A similar correlation between color removal and co-substrate supplementation was also observed for the treatment of real alkaline textile wastewater using pineapple wastewater as a co-substrate [19]. Their study revealed that elevating the co-substrate fraction was associated with the supply of more electron donors for catalyzing dye bond degradation [19]. A further increase in co-substrate fraction to 50% v/v insignificantly ($p > 0.05$) improved color R(%) to 89.46 ± 4.47% (see Figure 2a). It is suggested that a co-substrate fraction between 35% and 50% v/v could be suitable for supplying sufficient electron donors to enhance the dye decolorization efficiency. In a comparable study, glucose was used as an electron donor to improve MB biodegradation was initiated through the demethylation cleavage by an isolated bacterium [4]. However, this microbe's capacity to break down the heterocyclic and aromatic structure of MB dropped significantly ($p < 0.05$) to 28.31 ± 1.73% at 100/0 (% v/v) DW/WS mixing ratio. This pattern could be because of the limited substrate available to generate the required electron donors and the recalcitrant nature of the MB phenothiazine ring [5]. This inadequate electron transfer pattern suggests the occurrence of adsorption as another pathway for color removal. As such, some functional groups (e.g., —OH, —COOH, and —NH_2) on microbial cell surfaces could be involved in making complexations with the dye molecules [34]. This theory has been demonstrated previously [35], where the functional groups (e.g., C=O, —NH_2, and P—OH) on the *Klebsiella oxytoca* strains provided binding sites with MB cations to facilitate color removal.

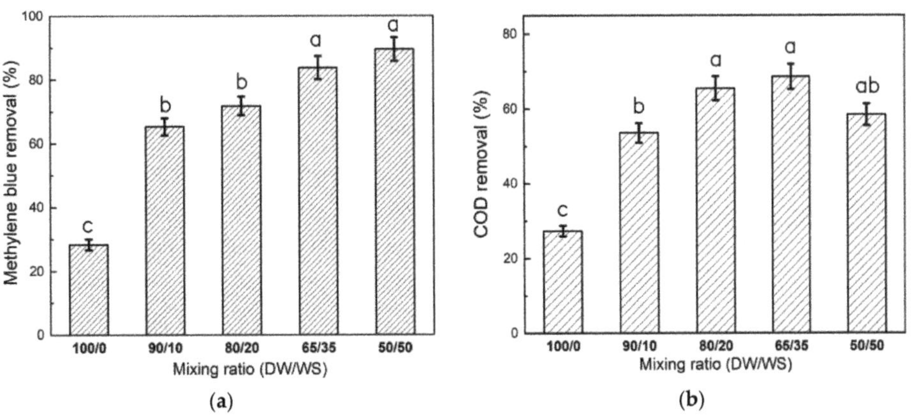

Figure 2. Effect of DW/WS mixing ratios on the removal efficiencies of (**a**) MB color and (**b**) COD. The statistical significance between DW/WS mixing ratios was derived by one-way ANOVA followed by Tukey's post hoc test estimated using MATLAB mathematical software. Groups sharing the common letter do not statistically differ from each other at a significant level of alpha = 0.05.

The reduction in COD value followed a similar pattern for the degradation of MB dye molecules, where the COD R(%) remained almost stable between 65.41 ± 3.27% and

68.52 ± 3.40% by adapting the co-substrate fractions in the 20–35% v/v range (Figure 2b). The enzymatic hydrolysis of feedstock straw could produce monosaccharides, like xylose, glucose, and galactose, utilized by microorganisms as electron donors to form colorless aromatic amines [17]. Moreover, some microbial species could utilize dye molecules and their metabolites as C/N sources [4]. However, the microbial's ability to convert COD dropped to 58.38 ± 2.92% (see Figure 2b) when the co-substrate fraction in the feed wastewater increased beyond 35% (v/v). Although the 50/50% (v/v) contained the largest fraction of readily biodegradable organics, microbial inhibition could occur due to altered nutrient balance in the culture. This condition showed a C/N ratio of 33/1, which was beyond the ratio of 30/1 reported during the treatment of textile wastewater [36]. Similarly, Malik et al. [15] noted that excessive carbon supplementation would disrupt the bacterial activity, e.g., phylum *Halobacteriota*, *Euryacrhaeota*, *Desulfobacteriota*, and *Actinobacteriota*, under anaerobic conditions to degrade dye Remazol blue because of nutrient imbalance in the mixtures. The lowest R(%) of 27.35 ± 1.37% was recorded for the batch run with no co-substrate addition (i.e., 100/0 v/v), thus elucidating microbial inhibition to degrade textile compounds at high dye concentrations. The COD removal pattern by this group reflects the microbial's ability to utilize organic matter from decaying biomass as a substrate via endogenous metabolism, a similar pathway reported previously [18]. It has also been reported that long-term exposure to textile-dyeing wastewater stress would encourage microorganisms to secrete extracellular polymeric substances (EPSs), e.g., exopolysaccharides, for soluble COD utilization [25]. In particular, using 35% v/v of WS in the DW/WS digestion process could provide a suitable environment for enabling microbial consortia to degrade the MB recalcitrant compound.

3.1.2. Effect of Mixing Ratio on Methane Production from Dye Wastewater and WS Substrate

The cumulative amount of bio-CH_4 for about 40 d was successfully fitted with the modified Gompertz model (Figure 3a), with R^2 values between 0.97 and 0.99. The onset of CH_4 production experienced an initial delay lasting 10–12 d (lag phase) required for microbial acclimatization to the dye-related environment. Following this adaptation phase, there was a gradual rise in CH_4 production, indicating an augmentation in microbial activity to facilitate substrate conversion into CH_4. The maximum cumulative CH_4 production of 113.08 ± 7.35 mL was achieved at the 65/35% v/v (DW/WS) mixing ratio. This finding was also consistent with the highest organic matter degradation (given as COD removal 68.52 ± 3.40%, as discussed in Section 3.1.1; see Figure 2b; Supplementary Table S1). This condition was also accompanied by exhibiting the maximum CH_4 production rate (R) of 5.20 ± 0.26 mL/d, with a lag period (λ) of 10.1 ± 0.1 days, indicating favorable physiological methanogenic activity for readily converting the substrate into CH_4. An extended λ was observed for the 100/0 v/v group, also corresponding to the lowest R of 1.30 ± 0.07 mL/d. This finding demonstrated that microbes needed a more extended period to acclimatize to the toxicity induced by textile-dyeing effluents.

Cumulative CH_4 production significantly increased from 65.64 ± 4.25 to 113.08 ± 7.33 mL by increasing the WS co-substrate fraction from 10% to 35%, respectively (Figure 3b). Raising the WS fraction in the digesters could offer more carbon sources to offset the nutrient limitation in DW, making the entire digestion process simpler for microbes. Digestion at 65/35 (v/v) DW/WS maintained a C/N ratio of 24.12, complying with the preferable range (C/N of 21–25) for maximizing biogas from textile wastewater with food waste and dairy effluent [37]. Similarly, the addition of 40% of food waste to garden waste and tofu waste digestion enhanced CH_4 productivity (370.54 mL/g VS) under a balanced C/N ratio (16.68–22.04), favoring the enrichment of genus *Methanosarcina archaea* species to utilize H_2/CO_2 to produce CH_4 [38]. Further, increasing the WS fraction beyond 35% showed a slight drop in P to 107.16 ± 7.25 mL observed at a 50/50 (v/v) DW/WS mixing ratio, indicating substrate underutilization (i.e., the amount of biodegradable organics could exceed the microbe utilization capacity). It is also suggested that elevating feeding substrate concentration over a certain threshold could lead to a shift in SIR to an unfavorable value, dominating hydrogenotrophic methanogenic and syntrophic bacteria [39]. Their study also demonstrated that the unpreferable SIR value caused a shift in the methanogenesis pathway (i.e., from acetoclastic methanogenesis to syntrophic pathways) with an about 60% reduction in the genus *Aminobacterium* species responsible for CH_4 production [39].

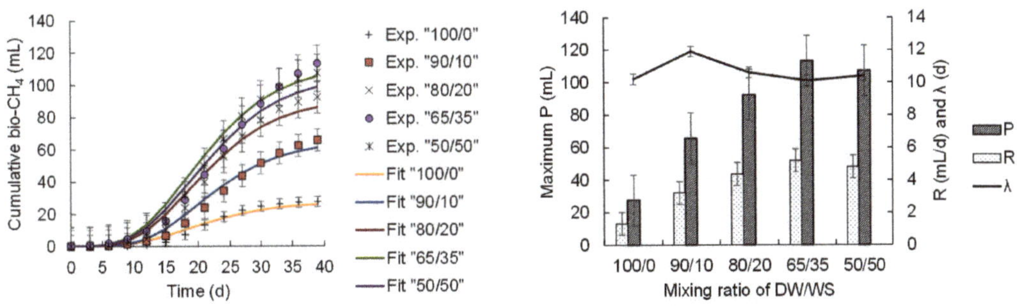

(**a**) (**b**)

Figure 3. Cumulative bio-CH_4 patterns at various DW/WS mixing ratios of 100/0, 90/10, 80/20, 65/35, and 50/50 (*v*/*v*), (**a**) experimental data "Exp." Compared with fitting the Gompertz model to data "Fit", and (**b**) Gompertz model parameters, where Maximum P is the cumulative methane production at the end of the anaerobic digestion period (mL), and R is the maximum CH_4 production rate (mL/d).

3.1.3. COD Mass Balance for Validation of Bio-CH_4 Production Data

The conversion of total influent COD into CH_4 and other end products showed a mass balance of over 85% (Table 2). Some portions of COD mass, such as non-degradable organic matter, COD stored in bacterial cells, and dissolved CH_4, were undetected in this mass balance model [12].

Table 2. COD mass balance at varying DW/WS mixing ratios of 100/0, 90/10, 80/20, 65/35, and 50/50 (*v*/*v*), showing the transformation of $COD_{initial}$ to CH_4 and biomass accumulation.

Mixing Ratios	100/0	90/10	80/20	65/35	50/50
SIR	0.50 ± 0.03	0.31 ± 0.02	0.30 ± 0.02	0.29 ± 0.01	0.28 ± 0.01
C/N	7.96 ± 0.54	11.6 ± 0.83	17.9 ± 1.22	24.12 ± 1.67	33.44 ± 2.39
Final pH	5.19 ± 0.23	6.38 ± 0.25	6.77 ± 0.29	6.85 ± 0.29	6.82 ± 0.30
$COD_{initial}$ total (mg/L)	1810 ± 131	2240 ± 155	2610 ± 180	3050 ± 218	3530 ± 252
COD_{final} total (mg/L)	1320 ± 94	1040 ± 76	902 ± 60	960 ± 69	1470 ± 109
$COD_{soluble}$ final (mg/L)	987 ± 71	865 ± 64	850 ± 63	848 ± 63	1260 ± 91
VFA/alkalinity	0.32 ± 0.02	0.27 ± 0.02	0.20 ± 0.01	0.19 ± 0.01	0.27 ± 0.02
CH_4 yield (mL/g COD)	75.90 ± 5.42	146.52 ± 10.37	176.73 ± 12.44	185.37 ± 12.58	151.78 ± 10.22
CH_4 yield (mL/g $COD_{removed}$)	280.36 ± 19.03	273.50 ± 19.25	270.06 ± 19.20	270.52 ± 19.14	260.09 ± 18.81
ΔVSS (mg/L)	125.44 ± 8.82	191.04 ± 13.07	252.21 ± 17.35	358.53 ± 25.80	369.15 ± 26.24
$COD_{biomass}/COD_{initial}$ (%)	9.84 ± 0.72	12.11 ± 0.82	13.72 ± 1.04	16.69 ± 1.37	14.85 ± 1.14
$COD_{CH_4}/COD_{initial}$ (%)	21.69 ± 1.58	41.86 ± 3.03	50.49 ± 3.31	52.96 ± 3.61	43.37 ± 2.84
$COD_{soluble}/COD_{initial}$ (%)	54.53 ± 3.95	38.62 ± 2.52	32.57 ± 2.19	27.80 ± 2.11	35.69 ± 2.71
COD mass balance (%)	86.06 ± 5.72	92.59 ± 6.82	96.78 ± 7.07	97.46 ± 7.22	93.91 ± 6.87

Increasing co-substrate concentration in the feed provided a favorable environment for microbial growth, probably due to the availability of easily degradable organics in WS, which also justifies the highest biomass growth ($COD_{biomass}/COD_{initial}$ = 16.69 ± 1.37%) at 65/35 mixing ratio. In particular, WS could be microbially consumed and fermented into glucose, VFAs, reducing sugars, soluble proteins, and other low molecular weight monomers to induce metabolic enzymes and co-enzymes [18]. Moreover, WS addition maintained a proper medium's buffering capacity (VFA/alkalinity = 0.19) with a pH condition of 6.85 ± 0.29, which is within a favorable range (pH 6.6–7.3) for acetoclastic methanogens growth [40]. Likely, methane production returned the highest yield ($COD_{CH_4}/COD_{initial}$) of 50–53% (Table 2) at 80/20 and 65/35 (*v*/*v*) DW/WS, probably due to the adaptation of methanogenic consortia to the balanced SIR range. Similarly, the methanogens grown under the optimum pH range (6.4–7.2) during garden waste (GW) digestion maintained better CH_4 potential, assigning to the quick adaptation of the genus *Methanosaeta archaea* species to the balanced SIR range [38].

For the group with no co-substrate (100/0 *v*/*v*), most of the feed COD remained in soluble form ($COD_{soluble}/COD_{initial}$ = 54.53 ± 3.95%). This poor COD conversion efficiency could be attributed to the high fraction of DW in the mixture and the lack of interaction between C—O as electron donors to the aromatic ring of MB (electron acceptor) [41]. Some microorganisms are encouraged to activate physiological antioxidative defense systems (release more metabolic secretions) during exposure to the toxic effects of textile dyes, increasing the COD soluble fraction in the final effluent [42]. Moreover, higher dye concentrations could be associated with faster cell lysis and death (i.e., biomass decay that inhibits methanogenesis), giving the lowest $COD_{biomass}/COD_{initial}$ of 9.84 ± 0.72%. This condition was also associated with the smallest $COD_{CH_4}/COD_{initial}$ (21.69 ± 1.58%) fraction, possibly due to substrate limitation to supply the required nutrients to microbial consortia. The insufficient $COD_{CH_4}/COD_{initial}$ percentage could also be validated by a pH drop to 5.19 ± 0.23, attributed to the accumulation of VFAs in the digester (i.e., a high portion of COD_{feed} was acidified and not converted to CH_4). Similarly, lower methane conversion efficiency was observed under pH-stressed AD (pH< 6.0) due to the dominance of acidogenesis over methanogenesis responsible for VFA accumulation and subsequently raising $COD_{soluble}$ in the final supernatant [43].

3.2. Biochar Production (Results of the Second Experiment)

3.2.1. Morphology and Elemental Composition of Raw Sludge, Exhausted Sludge, and Biochar

The pyrolysis of textile sludge yielded 61.3% biochar (0.613 g/g sludge), complying with the 0.40–0.73 g/g dry sludge range reported earlier [44]. Moreover, this value closely aligns with the biochar yield (0.60 g/g sludge) obtained from the pyrolysis of petrochemical digestate [12]. The percentage loss in weight could be associated with the dehydration (release of moisture) and volatilization of some organic compounds [45]. The main elements in the ash were Si, Al, Fe, P, Ca, Mg, and K, showing ash contents of 28.38%wt for sludge and 47.78%wt for biochar. These values agree with the enrichment of inorganics (ash) in the biochar by the order of 1.4–2.4 times the ash fraction in the raw sludge [44].

The SEM morphology of raw sludge was relatively compact, with well-arranged layers and almost vacant pores (Figure 4a). Some bright white spots appeared on the exhausted sludge surface, probably because various bacterial species were more metabolically active within the surface layer to form granular structures [46]. In a comparable study, granular sludge containing decolorization and degradation microbial communities was responsible for dye mineralization, generating aromatic amines [47]. The folding structure on the sludge surface subsequently disappeared in biochar, depending on the decomposition and carbonization degree of organic matter at higher temperatures. The gradual increase in the concentration of inorganic components (e.g., minerals and metals) during pyrolysis could make the biochar pores smaller, further increasing the associated surface area [22]. This pattern, in addition to the increased decomposition of organic substances after pyrolysis, could increase the ash content and other residues occupying the pores initially present in the digestate.

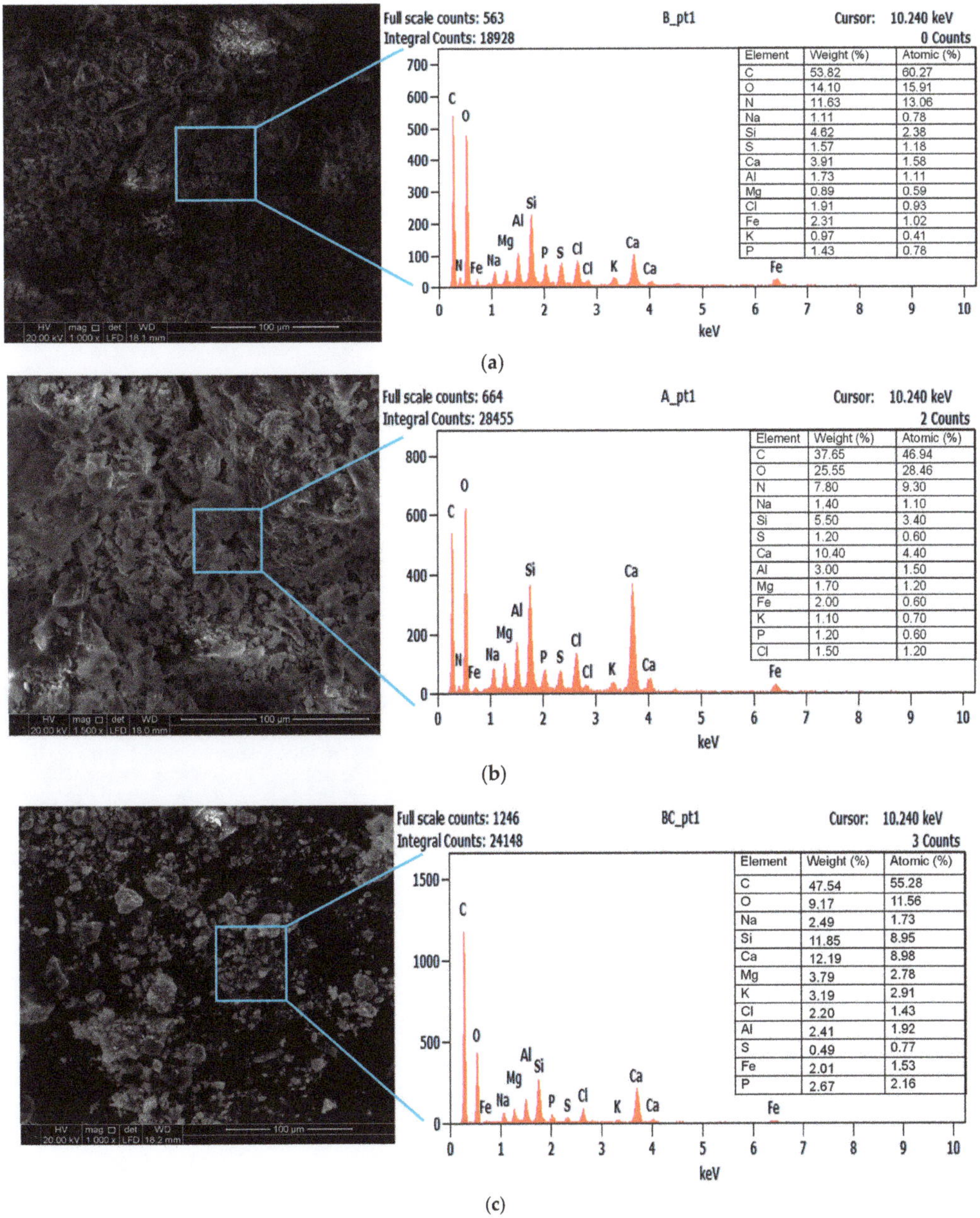

Figure 4. SEM/EDX characterization of (**a**) raw sludge, (**b**) exhausted sludge, and (**c**) anaerobic sludge biochar.

The raw sludge particles contained elements, e.g., carbon (C: 53.82%), oxygen (O: 14.10%), and nitrogen (N: 11.63%). It also compressed various inorganic microelements, e.g., Al, Si, Na, Cl, Ca, Mg, P, K, and Fe (Figure 4a). The elemental contents of C and N reduced by 30% and 33%, respectively, after the AD process (see exhausted sludge in Figure 4b). These elements could have been utilized by the methanogenic microorganisms.

Pyrolysis temperatures enriched most of the inorganic elements present in the sludge (see biochar in Figure 4c), possibly due to the gradual loss of volatile organic contents (C, H, N, and O) during the sludge thermal treatment. These features are comparable with the characteristics of biochar prepared from the fermenter digestate, showing enrichment of the Ca, Mg, Na, Si, S, P, K, and Fe elements [48]. The increased amount of cationic elements (e.g., Na^+, Mg^{2+}, Ca^{2+}, and K^+) supports the use of biochar for heavy metal adsorption from wastewater via the ion exchange mechanism [12]. The presence of K, Mg, Ca, S, and P elemental composition makes biochar suitable for soil amendment applications. The reduction in N and S contents suggests their volatilization, releasing gaseous products as NO_X, NH_3, and SO_2 [45]. The C content presented a significant increase of 26%, and the O/C ratio reached ≈0.2 after pyrolysis (Figure 4c). It is supposed that the amorphous regions of digestate were partially disintegrated, and the degree of carbonization and aromatization was enhanced, yielding a stable biochar material against the biological decomposition in the soil environment [16]. This low O/C ratio could also be linked to the decomposition of oxygen functional groups, e.g., –C=O, –COOH, and O–H, in the digestate during the decarbonylation and decarboxylation processes [49]. Based on the aforementioned properties, the prepared biochar possesses a high carbon sequestration potential (O/C atomic ratio < 0.2), with an estimated half-life of over 1000 years [50].

3.2.2. Crystallographic Structure of Raw Sludge, Exhausted Sludge, and Biochar

The XRD of the raw sludge (Figure 5a) showed a broad non-crystalline peak at $2\theta = 20.93°$, corresponding to the (002) plane. Other diffraction peaks appeared at $2\theta \approx 26.67°$ (101), 29.39° for calcite [51], and 42.65° for quartz [52], representing the sludge major component. However, more peaks appeared in the exhausted sludge, possibly explaining the reduction in the amorphous cellulose content [53]. Incrementing the crystalline cellulose was further validated by having a greater CrI value of 44.5% for exhausted sludge compared with 32.5% for raw sludge. As such, some microbes tended to cleave the amorphous fraction of cellulose during the AD process. Disrupting the amorphous cellulose through the enzymatic hydrolysis of water hyacinth was also accompanied by increasing CrI from 50.13% to 60.21% [3].

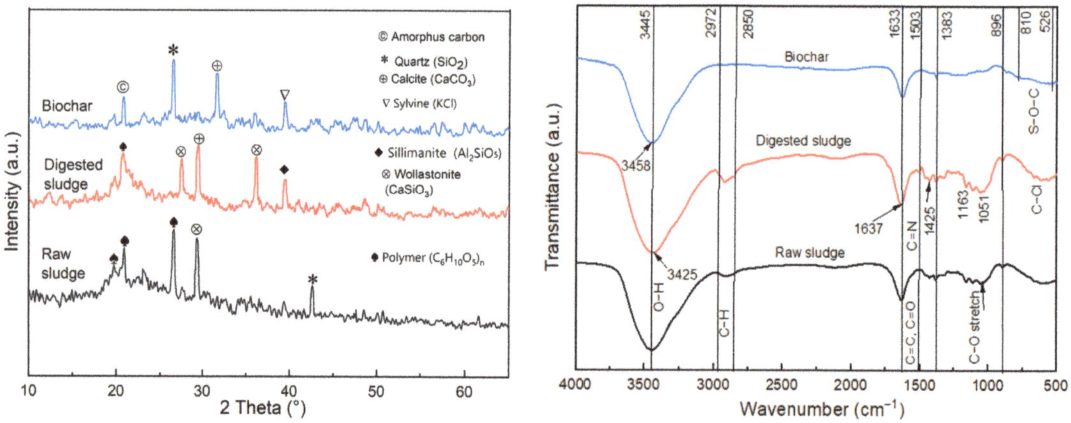

(**a**) (**b**)

Figure 5. Characterization of raw sludge, exhausted sludge, and anaerobic sludge biochar for (**a**) crystallinity and (**b**) main functional groups.

The crystalline phases of the digestate denoted the presence of silicates, aluminosilicates, and carbonates, e.g., wollastonite ($CaSiO_3$, PDF#00-001-0720), sillimanite (Al_2SiO_5), and calcium carbonate as calcite ($CaCO_3$, PDF#01-072-1650) [54], due to sufficient Si, Ca, and Al contents of WS. The intensity of these crystalline peaks decreased in biochar, indicating a shift toward a more amorphous structure due to the decomposition of crystalline domains within the solid digestate under pyrolysis temperatures [55]. The CrI reduced from 44.51% to 35.65% after pyrolysis, elucidating the formation of more amorphous carbon [56].

The mineral phases of the biochar could be monitored by the diffraction peaks at $2\theta = 26.61°$ for SiO_2 (silica quartz structure) [57], confirming the high ash content of biochar. Other peaks at $2\theta = 31.65°$ (dellaite) and 39.47° (crystalline components of sylvine, KCl) [58] were noticed for the biochar XRD pattern. This pattern also comprised other inorganic minerals related to magnetite (Fe_3O_4 crystalline, PDF#01-076-0955) and aluminum oxide II (Al_2O_3, PDF#01-075-0278). In a comparable study [59], sludge-derived biochar also contained quartz and kaolinite as the major crystalline minerals and major elements of Ca, Al, Fe, and Si.

3.2.3. Main Functional Groups of Raw Sludge, Exhausted Sludge, and Biochar

FTIR spectra of the raw sludge exhibited dominant peaks at 3445 cm^{-1} (–OH) for the hydroxyl group and 2970–2850 cm^{-1} (C–H) in the aliphatic CH_2 and CH groups (Figure 5b). The intensity of the sludge FTIR peaks showed a slight shift after AD, assigned to the microbial degradation of the substrate structure into CH_4, CO_2, and other metabolites. For instance, the shift in band for the O–H group around 3425 cm^{-1} could be assigned to the binding of MB with the sludge biomass due to N–H stretching [60]. The hydrogen atoms on the adsorbent hydroxyl groups could form hydrogen bonds with nitrogen atoms in MB. A previous article also demonstrated that a shift in transmittance peak from 3407 to 3411 cm^{-1} could describe the occurrence of hydrogen bonding between the nitrogen atoms (N^{+}) in the MB molecule and OH^{-} (as negatively charged hydroxyl groups) [3]. The absorption peaks between 2850 and 2972 cm^{-1} for the aliphatic C–H stretching bonds were slightly widened and shifted, probably due to the activity of bacterial species able to decompose aliphatic fraction contents accompanied by the conversion of VFAs into CH_4 and CO_2 [61]. The peak at ≈2900 cm^{-1} could represent the C–H stretching vibration in the methylene group [62], suggesting that MB biodegradation initiated through the demethylation cleavage by some bacterial species [4]. Some peaks appeared in the treated sample, as compared with the raw sludge, suggesting the dye degradation scheme (e.g., the C=N and C—S bonds in the MB molecular structure could be oxidized to generate small molecular substances [9]). The appearance of a new vibration band at 1503 cm^{-1} agrees with the bending vibrations of MB molecule adsorbed into sludge, while narrowing the band characteristic of OH vibration suggests MB molecule fixation [63]. In a similar study, the adsorption of MB onto the bacterial cell wall of sludge could be supported by C=O electrostatic attraction, $-NH_2$, P-OH hydrogen bonding, and van der Waals forces [35]. Similarly, the appearance of peaks in the 820–557 cm^{-1} region, describing the C-Cl stretching [64], could justify the interaction of MB molecules with the sludge matrix.

Some peaks at 2972–2850 cm^{-1} and 1503 cm^{-1} disappeared after sludge pyrolysis, indicating the thermal decomposition of the organic compounds into gaseous products. The decrease in the intensity of multiple peaks at 1383–1051 cm^{-1} could also indicate the transformation of the remaining organic fatty hydrocarbons of sludge into volatiles under the thermal treatment process. The peaks in the 1633–1503 cm^{-1} range could represent the conjugated aromatic ring stretching, where the C=C bonds in biochar suggested the formation of stable aromatic ring systems [65]. The bands at the 1163–900 cm^{-1} region could be related to the polysaccharide components [60]. The weak peaks between 810–526 cm^{-1} could reveal the existence of Si–O–Si and Si–C, consistent with the presence of quartz in the as-prepared biochar (see Figure 5a). These functional groups would promote the cation-exchange capacity of the soil and improve crop nutrition during the land application of biochar [49].

3.3. Economic Analysis for the Anaerobic Co-Digestion System Treating Dye Wastewater

The economic feasibility of the biogas and biogas/biochar scenarios was performed based on the optimized mixing ratio of 65/35 (v/v) DW/WS for a medium-sized scheme treating 35 m^3/day of DW (Table 3). The operational parameters of sludge retention time (SRT) = 14.6–16.2 d and food-to-microorganism (F/M) ratio of 0.68–0.71 g COD/g VSS/d complied with a previous study [66], avoiding VFA accumulation and maintaining balance between the faster-growing bacteria and the slower-growing groups. The computation of capital costs was based on elements like construction facilities, mechanical and electrical apparatus, piping and fittings, and site footprint, following the previous study [32]. The total capital cost of scenario#2 was estimated at USD 28,500 and reduced by ≈22% due to the exclusion of the pyrolysis unit. Contingencies maintained USD 2400, which was estimated at ≈9% of the capital costs. The prices of construction materials for digesters with a volume of 15 m^3 (hydraulic retention time of 10.29 h) were based on manufacturer specifications, local contractors, and supplier quotes.

Electrical energy demands varied among the two scenarios and dominated most of the operational costs, accounting for 34.02% and 40.84% for scenario#1 and scenario#2, respectively. This cost comprised the energy consumed by electrical appliances and a source of heat to maintain mesophilic conditions for the effective digestion of DW. It also considered the power required to operate a screw press for sludge dewatering, further facilitating the pyrolysis of dry sludge [67]. The energy consumed by drying sludge (assuming 105 °C for 24 h) was computed based on the previous study [68]. Energy demand for pyrolysis at 550 °C was estimated from 150 kWh/t total solids [68], assuming 4 h of daily operation. Elevating the electrical energy costs for scenario#2 was due to additional energy demands for maintaining pyrolysis temperatures.

In both scenarios, the methanogenic microbes could convert 52.96 ± 3.61% of the feed COD into CH_4 (see COD mass balance in Table 2), giving a daily biogas production of 19.79 m^3. CH_4 production would positively reduce the carbon footprint from the uncontrolled decomposition of waste organics. Thus, the carbon credit earned from biogenic CH_4 was estimated at 355 kg CO_2/d, contributing 1277 USD/y in environmental benefits. This system was also responsible for reducing COD pollution up to 68.52 ± 3.40% to give a yearly benefit of USD 2107, expressed as a profit of 0.08 USD/kg $COD_{removed}$ [12]. This shadow profit reflects the benefit of avoiding environmental damages from the uncontrolled disposal of untreated DW. Although installing a pyrolysis unit raised capital and operating costs of scenario#2, biochar production could add extra revenues to the cash flow. Sludge accumulation was determined from the COD conversion efficiency ($COD_{biomass}$ = 16.69 ± 1.37% of $COD_{initial}$; see Table 2), which could yield 3.76 kg/d of solid digestate (estimated from 1.42 g COD/g biomass conversion ratio). The pyrolysis scheme would increase the profitability of scenario#2, where the accumulated sludge would produce 2.31 kg biochar/d estimated from 3.76 kg solid digestate/d multiplied by a biochar yield of 0.163 kg/kg dry sludge. The GHG offset benefit was estimated from the amount of carbon stored in biochar, giving 1.11 kg/d (using the C fraction of 48% in EDX; Figure 4c). With the carbon trading price of 10 USD/ton CO_2 and the biochar price of 1.5 USD/kg, the pyrolysis scheme could deliver an extra 2616 USD/y to the cash flow of scenario#2. The biochar price is within the comparable range of the global biochar price (0.09–8.85 USD/kg), as reported elsewhere [69]. Conversely, the net profits of scenario#1 suffered from sludge disposal expenses compared with scenario#2.

Incorporating a pyrolysis unit would recover the initial investment capital in 6.5 years compared with 8.4 years for the AD alone scenario. Moreover, the profits of scenario#2 would maintain a positive net present value (NPV) with an internal rate of return greater than the project discount rate (8%). These findings demonstrated that the study outputs could pave the way toward the circular economy of recalcitrant textile dyes and residual sludge, meeting several sustainable development goal (SDG) targets (see Supplementary Figure S1). Further studies are required to consider the liquid (bio-oil) and gases (e.g., non-condensable gases with a high-calorie value) generated from the co-pyrolysis of sludge with wood biomass and estimate the profitability criteria with the NPV.

Table 3. Economic analysis for the AD and AD/pyrolysis scenarios treating dye wastewater.

Item	Scenario#1 AD	Scenario#2 AD/Pyrolysis	Justification
		Capital costs (USD)	
Construction	10,000	10,000	Included site footprint, screening unit, receiving tanks, anaerobic unit (assumed as 15 m^3 of lightweight structural concrete with a foundation), piping, and fitting. Maintained 35–45% of the total capital costs
Mechanical equipment	9800	11,100	Included the cost of pumps for lifting wastewater and sludge, a screw press for sludge dewatering, a heat exchanger, and a mixer. Maintained 40–44% of the capital costs
Pyrolysis unit	0	5000	Purchase price and shipping of a pyrolysis unit. The price included an additional condenser, furnace, and gas cleaning equipment.
Contingency	2400	2400	About 9% of the capital costs
Total	22,200	28,500	The extra cost of purchasing and installing a pyrolysis unit increases the capital costs of scenario#2 by about 30% compared with scenario#1
		Operating costs (USD/y)	
Electrical energy consumption	1218	1819	Energy consumption for maintaining mesophilic conditions (37 °C), drying, pyrolysis, and operating equipment (motors, pumps, lighting) computed by adding up individual energy demands assuming 0.05 USD/kWh electricity tariff
Chemical utilization	962	1062	Chemicals and reagents, e.g., for pH adjustment, buffering, nutrients, for effective microbial degradation assuming a unit cost of 0.03 USD/kg
Water utilization	485	533	Water for mixing substrates, cleaning assuming a unit cost of water 0.02 USD/m^3
Maintenance	380	420	About 10% of the operating costs
Overhead and Staff salaries	535	620	Taxes, process control, legal fees, insurance, and office supplies
Total operating costs	3580	4454	Extra electrical energy consumption for scenario#2 for pyrolysis
		Profits (USD/y)	
Pollution reduction (COD removal)	2107	2107	Calculated from an organic load of 106.75 kg COD/d, COD removal efficiency of 68.52%, and 0.08 USD/kg COD removed. These values correspond to the design criteria of an organic loading rate of 7.12 kg/m^3/d and an F/M ratio of 0.69 d^{-1}
Biogas selling	2850	2850	Computed from the COD mass balance of COD_{CH4}/COD_{in} = 52.96% and assuming a unit cost of 0.4 USD/m^3 of biogas
Carbon credit (biogas)	1277	1277	The biogenic carbon emissions were calculated from CO_2 trading of 10 USD/ton CO_2
Biochar selling	0	1246	Derived from $COD_{biomass}/COD_{initial}$ of 16.69%, biochar yield of 0.613 g/g, and 1.5 USD/kg biochar selling price
Carbon credit (biochar)	0	15	Calculated from carbon content in biochar = 47.54% (see EDX characteristics in Figure 4c), and CO_2/C molar ratio of 44/12.
Sludge disposal	0	1355	Denotes transportation costs and fees paid for sludge disposal (without pyrolysis) in a landfill. The amount of sludge generated (10.04 kg/d) was based on an SRT of 15.49 d, which is considered suitable to avoid VFAs accumulation and maintain a balance between the faster-growing bacteria and the slower-growing groups [66]
Total profits	6233	8849	Additional profits for scenario#2 from biochar selling and carbon credit from biochar

Table 3. *Cont.*

Item	Scenario#1 AD	Scenario#2 AD/Pyrolysis	Justification
		Profitability criteria	
Net profit (USD/y)	2653	4395	Computed as the difference between the total annual revenues and the annual operating costs
Payback period (PBP in y)	8.4	6.5	This time should be compared with the project lifetime (10 y)
NPV (USD)	−4396	989	Net present value is estimated using an 8% discount rate and a project lifetime of 10 years
IRR (%)	3.4%	8.8%	Internal rate of return should be greater than or equal to the project discount rate (8%)

4. Conclusions

This study successfully developed a sustainable strategy that could cost-effectively manage recalcitrant dye wastewater (DW) based on biological and thermal treatment processes. Adding extracts of wheat straw (WS) to DW under anaerobic conditions enhanced the methylene blue (MB) degradation efficiency by ≈3-fold higher than the mono-digestion assays (only DW). The 35% v/v WS supplementation showed a better synergistic interaction with DW, in which 52.96 ± 3.61% of the influent COD was transformed to bio-CH_4 production. Moreover, the thermal treatment of solid digestate depicted a promising route for sludge management, getting biochar characterized by a high degree of aromaticity (O/C ≈ 0.2). The profits related to pollution reduction, biogas, biochar, and carbon trading showed that the integrated digestion/pyrolysis scenario would overcome capital costs in a time less than the project lifetime (<10 y). The profitability scenario of the digestion/pyrolysis project also revealed a positive net present value, with an internal rate of return greater than the discount rate (8%). Further studies are required to determine the best post-treatment phase that could achieve complete mineralization of the aromatic compounds to meet the discharge standards. Moreover, some future studies should aim to identify the microbial strains and enzymatic activities involved in the degradation processes and carry out phytotoxicity analysis of the biodegradation products.

Supplementary Materials: The following supporting information can be downloaded at https://www.mdpi.com/article/10.3390/w16142025/s1, Table S1. Data of CH_4 recovery from the anaerobic digestion of dye wastewater compared with the current study; Figure S1: Correlating the study output from the anaerobic digestion (scenario-1) and anaerobic digestion/pyrolysis (scenario-2) treating dye wastewater to the sustainable development goals (SDGs).

Author Contributions: Conceptualization, M.N., M.F. and M.G.I.; methodology, A.T.; software, A.T.; formal analysis, A.T.; writing—original draft preparation, A.T.; writing—review and editing, M.N., M.F. and M.G.I.; visualization, M.N., M.F. and M.G.I.; supervision, M.N., M.F. and M.G.I. All authors have read and agreed to the published version of the manuscript.

Funding: This research was funded by TICAD7, the Egypt-Japan University of Science and Technology (E-JUST), and the Japanese International Corporation Agency (JICA).

Data Availability Statement: The data that support the findings of this study are within the article (and or) its Supplementary Materials.

Acknowledgments: The author is very grateful to TICAD7 for providing financial support in the form of an MSc. scholarship. Also, thanks to JICA—Japan International Corporation Agency for providing all the facilities and equipment needed to conduct this study.

Conflicts of Interest: The authors declare no conflicts of interest.

References

1. Nnaji, C.; Ume, S.; Obasi, U.; Anadebe, C.; Ezemagu, G.; Okeke, U.; Onukwuli, D. Machine learning-based performance evaluation and sludge characterization studies of oxidized starch-aluminum electrode assisted by direct current treatment of dye laden wastewater. *Results Eng.* **2023**, *20*, 101576. [CrossRef]
2. Saratale, G.R.; Gandhi, S.A.; Purankar, V.M.; Kurade, B.M.; Govindwar, P.S.; Oh, E.S.; Saratale, D.G. Decolorization and detoxification of sulfonated azo dye C.I. Remazol Red and textile effluent by isolated *Lysinibacillus* sp. *RGS J. Biosci. Bioeng.* **2013**, *115*, 658–667. [CrossRef] [PubMed]
3. Msemwa, G.; Ibrahim, M.; Manabu, F.; Nasr, M. Phytomanagement of textile wastewater for dual biogas and biochar production: A techno-economic and sustainable approach. *J. Environ. Manag.* **2022**, *322*, 116097. [CrossRef] [PubMed]
4. Kishor, R.; Saratale, D.; Saratale, G.; Ferreira, R.; Bilal, M.; Iqbal, N.; Bharagava, N. Efficient degradation and detoxification of methylene blue dye by a newly isolated ligninolytic enzyme producing bacterium *Bacillus albus* MW407057. *Colloids Surf. B Biointerfaces* **2021**, *206*, 111947. [CrossRef] [PubMed]
5. Ogunlaja, A.; Nwankwo, N.; Omaliko, E.; Olukanni, D. Biodegradation of Methylene Blue as an Evidence of Synthetic Dyes Mineralization during Textile Effluent Biotreatment by *Acinetobacter pittii*. *Environ. Process.* **2020**, *7*, 931–947. [CrossRef]
6. Kalengyo, B.R.; Ibrahim, G.M.; Fujii, M.; Nasr, M. Utilizing orange peel waste biomass in textile wastewater treatment and its recyclability for dual biogas and biochar production: A techno economic sustainable approach. *Biomass Convers. Biorefin.* **2023**. [CrossRef]
7. Demarema, S.; Nasr, M.; Ookawara, S.; Abdelhaleem, A. Enhanced synergistic system for the persulfate activation under visible light using novel N–ZnO photocatalyst supported on Lantana camara-based biochar. *Chemosphere* **2024**, *349*, 140840. [CrossRef] [PubMed]
8. Sun, Y.; Cheng, S.; Lin, Z.; Yang, J.; Li, C.; Gu, R. Combination of plasma oxidation process with microbial fuel cell for mineralizing methylene blue with high energy efficiency. *J. Hazard. Mater.* **2020**, *384*, 121307. [CrossRef] [PubMed]
9. Wu, K.; Shi, M.; Pan, X.; Zhang, J.; Zhang, X.; Shen, T.; Tian, Y. Decourization and biodegradation of methylene blue dye by ligninolytic enzyme-producing *Bacillus thuringiensis*: Deradation products and pathway. *Enzym. Microb. Technol.* **2022**, *156*, 109999. [CrossRef]
10. Labayen, J.J.; Yanac, K.; Yuan, Q. Effect of substrate-to-inoculum ratio on anaerobic digestion of treated and untreated cotten textile waste. *Int. J. Environ. Sci. Technol.* **2021**, *18*, 287–296. [CrossRef]
11. Kumar, V.; Thakur, C.; Chaudhari, K. Anaerobic biological treatment of dye bearing water in anaerobic sequencing batch reactor: Performance and kinetics studies. *J. Indian Chem. Soc.* **2022**, *99*, 10. [CrossRef]
12. Atukunda, A.; Ibrahim, G.M.; Fujii, M.; Ookawara, S.; Nasr, M. Dual biogas/biochar production from anaerobic co digestion of petrochemical and domestic wastewater: A techno economic and sustainable approach. *Biomass Convers. Biorefin.* **2024**, *14*, 8793–8803. [CrossRef]
13. Rahman, U.W.; Khan, D.M.; Khan, Z.M. Anaerobic biodegradation of benzene-laden wastewater under mesophilic environment and simultaneous recovery of methane-rich biogas. *J. Environ. Chem. Eng.* **2018**, *6*, 2957–2964. [CrossRef]
14. Ikram, M.M.; Zahoor, M.; Hanafiah, M.; Oyekanmi, A.; Ullah, R.; Farraj, D.; Gulfam, N. Biological Degradation of the Azo Dye Basic Orange 2 by *Escherichia coli*: A Sustainable and Ecofriendly Approach for the Treatment of Textile Wastewater. *Water* **2022**, *14*, 2063. [CrossRef]
15. Malik, A.R.; Vistanty, H.; Suhardi, H.S. Performance of anaerobic co-digestion with honey processing wastewater as co-substrate for treating synthetic wastewater containing commercial anthraquinone dye Remazol blue RSP: Effect of C:N ratio and HRT. *Bioresour. Technol. Rep.* **2022**, *19*, 101157. [CrossRef]
16. Liu, Y.; Ran, C.; Siyal, A.A.; Song, Y.; Jiang, Z.; Dai, J.; Zhang, T. Comperative study for fluidized bed pyrolysis of textile dyeing sludge and municipal sewage sludge. *J. Hazard. Mater.* **2020**, *396*, 122619. [CrossRef] [PubMed]
17. Rasool, K.; Mahmoud, A.K.; Lee, S.D. Influence of co-substrate on textile wastewater treatment and microbial community changes in the anaerobic biological sulfate reduction process. *J. Hazard. Mater.* **2015**, *299*, 453–461. [CrossRef] [PubMed]
18. Huang, J.; Chu, S.; Chen, J.; Chen, Y.; Xie, Z. Enhanced reduction of an azo dye using henna plant biomass as a solid-phase electron donor, carbon source, and redox mediator. *Bioresour. Technol.* **2014**, *161*, 465–468. [CrossRef] [PubMed]
19. Nuid, M.; Aris, A.; Krishnen, R.; Chelliapan, S.; Muda, K. Pineapple wastewater as co-substrate in treating real alkaline, non-biodegradable textile wastewater using biogranulation technology. *J. Environ. Manag.* **2023**, *344*, 118501. [CrossRef] [PubMed]
20. Tugume, M.; Ibrahim, G.; Fujii, M.; Nasr, M. Management of Cheese Whey Wastewater and Greywater for Dual Biogas and Biochar Production: A Techno Economic and Sustainable Approach. *Waste Biomass Valorization* **2024**, *15*, 4373–4393. [CrossRef]
21. Elreedy, A.; Fujii, M.; Tawfik, A. Factors affecting on hythane bio-generation via anaerobic digestion of mono-ethylene glycol contaminated wastewater: Inoculum-to-substrate ratio, nitrogen-to-phosphorus ratio and pH. *Bioresour. Technol.* **2017**, *223*, 10–19. [CrossRef] [PubMed]
22. Alghashm, S.; Song, L.; Liu, L.; Ouyang, C.; Zhou, J.; Li, X. Improvement of Biogas Production Using Biochar from Digestate at Different Pyrolysis Temperatures during OFMSW Anaerobic Digestion. *Sustainability* **2023**, *15*, 11917. [CrossRef]
23. Khan, R.; Shukla, S.; Kumar, M.; Zuorro, A.; Pandey, A. Sewage sludge derived biochar and its potential for sustainable environment in circular economy: Advantages and challenges. *J. Chem. Eng.* **2023**, *471*, 144495. [CrossRef]

24. Tawfik, A.; Ali, M.; Danial, A.; Zhao, S.; Meng, F.; Nasr, M. 2-Biofuels (H_2 nad CH_4) production from anaerobic digestion of biscuits wastewater: Experimental study and techno-economic analysis. *J. Water Process Eng.* **2021**, *39*, 101736. [CrossRef]
25. Barathi, S.; Aruljothi, N.; Karthik, C.; Padikasan, A.; Ashokkumar, V. Biofilm mediated decolorization and degradation of reactive red 170 dye by the bacterial consortium isolated from the dyeing industry wastewater sediments. *Chemosphere* **2022**, *286*, 131914. [CrossRef] [PubMed]
26. Rahmani, M.A.; Tyagi, K.V.; Gunjyal, N.; Kazmi, A.A.; Ojha, P.S.; Moustakas, K. Hydrothermal and Thermal-alkali pretreatments of wheat straw: Co-digestion, subtrate solubilization, biogas yield and kinetic study. *Environ. Res.* **2023**, *216*, 114436. [CrossRef]
27. Jadhav, U.U.; Dawkar, V.V.; Telke, A.A.; Govindwar, P.S. Decolorization of Direct Blue GLL with enhanced lignin peroxidase enzyme production in Comamonas sp UVS. *J. Chem. Technol. Biotechnol.* **2009**, *84*, 126–132. [CrossRef]
28. Liu, C.; Li, H.; Ni, Q.J.; Zhuo, G.; Chen, W.; Zheng, Y.; Zhen, G. Effect of municipal sludge-based biochar produced at different pyrolysis temperatures on humification and oxytetracycline degradation of pig manure composting. *Sci. Total Environ.* **2024**, *906*, 167816. [CrossRef] [PubMed]
29. APHA. *Standard Methods for the Examination of Water and Wastewater*, 23rd ed.; American Public Health Association: Washington, DC, USA, 2017.
30. Bashiri, B.; Fallah, N.; Bonakdarpour, B.; Elysai, S. The development of aerobic granules from slaughterhouse wastewater in treating real dyeing wastewater by Sequencing Batch Reactor (SBR). *J. Environ. Chem. Eng.* **2018**, *6*, 5536–5543. [CrossRef]
31. Segal, L.; Creely, J.J.; Martin, A.E., Jr.; Conrad, C.M. An Empirical Method for Estimating the Degree of Crystallinity of Native Cellulose Using the X-ray Diffractometer. *Text. Res. J.* **1959**, *29*, 786–794. [CrossRef]
32. Ansari, A.; Ravindran, B.; Gupta, K.; Nasr, M.; Rawat, I.; Bux, F. Techno-economic estimation of wastewater phycoremediation and environmental benefits using Scenedesmus obliquus microalgae. *J. Environ. Manag.* **2019**, *240*, 293–302. [CrossRef] [PubMed]
33. Tawfik, A.; Bakr, H.M.; Nasr, M.; Haider, J.; k.Al Mesfer, M.; Lim, H.; Qyyum, M.A.; Lam, S.S. Economic and environmental sustainability for anaerobic biological treatment of wastewater from paper and cardboard manufacturing industry. *Chemosphere* **2022**, *289*, 133166. [CrossRef] [PubMed]
34. Danouche, M.; Arroussi, E.H.; Bahafid, W.; Ghachtouli, E.N. An overview of the biosorption mechanism for the bioremediation of synthetic dyes using yeast cells. *Environ. Technol. Rev.* **2021**, *10*, 58–76. [CrossRef]
35. He, T.; Hua, J.; Chen, R.; Yu, L. Adsorption characteristics of methylene blue by a dye-degrading and extracellular polymeric substance -producing strain. *J. Environ. Manag.* **2021**, *288*, 112446. [CrossRef] [PubMed]
36. Gadow, I.S.; Estrada, L.A.; Li, Y.Y. Characterization and potential of two different anaerobic mixed microflora for bioenergy recovery and decolorization of textile wastewater: Effect of C/N ratio, dye concentration and pH. *Bioresour. Technol. Rep.* **2022**, *17*, 100886. [CrossRef]
37. Ghosh, S.; Kunnoth, B.; Pilli, S.; Rao, V.; Tyagi, D. Novel hybrid system for organic matter removal and energy production from dairy and textile wastewaters: Anaerobic digestion and electrocoagulation approach. *Biomass Convers. Biorefin.* **2024**. [CrossRef]
38. Song, Y.; Meng, S.; Chen, G.; Yan, B.; Zhang, Y.; Tao, J.; Li, J. Co-digestion of garden waste, food waste, and tofu residue: Effects of mixing ratio on methane production and microbial community structure. *J. Environ. Chem. Eng* **2021**, *9*, 105901. [CrossRef]
39. Xiao, Y.; Zan, F.; Zhang, W.; Hao, T. Alleviating nutrient imbalance of low carbon-to-nitrogen ratio food waste in anaerobic digestion by controlling the inoculum-to-substrate ratio. *Bioresour. Technol.* **2022**, *346*, 126342. [CrossRef] [PubMed]
40. Ciotola, J.; Martin, F.; Tamkin, A.; Castaño, M.; Rosenblum, J.; Bisesi, S.; Lee, J. The Influence of Loading Rate and Variable Temperatures on Microbial Communities in Anaerobic Digesters. *Energies* **2014**, *7*, 785–803. [CrossRef]
41. Anfar, Z.; El Haouti, R.; Lhanafi, S.; Benafqir, M.; Azougarh, Y.; El Alem, N. Treated digested residue during anaerobic co-digestion of Agri-food organic waste: Methylene blue adsorption, mechanism and CCD-RSM design. *J. Environ. Chem. Eng.* **2017**, *5*, 5857–5867. [CrossRef]
42. Ramzan, U.; Shakoori, R.; Shakoori, R.; Abbas, Z.; Wabaidur, M.; Eldesoky, E.; Rafatullah, M. Biodegradation and decolorization of Reactive Red 2 azo dye by Paramecium jenningsi and Paramecium multimicronucleatum in industrial wastewater. *Biomass Convers. Biorefin.* **2024**, *14*, 7753–7761. [CrossRef]
43. Lackner, N.; Wagner, O.; Markt, R.; Illmer, P. pH and Phosphate Induced Shifts in Carbon Flow and Microbial Community during Thermophilic Anaerobic Digestion. *Microorganisms* **2020**, *8*, 286. [CrossRef] [PubMed]
44. Tarelho, L.; Hauschild, T.; Vilas-Boas, A.; Silva, D.; Matos, M. Biochar from pyrolysis of biological sludge from wastewater treatment. *Energy Rep.* **2020**, *6* (Suppl. S1), 757–763. [CrossRef]
45. Liu, J.; Huang, S.; Chen, K.; Wang, T.; Mei, M.; Li, J. Preparation of biochar from food waste digestate: Pyrolysis behavior and product properties. *Bioresour. Technol.* **2020**, *302*, 122841. [CrossRef] [PubMed]
46. Krysiak-Baltyn, K.; Cavalida, R.; Thwaites, B.; Reeve, P.; Scales, P.; Van den Akker, B.; Gras, S. Comparison of physical characteristics and dewatering behaviour between granular and floccular sludges generated from the same sewage source. *J. Water Process Eng.* **2019**, *29*, 100785. [CrossRef]
47. Zhu, Y.; Wang, W.; Ni, J.; Hu, B. Cultivation of granules containing anaerobic decolorization and aerobic degradation cultures for the complete mineralization of azo dyes in wastewater. *Chemosphere* **2020**, *246*, 125753. [CrossRef] [PubMed]
48. Eraky, M.; Nasr, M.; Elsayed, M.; Ai, P.; Tawfik, A. Synergistic interaction of tween 20 and magnesium@ functionalized graphene oxide nano-composite for dual productivity of biohydrogen and biochar from onion peel waste. *Renew. Energy* **2023**, *216*, 119082. [CrossRef]

49. Nasr, M.; Tawfik, A.; Awad, H.M.; Galal, A.; El-Qelish, M.; Qyyum, M.A.; Ali Khan, M.M.; Rehan, M.; Nizami, A.-S.; Lee, M. Dual production of hydrogen and biochar from industrial effluent containing phenolic compounds. *Fuel* **2021**, *301*, 121087. [CrossRef]
50. Ullah, H.; Abbas, Q.; Ali, U.M.; Cheema, I.A.; Yousaf, B.; Rinklebe, J. Synergistic effects of low-/medium-vacuum carbonization on physicochemical properties and stability characteristics of biochars. *Chem. Eng. J.* **2019**, *373*, 44–57. [CrossRef]
51. Li, M.; Ren, Z.; Gao, H.; Zhang, A.; Sun, Z. Preparation of tobermorite onto flux-calcinated diatomite surface and the adsorption properties and mechanism of methylene blue. *Desalin. Water Treat.* **2020**, *188*, 247–256. [CrossRef]
52. Fan, H.; Li, L.; Li, Z.; Shang, S. Structure of Sewage Sludge-Clay Multiscale Composite Particles to Control the Mechanism of SO_2 and H_2S Gas Release. *Materials* **2022**, *15*, 1855. [CrossRef] [PubMed]
53. Vasić, K.; Knez, Ž.; Leitgeb, M. Bioethanol Production by Enzymatic Hydrolysis from Different Lignocellulosic Sources. *Molecules* **2021**, *26*, 753. [CrossRef] [PubMed]
54. Chahal, H.; Matthews, S.; Jones, M. Fabrication of Calcium Phosphate Coatings by the Inflight Reaction of Precursor Feedstocks Using Plasma Spraying. *J. Therm. Spray Technol.* **2023**, *32*, 1465–1481. [CrossRef]
55. Ma, Q.; Song, W.; Wang, R.; Zou, J.; Yang, R.; Zhang, S. Physicochemical properties of biochar derived from anaerobically digested dairy manure. *Waste Manag.* **2018**, *79*, 729–734. [CrossRef] [PubMed]
56. Das, K.S.; Ghosh, K.G.; Avasthe, R.; Sinha, K. Morpho-mineralogical exploration of crop, weed and tree derived biochar. *J. Hazard. Mater.* **2021**, *407*, 124370. [CrossRef] [PubMed]
57. Wan, D.; Cheng, X.; Shi, Y.; Chen, Z.; Liu, Y.; Han, X.; Liu, Y.; Zhang, Z. Insights into lead removal in water using a novel carbonized material derived from the by-product of oil refining: Action mechanism and performance optimization. *J. Chem. Technol. Biotechnol.* **2021**, *96*, 3224–3236. [CrossRef]
58. Kumar, A.; Bhattacharya, T.; Shaikh, W.; Chakraborty, S.; Owens, G.; Naushad, M. Valorization of fruit waste-based biochar for arsenic removal in soils. *Environ. Res.* **2022**, *213*, 113710. [CrossRef] [PubMed]
59. Shen, T.; Tang, Y.; Lu, X.-Y.; Meng, Z. Mechanisms of copper stabilization by mineral constituents in sewage sludge biochar. *J. Clean. Prod.* **2018**, *193*, 185–193. [CrossRef]
60. Martínez, E.; Fierro, J.; Rosas, J.; Lobato, A.; Otero, M.; Gómez, X. Assessment of Cationic Dye Biosorption onto Anaerobic Digested Sludge: Spectroscopic Characterization. *Environ. Prog. Sustain. Energy* **2016**, *35*, 1330–1337. [CrossRef]
61. Provenzano, M.; Malerba, A.; Pezzolla, D.; Gigliotti, G. Chemical and spectroscopic characterization of organic matter during the anaerobic digestion and successive composting of pig slurry. *Waste Manag.* **2014**, *34*, 653–660. [CrossRef]
62. Wang, J.; Kou, L.; Huang, Z.; Zhao, L. One-pot preparation of MnOx impregnated cotton fibers for methylene blue dye removal. *RSC Adv.* **2018**, *8*, 21577–21584. [CrossRef] [PubMed]
63. Ettahiri, Y.; Bouna, L.; Hanna, J.; Benlhachemi, A.; Pilsworth, H.; Bouddouch, A.; Bakiz, B. Pyrophyllite clay-derived porous geopolymers for removal of methylene blue from aqueous solutions. *Mater. Chem. Phys.* **2023**, *296*, 127281. [CrossRef]
64. Chen, Z.; Chen, H.; Pan, X.; Lin, Z.; Guan, X. Investigation of Methylene Blue Biosorption and Biodegradation by *Bacillus thuringiensis* 016. *Water Air Soil Pollut.* **2015**, *226*, 146. [CrossRef]
65. Feng, D.; Zhao, Y.; Zhang, Y.; Zhang, Z.; Zhang, L.; Gao, J.; Sun, S. Synergetic effects of biochar structure and AAEM species on reactivity of H_2O-activated biochar from cyclone air gasification. *Int. J. Hydrogen Energy* **2017**, *42*, 16045–16053. [CrossRef]
66. Chen, Y.; Xiao, K.; Jiang, X.; Shen, N.; Zeng, R.; Zhou, Y. Long solid retention time (SRT) has minor role in promoting methane production in a 65 °C single-stage anaerobic sludge digester. *Bioresour. Technol.* **2018**, *247*, 724–729. [CrossRef] [PubMed]
67. Li, H.; Feng, K. Life cycle assessment of the environmental impacts and energy efficiency of an integration of sludge anaerobic digestion and pyrolysis. *J. Clean. Prod.* **2018**, *195*, 476–485. [CrossRef]
68. Monlau, F.; Sambusiti, C.; Antoniou, N.; Barakat, A.; Zabaniotou, A. A new concept for enhancing energy recovery from agricultural residues by coupling anaerobic digestion and pyrolysis process. *Appl. Energy* **2015**, *148*, 32–38. [CrossRef]
69. Saharudin, M.D.; Jeswani, K.H.; Azapagic, A. Biochar from agricultural wastes: Environmental sustainability, economic viability and the potential as a negative emissions technology in Malaysia. *Sci. Total Environ.* **2024**, *919*, 170266. [CrossRef]

water

MDPI

Article

Green Regeneration and Resource Recovery of Nickel-Plating Waste Solution: A Synergistic Study of Electrodialysis and Advanced Oxidation

Xiaolong Xiong [1,†], Kangping Cui [1,*,†], Haiyang Li [1] and Wenming Wu [2]

1 School of Resources and Environmental Engineering, Hefei University of Technology, Hefei 230000, China; 2023171157@mail.hfut.edu.cn (X.X.); lhy@sc-link.com.cn (H.L.)
2 Zhongxin Link Environmental Technology (Anhui) Co., Ltd., Lu'an 237000, China; wwm@sc-link.com.cn
* Correspondence: cuikangping@hfut.edu.cn; Tel.: +86-186-5595-3355
† These authors contributed equally to this work.

Abstract: Electroless nickel plating is a chemical deposition process in which nickel ions within a plating solution are reduced by a chemical reducing agent and subsequently deposited onto the surface of a solid substrate. Chemical nickel-plating wastewater contains substantial amounts of phosphorus as well as abundant nickel resources. In this study, electrodialysis coupled with advanced oxidation techniques was utilized for the efficient recovery of nickel and phosphorus from spent nickel-plating solutions. The end-of-life tank solution from chemical nickel plating was treated via electrodialysis to remove harmful phosphite and sulfate ions, enabling the purified solution to be reused in plating production by supplementing it with appropriate amounts of sodium hypophosphite and nickel sulfate. Subsequently, the concentrate generated from electrodialysis was treated using peroxydisulfate (PDS)-based advanced oxidation technology to break nickel complexation and simultaneously promote the oxidation of hypophosphite and phosphite ions. Finally, $Ca(OH)_2$ was employed as a precipitating agent to effectively recover phosphorus from the treated concentrate. From an economic perspective, optimal process conditions were determined as follows: a current density of 20 mA/cm^2, concentrate-to-dilute water volume ratio of 1:1, current speed of 1.0 m^3/h, and a sodium sulfate concentration in concentrate of 20 g/L. Under these conditions, the migration rates of $H_2PO_2^-$ and HPO_3^{2-} ions reached 67.3% and 62.53%, respectively, whereas Ni^{2+} exhibited significantly lower mobility at only 6.77%. The purified wastewater recovered approximately 60% of its initial plating activity. Regarding the concentrate—which is a by-product of electrodialysis—the hypophosphite ions were nearly completely oxidized using a PDS dosage of 0.3 mol/L. Furthermore, when the Ca/P molar ratio was adjusted to 2.0, total phosphorus (TP) and nickel (Ni) removal efficiencies exceeded 98% and 93%, respectively.

Keywords: electrodialysis; hypophosphite; green regeneration; phosphorus recovery

1. Introduction

Chemical nickel plating has been widely applied in the surface treatment industry due to its relatively simple equipment requirements, the elimination of an external power supply, auxiliary electrodes, and transmission systems, and advantages such as uniform plating thickness, excellent adhesion, superior mechanical properties, and high coating hardness [1,2]. Hypophosphite-based electroless nickel-plating solutions represent the predominant industrial choice, typically employing nickel sulfate as the main metal salt, sodium hypophosphite as a reducing agent, citric acid or its salts as complexing agents, sodium acetate as a buffering agent, and minor additives to facilitate nickel deposition through redox reactions [3,4]. However, this process has significant drawbacks. With ongoing plating reactions, by-products such as phosphite continuously accumulate, causing the performance deterioration of the plating bath. When the phosphite concentration reaches a certain threshold, the plating solution loses effectiveness and becomes a waste solution [5]. Typically, such waste solutions contain 10–200 g/L phosphorus, 2–7 g/L nickel, more than 20 g/L sodium sulfate, trace organic substances, and impure metal ions [6]. It has been reported that the service life of existing chemical nickel-plating baths typically does not exceed 10 Mean Times of Solution (MTO_S) [7]. Additionally, the disposal cost of chemical nickel-plating wastewater is notably high; previous studies indicated that the treatment cost for 1 m^3 of the waste solution ranges from approximately 2700 to 4000 RMB, significantly burdening enterprises in the competitive surface-treatment industry [8]. Due to its toxic constituents, chemical nickel-plating wastewater poses severe environmental risks. Nickel and nickel compounds significantly threaten aquatic ecosystems and human health, while phosphorus contributes to eutrophication, potentially triggering algae blooms, cyanobacteria proliferation, and red tides [9–11]. Moreover, nickel, a valuable yet expensive metal, and phosphorus, a non-renewable resource, have economic importance that mandates their recovery. Traditional nickel recovery methods, such as sludge formation, fail to achieve precise resource recovery and maximal utilization [12–14].

Researchers have long investigated various treatment technologies for chemical nickel-plating wastewater, including chemical precipitation [15], advanced oxidation processes [16,17], electrochemical methods [18], ion exchange [19], adsorption [20], membrane separation [21], and biological treatments [22]. The key achievements and shortcomings of existing processing technologies are summarized in Table 1. Previous studies also indicate that conventional precipitation is insufficient for complexed nickel removal, and hypophosphite treatment is problematic by traditional means. Only limited studies have explored the decomposition of complexed nickel species by hydrogen peroxide (H_2O_2). For example, Liang et al. developed CuO-CeO_2-CoOx composite nanocatalysts for a heterogeneous Fenton-like process, achieving over 99.9% removal of nickel(II)–citrate complexes, but this utilized a relatively low initial nickel concentration (1.0 mM) without addressing the oxidation of hypophosphite and phosphite [23]. In recent years, numerous researchers have explored regeneration and reuse technologies for electroless nickel-plating wastewater. Among these, electrodialysis (ED) has gained widespread application due to its excellent ion selectivity. ED is a membrane separation technology that utilizes an electric field and ion-exchange membranes to selectively transport and separate ions from a solution [24]. It plays a crucial role in desalination, ion removal, and concentration-based separation processes. Song et al. employed ED under conditions of 65 mA/cm^2 current density, 37 °C temperature, 5 cm membrane spacing, and 0.02 L/s current speed for 48 h, achieving approximately 60.86% removal of phosphite, but incurring losses of 63.13% hypophosphite and 1.2% nickel ions [25]. Moreover, the concentrated wastewater generated by ED treatment remained inadequately addressed, limiting its practical application.

Currently, advanced oxidation methods employing strong oxidants have emerged as effective solutions for high-concentration hypophosphite and phosphite wastewater [26–28]. Several studies report the successful oxidation of hypophosphite and phosphite using peroxymonosulfate (PMS)-based processes. For instance, Dong et al. achieved hypophosphite ($H_2PO_2^-$) and phosphite (HPO_3^{2-}) conversion rates of 100% and 99.69%, respectively, via PMS activation combined with alkaline and thermal conditions [29]. Similarly to PMS and

H_2O_2, PDS has recently attracted significant attention as a powerful oxidant [30]. Commonly employed activation methods for PDS include photocatalytic activation [31,32], thermal activation [33,34], transition metal catalysis [35–37], carbon–material catalysis [38–40], electrical activation [41], and alkaline activation [42,43]. However, each activation technique still faces challenges that necessitate further exploration and optimization.

Table 1. Comparative summary of existing technologies.

Processing Technology	Key Achievements	Drawbacks
chemical precipitation	Remarkable treatment effect; mature process; simple operation; wide range of applications	Produces large amounts of sludge; difficult to completely remove complexed nickel; stringent pH control requirements
advanced oxidation processes	Strong broad-spectrum degradation ability; highly efficient complex-breaking and destabilization; suitable for deep treatment	Higher operating costs; harsh reaction conditions; complex equipment and high maintenance requirements
electrochemical methods	High degree of automation control; green environmental protection; no secondary pollution; suitable for the recovery of valuable metals	Higher equipment and power costs; electrodes are susceptible to passivation or corrosion; pH- and conductivity-sensitive
ion exchange	Stable and clear effluent quality; resource recycling; easy to operate; can be automated	The resin easily becomes poisoned or clogged; the scope of application is limited; there are difficulties in the treatment of regeneration fluid.
adsorption	Highly efficient removal of nickel ions (free and complexed); high green potential; suitable for low concentrations of wastewater treatment	Limited adsorption capacity, which needs to be replaced or regenerated frequently; regeneration operation is complicated and may lead to secondary pollution; poor treatment of high concentrations of wastewater
membrane separation	Compact process with small footprint; no additives; no chemical reaction; adaptable and applicable to a variety of pollutants	Serious membrane contamination problems; high investment and operating costs; high requirements for influent water quality
biological treatments	Green environmental protection with no secondary pollution; produces a wide source of materials, making it renewable; low concentration of nickel has a good removal ability	Difficulty to achieve rapid industrialization and diffusion; stringent requirements for environmental conditions; regeneration and longevity of biomaterials

In light of these challenges, the optimal treatment strategy involves selectively removing harmful sulfate and phosphite species while preserving other valuable components in the waste solution. The purified solution can then be reused after appropriate supplementation, significantly extending its operational lifespan, reducing environmental impacts, and lowering wastewater treatment costs. Therefore, this study proposes an innovative treatment strategy that couples electrodialysis with PDS-based advanced oxidation, maximizing nickel and phosphorus recovery from chemical nickel-plating wastewater. Specifically, the effects of key ED operating parameters on hypophosphite and phosphite ion transport were investigated to determine optimal conditions. Subsequently, the influences of PDS dosage and reaction conditions on hypophosphite oxidation, as well as the impact of Ca/P molar ratios on final effluent quality, were explored. Consistent with cleaner production principles, this study achieved the purification, reuse, and near-zero discharge of nickel-plating wastewater, providing practical and cost-effective solutions for enhanced resource recycling, reduced pollutant emissions, and lower wastewater treatment costs within the surface treatment industry.

2. Materials and Methods

2.1. Materials and Composition of Wastewater

The high-concentration hypophosphite and phosphite wastewater used in this study was obtained from an industrial chemical nickel-plating process. Wastewater composition analysis showed significant concentrations of phosphorus and nickel, accompanied by other constituents such as sulfate ions, organic impurities, and trace metal ions. The ED system and associated membranes were procured from Shandong Tianwei Membrane Technology Co., Ltd. (Weifang, China) Analytical chemicals, including sodium thiosulfate standard solution, iodine standard solution, starch indicator, and sodium bicarbonate, were purchased from Shanghai National Pharmaceutical Group Chemical Co., Ltd. (Shanghai, China) Other chemicals employed in this study were of analytical grade. The specific composition of the chemical nickel-plating wastewater is detailed in Table 2. Deionized water was used throughout the experimental procedures.

Table 2. Composition of chemical nickel-plating wastewater.

Items	**pH**	**Ni^{2+} (g/L)**	**$H_2PO_2^-$ (g/L)**	**HPO_3^{2-} (g/L)**	**PO_4^{3-} (g/L)**	**TP (g/L)**
Value	4.56 ± 0.2	5.82 ± 0.1	27.91 ± 2	124.96 ± 5	0.1687 ± 0.2	190.38 ± 5

2.2. Experimental Steps

2.2.1. Regeneration of Wastewater by ED

A volume of 1 L nickel-plating wastewater was introduced into the dilute chamber of the ED apparatus, while the 1 L sodium sulfate solution at a concentration of 20 g/L was prepared and added into both the concentrate and electrode chambers. After the power supply was activated, the conductivity of the solution in the concentration chamber was continuously monitored. The electrodialysis process was terminated once the conductivity reached approximately 78 ms/cm, indicating the completion of the ion migration. The overall reaction duration was approximately 120 min. Each set of experiments was repeated three times to minimize experimental errors.

2.2.2. Reuse of Regenerated Wastewater

The regenerated electroless nickel-plating wastewater (700 mL) was adjusted to contain a nickel concentration of approximately 6.0 g/L and a hypophosphite concentration of 12 g/L by supplementation with nickel sulfate and sodium hypophosphite solutions. The solution was then diluted to 1 L with purified water, and ammonia was added dropwise to adjust the solution's pH to approximately 4.6. The plating bath was maintained at a nickel concentration range of 5–6 g/L and a hypophosphite concentration between 18.75 and 22.5 g/L. After standard pretreatment procedures—including degreasing, rinsing, acid activation, and washing—the nickel-plating solution was heated to between 85 °C and 91 °C, and electroless nickel plating was performed on iron sheet samples. The plating experiments were carried out for four cycles, with each cycle consisting of three consecutive plating runs, each lasting 1 h. The performance of the regenerated plating bath was evaluated by comparing the plated samples against those obtained using a freshly prepared standard plating solution.

2.2.3. PDS Oxidation Experiment

The concentrated wastewater obtained from the ED process (100 mL) was subjected to advanced oxidation treatment using PDS. Different dosages of PDS were added to investigate the optimal oxidant concentration. The solution was continuously stirred during oxidation, and samples were collected every 10 min to evaluate the oxidation efficiency. Additionally, the effects of the reaction temperature and initial solution pH on the oxidation performance were systematically studied. The oxidation efficiencies of different oxidants were also compared under identical experimental conditions. Each set of experiments was repeated three times to minimize experimental errors.

2.2.4. Calcium Salt Dosing Experiment

Calcium hydroxide ($Ca(OH)_2$) was employed as the precipitant to recover phosphorus from the oxidized wastewater. Experiments were conducted at room temperature (approximately 25 °C) to determine the optimal calcium-to-phosphorus (Ca/P) molar ratio. Various Ca/P ratios (1.0, 1.5, 2.0, 2.5, and 3.0) were tested to examine their effects on the removal efficiencies of phosphorus and nickel ions. Additionally, the influence of reaction temperature on phosphorus recovery was investigated, and comparative studies were carried out using different calcium salts to assess their effectiveness in removing phosphorus and other residual ions from wastewater. Each set of experiments was repeated three times to minimize experimental errors.

2.3. Analytical Methods

The concentrations of hypophosphite and phosphite ions were determined by standard titration methods [44]. Orthophosphate (PO_4^{3-}) and TP concentrations were measured separately using ammonium molybdate spectrophotometry at a wavelength of 700 nm with a UV-Vis spectrophotometer (Model 752N, Shanghai Yidian Instrument Co., Ltd., Shanghai, China), following the protocols described in the Chinese national standard methods (State Environmental Protection Administration, Beijing, China, 1989) [45]. The concentrations of metal ions in the solution were analyzed by inductively coupled plasma optical emission spectroscopy (ICP-OES). Electron paramagnetic resonance (EPR) was used to detect the free radicals generated by the oxidation process. The specific operation was as follows: place a 2 mmol/L DMPO solution in the reaction system, add PDS for 30 s and then take a sample and cross the membrane; contain the filtrate in a 1.5 mL centrifuge tube; and then a small amount of filtrate is sucked up by capillary tube and put into the detector for detection. Test conditions: receiver gain 1×10^5; modulation amplitude 1 G, microwave power 10 Mw, modulation frequency 100 kHz. Additionally, solid precipitates obtained after the calcium precipitation experiments were characterized using X-ray diffraction (XRD, Rint2200, Rigaku Corporation, Tokyo, Japan) and scanning electron microscopy (SEM, JSM-6400, Jeol, Tokyo, Japan).

3. Results and Discussion

3.1. ED Regeneration of Wastewater

Since phosphite is the primary harmful contaminant in electroless nickel-plating wastewater, the purification efficiency of the ED process strongly depends on the mobility of phosphite ions. To optimize the process parameters, the present study investigated the effects of current density, current speed, sodium sulfate concentration in the concentrate chamber, and the volume ratio of concentrate with which to dilute solutions on the mobility of hypophosphite, phosphite, and nickel ions. In theory, nickel ions should remain in the dilute chamber due to the barrier effect of the anion–exchange membrane. However, since nickel ions in the plating wastewater mainly exist in complexed forms, a small fraction of nickel ions migrate across the membrane due to complexation with organic acids and other organics present in the wastewater [46].

It was observed that the current density was the predominant factor affecting ion mobility during the ED process. As illustrated in Figure 1a, increasing the current density significantly enhanced the mobility of hypophosphite and phosphite ions in the solution. When the current density was set at 20 mA/cm^2, the mobility of hypophosphite and phosphite reached 67.3% and 62.5%, respectively. Further increases in the current density led to only marginal improvements in ion mobility. However, higher current densities substantially reduced the reaction time required for wastewater purification. Specifically, at a current density of 10 mA/cm^2, approximately 120 min was required to complete the purification of 1 L of wastewater, whereas increasing the current density to 30 mA/cm^2 shortened this time to 80 min. Nevertheless, excessive current densities may lead to overheating and the potential shutdown of the power supply. Moreover, elevated current densities increased the temperature of the plating solution, which subsequently lowered the migration efficiency of harmful ions, such as phosphite and sulfate.

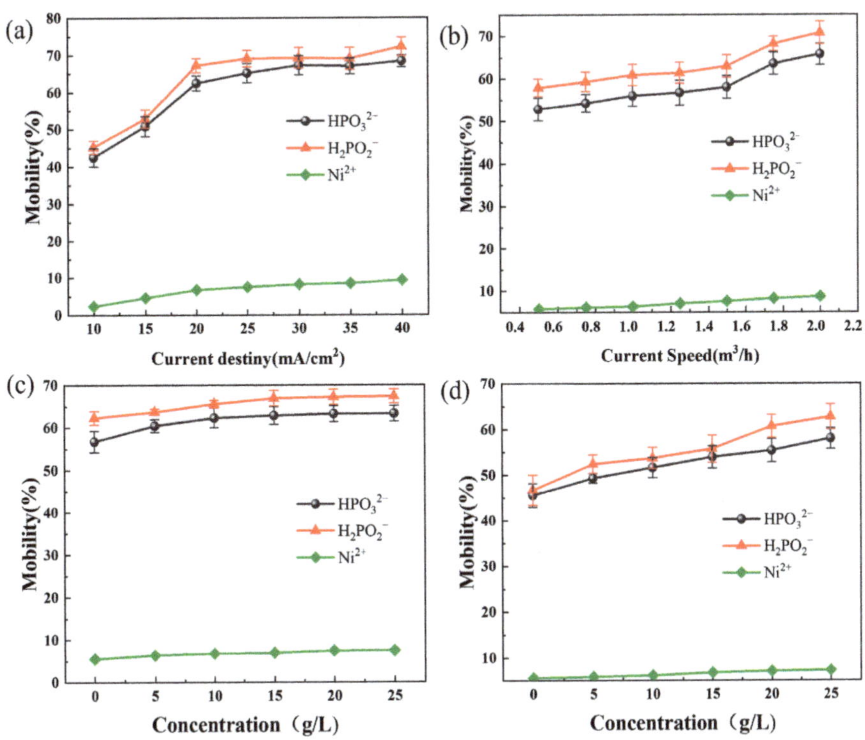

Figure 1. Effects of various parameters on ion mobility in the solution: ((**a**) effect of current density; (**b**) effect of current speed; (**c**) effect of sodium sulfate concentration in the concentrate chamber (concentrate-to-dilute ratio = 1:1); and (**d**) effect of sodium sulfate concentration in the concentrate chamber (concentrate-to-dilute ratio = 0.5:1)).

As illustrated in Figure 1a, the mobility of Ni^{2+} ions increased from 2.37% to 9.4% with an increase in current density. Considering the practical circumstances and overall economic efficiency, the optimal current density was determined to be 20 mA/cm^2. Figure 1b demonstrates that increasing the current speed gradually enhances the ion mobility, eventually reaching a stable plateau. At a current speed of 1.0 m^3/h, ion mobility achieved its maximum stable value, and further increases in the current speed did not lead to significant improvements. Therefore, a current speed of 1.0 m^3/h was recommended to maximize phosphite migration efficiency.

From Figure 1c,d, it can be observed that increasing the sodium sulfate concentration in the concentrate chamber slightly enhanced ion mobility; however, the improvement was not significant. Nevertheless, a higher sodium sulfate concentration effectively increased the initial conductivity in the concentrate chamber, thus shortening the reaction time. A comparative analysis of Figure 1c,d indicates that reducing the volume ratio of the concentrate to dilute water negatively affects ion mobility and significantly prolongs the reaction duration. Hence, considering these comprehensive factors, the recommended process parameters are a sodium sulfate concentration of 20 g/L in the concentrate chamber and a concentrate-to-dilute volume ratio of 1:1.

3.2. Reuse of Regenerated Wastewater

In the chemical nickel-plating process, the plating performance is significantly influenced by the phosphite concentration. Typically, plating performance deteriorates with the increasing phosphite concentration. Specifically, when the phosphite concentration reaches approximately 120 g/L, defects such as reduced coating adhesion and leakage become evident during thick plating, necessitating the termination of the plating process. In comparison, the thin plating processes can tolerate higher phosphite concentrations (up to approximately 200 g/L) before similar defects occur. In this study, considering that thick plating operations generally exhibit bonding issues and plating defects at lower phosphite concentrations (>120 g/L), thick plating was selected to evaluate the regeneration and reuse potential of the wastewater after ED treatment.

Figure 2 compares the performance of plated parts produced using the freshly formulated plating solution and the regenerated plating solution. As illustrated, parts plated using the regenerated solution exhibited slightly inferior performance relative to those plated with the freshly prepared bath, although the overall differences remained within an acceptable range. During the first three plating cycles, the regenerated plating bath consistently maintained coating thicknesses above 16.0 μm and deposition rates greater than 0.2 μm/min, fully meeting industrial production requirements. However, in the fourth cycle, plating quality decreased significantly, corresponding to an elevated phosphite concentration (>120 g/L) within the plating bath, indicating that the bath had accumulated excessive impurities and required disposal. Compared with the freshly prepared plating bath, which maintained stable performance for a longer duration, the regenerated plating bath effectively restored approximately 60% of the original activity.

In summary, the electrodialysis-based regeneration of aged nickel-plating solutions is technically viable, presenting substantial practical value and broad prospects for industrial application.

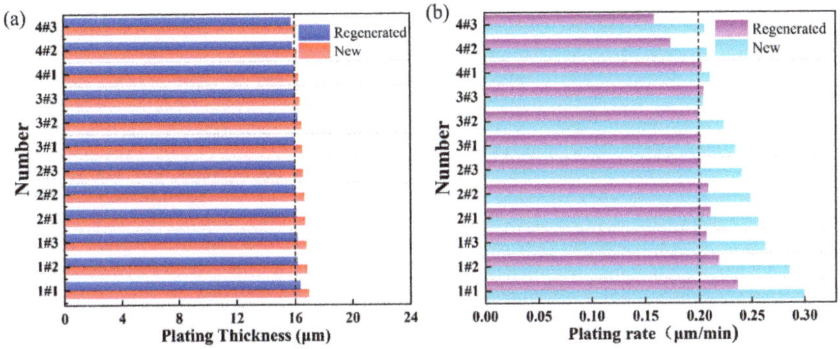

Figure 2. Plating performance test. ((**a**) Plating thickness; (**b**) plating rate).

3.3. Recovery of Phosphorus from Concentrated Wastewater

3.3.1. Phosphorus Speciation Changes During Oxidation

Figure 3 illustrates the influence of the PDS concentration on phosphorus speciation during the oxidation reaction. As shown, increasing the PDS dosage resulted in significant changes in phosphorus speciation. Specifically, the hypophosphite concentration decreased continuously, accompanied by a corresponding increase in the orthophosphate concentration, consistent with expected oxidation trends. When observing the phosphite concentration, an initial increase followed by stabilization was noted. This phenomenon was attributed to the relatively slower oxidation kinetics of phosphite compared with hypophosphite. Initially, hypophosphite rapidly oxidized into phosphite, leading to an initial rise in the phosphite concentration, followed by a plateau phase. Subsequently, as hypophosphite oxidation neared completion, the oxidation of phosphite commenced at a comparatively slower rate at higher oxidant dosages and prolonged reaction times.

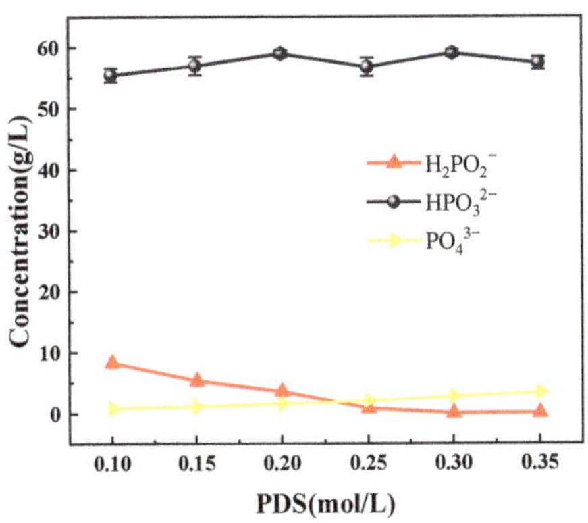

Figure 3. Effect of PDS concentration on changes in phosphorus morphology.

Although the complete oxidation of low-valent phosphorus (hypophosphite and phosphite) to orthophosphate was not fully achieved, the phosphorus recovery efficiency remained unaffected. This is attributed to the fact that both calcium phosphite and calcium phosphate are water-insoluble precipitates, effectively removable by calcium hydroxide precipitation. Therefore, the incomplete oxidation of phosphite does not impede phosphorus recovery, as both calcium phosphite and calcium phosphate can be effectively precipitated and removed from the solution.

3.3.2. Effects of Reaction Parameters

As shown in Figure 3, the oxidation efficiency of hypophosphite increased significantly with the increasing PDS dosage. Specifically, the oxidation efficiency rose from 48.04% to 99.8% when the PDS dosage was increased from 0.1 mol/L to 0.3 mol/L. A further increase in the PDS concentration resulted in negligible improvement, as the hypophosphite was already nearly completely oxidized at 0.3 mol/L. Hence, considering both cost and practical application, a PDS dosage of 0.3 mol/L is recommended for hypophosphite oxidation.

Figure 4b illustrates that the oxidation performance of various oxidants toward hypophosphite was similar under identical experimental conditions, with all tested oxidants eventually achieving near-complete oxidation. Figure 4c demonstrates the influence of temperature on the oxidation process. Considering industrial practicalities, the experiments were carried out within a common operational temperature range (10–40 °C). The results indicate that temperature primarily influences reaction kinetics; specifically, when the temperature exceeded 30 °C, hypophosphite was completely oxidized within 50 min.

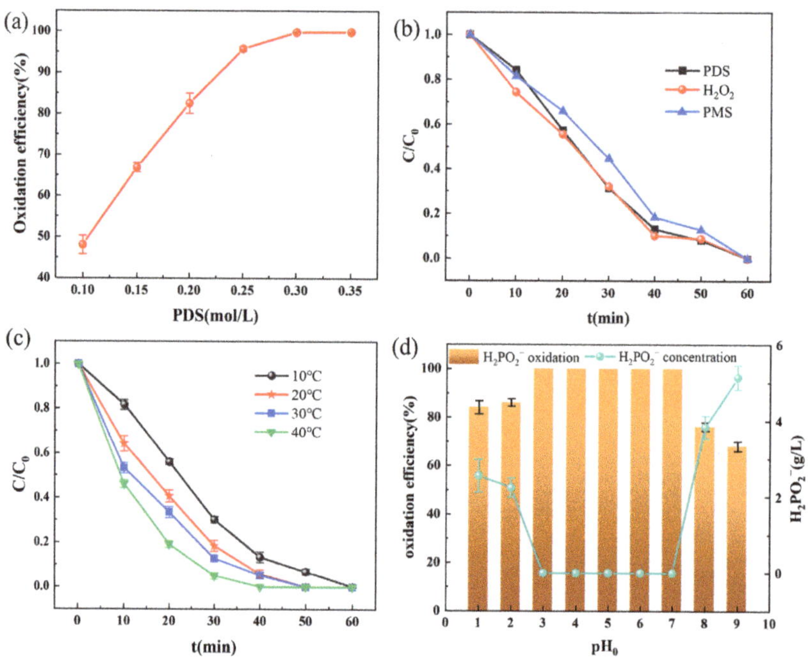

Figure 4. Effects of various reaction parameters on the oxidation efficiency of hypophosphite: ((**a**) effect of PDS concentration; (**b**) comparison of different oxidants; (**c**) effect of reaction temperature; and (**d**) effect of initial solution pH).

Figure 4d depicts the effect of initial solution pH on hypophosphite oxidation. The results show that complete oxidation of hypophosphite occurred within approximately 1 h when the initial pH ranged between 3 and 7 at a PDS dosage of 0.3 mol/L. However, oxidation efficiency decreased under strongly acidic conditions (pH < 3), with an oxidation efficiency of only 84.15% at pH 1.0, indicating the significant inhibition of PDS activity. Furthermore, oxidation efficiency markedly decreased in alkaline conditions, with only 68.18% oxidation at pH 9.0. These findings confirm that the oxidation of hypophosphite by

PDS is highly sensitive to solution pH, with optimal performance achieved in near-neutral conditions.

It was observed that solution pH significantly influenced the oxidation mechanism and efficiency of PDS. Electron paramagnetic resonance (EPR) experiments were performed to clarify the radical species involved during the oxidation process at different initial pH conditions (pH_0 = 3.0 and pH_0 = 11), and the results are presented in Figure 5. Signals corresponding to both DMPO–·OH and DMPO–$SO_4^-\cdot$ adducts were detected in both acidic (pH_0 = 3.0) and alkaline (pH_0 = 11) conditions, indicating the coexistence of $SO_4^-\cdot$ and ·OH radicals during PDS oxidation. However, under alkaline conditions (pH_0 = 11), a notably stronger signal for DMPO–·OH was detected, suggesting that $SO_4^-\cdot$ radicals were largely converted to ·OH radicals.

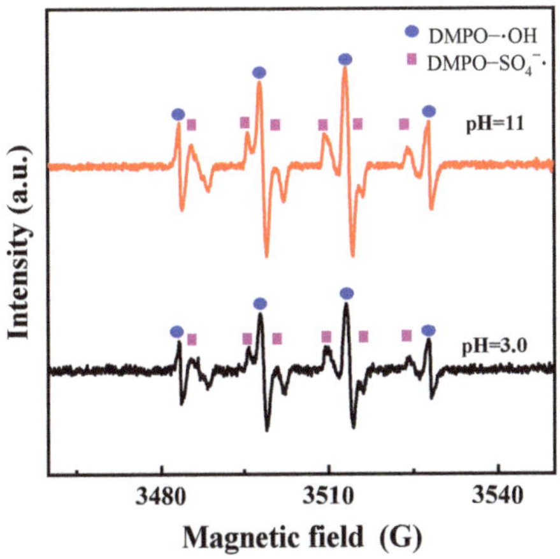

Figure 5. EPR spectra under different initial pH conditions.

Considering the redox potentials of these two radicals (E_{OH} = 2.80 eV, $E_{SO_4^-\cdot}$ = 2.5–3.1 eV), the conversion of $SO_4^-\cdot$ to ·OH may decrease the overall oxidation capability of PDS under alkaline conditions [47,48]. Furthermore, the half-life of the ·OH radical ($t_{1/2} < 1$ μs) is significantly shorter than that of the $SO_4^-\cdot$ radical ($t_{1/2} \approx 4$ s) [49,50], resulting in a substantial reduction in the radical lifetime, thus limiting any effective interaction with hypophosphite ions.

3.3.3. Effect of Calcium Dosage on Phosphorus Recovery

After oxidation treatment, phosphite and orthophosphate are present in the concentrated wastewater, and these phosphorus species can be precipitated by adding calcium salts to facilitate phosphorus recovery. To evaluate phosphorus recovery efficiency and optimize calcium salt selection, four different calcium salts were tested at a fixed Ca/P molar ratio of 2.0, as illustrated in Figure 6. The results indicate that CaO exhibited the lowest removal efficiency for TP and Ni^{2+} ions. In contrast, calcium hydroxide ($Ca(OH)_2$) demonstrated superior removal efficiency for both TP and Ni^{2+}. Although calcium chloride and calcium carbonate achieved moderate removal efficiencies, their addition introduced impure ions, complicating the subsequent wastewater treatment processes. In contrast, the use of $Ca(OH)_2$ did not introduce extraneous ions, thereby simplifying downstream processing. Additionally, the $Ca(OH)_2$ supernatant, which is strongly alkaline, can effectively neutralize the acidic electroplating wastewater commonly generated in industrial wastewater treatment facilities.

Thus, considering phosphorus and nickel removal efficiency, the absence of impure ions, and practical treatment convenience, $Ca(OH)_2$ is recommended as the optimal calcium salt for phosphorus recovery.

Figure 7a illustrates the effect of the Ca/P molar ratio on TP removal from the concentrated wastewater. At low Ca/P ratios, the solution was weakly alkaline, which was not conducive to the effective precipitation of phosphite and phosphate species, resulting in a relatively lower phosphorus removal rate. As the Ca/P molar ratio increased, the phosphorus removal rate improved significantly, reaching an optimum of 98.75% at Ca/P = 2.0. Beyond this optimal ratio, further increases in calcium dosage led to minimal improvements in the phosphorus removal rate and even slightly reduced effectiveness. This reduction occurred because excessively high solution pH favored the formation of $Ca(OH)_2$ precipitates over calcium phosphate and calcium phosphite, thus inhibiting phosphorus precipitation [51,52].

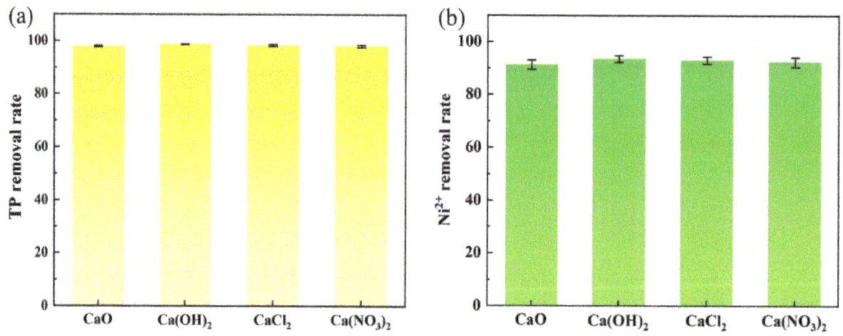

Figure 6. Comparison of ion removal efficiencies using different calcium salts: ((**a**) TP removal rate; (**b**) Ni^{2+} removal rate).

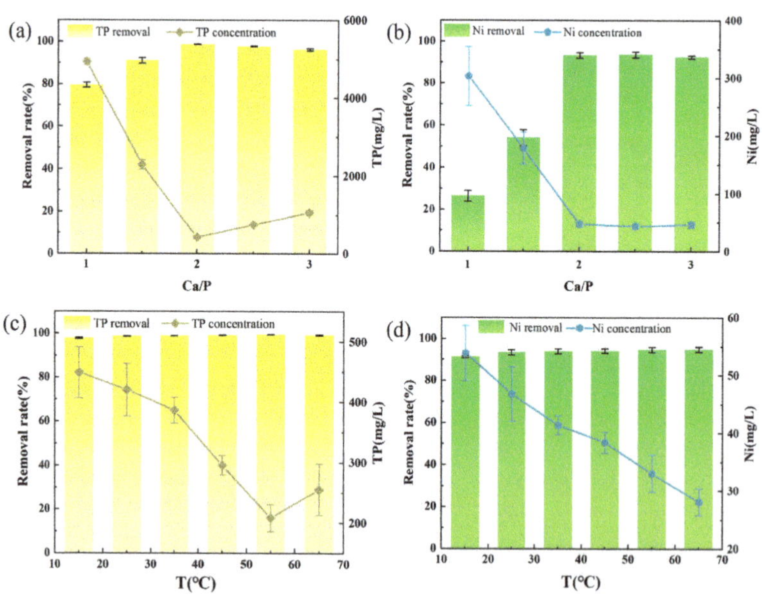

Figure 7. Effect of different parameters on ion removal efficiencies: ((**a**) effect of Ca/P molar ratio on TP removal rate; (**b**) effect of Ca/P molar ratio on Ni removal rate; (**c**) effect of reaction temperature on TP removal rate; and (**d**) effect of reaction temperature on Ni removal rate).

As shown in Figure 7b, the optimal Ni removal rate (93.42%) was also achieved at a Ca/P ratio of 2.0. Further increases in calcium dosage did not significantly enhance the nickel removal rate, and a residual nickel concentration (~40 mg/L) remained difficult to remove completely. Considering the combined removal rate of phosphorus and nickel, the Ca/P molar ratio of 2.0 was identified as optimal.

The impact of the reaction temperature on phosphorus removal is presented in Figure 7c. With increasing temperature, the phosphorus removal rate initially increased, peaking at 99.53% at approximately 55 °C, and subsequently declined slightly. Elevated temperature enhanced molecular diffusion, accelerating the precipitation reaction. However, excessively high temperatures caused the partial dissolution of loosely structured calcium phosphate crystals, releasing phosphate ions back into the solution and, thus, lowering the overall phosphorus removal rate. Even at room temperature (25 °C), the phosphorus removal rate reached 98.75%, indicating the sufficient availability of phosphorus species in the solution to facilitate rapid precipitation.

Figure 7d demonstrates the effect of temperature on the nickel removal rate. Increasing temperature gradually improved nickel removal, ultimately reaching a maximum removal rate of 94.77%, corresponding to a residual nickel concentration of approximately 20 mg/L.

3.3.4. Analysis of Precipitated Products

Figure 8 shows XRD patterns and scanning SEM images of the precipitates obtained after phosphorus recovery. The XRD analysis revealed that the precipitated products predominantly consisted of $CaHPO_3 \cdot H_2O$, along with minor amounts of hydroxyapatite ($Ca_5(PO4)_3OH$) and $CaSO_4 \cdot 2H_2O$. Although $Ni(OH)_2$ was theoretically expected to form, it was not detected by XRD, indicating that its concentration in the precipitate was very low. Considering the agricultural importance of phosphorus as a nutrient, the obtained precipitate has potential applications as a phosphorus fertilizer supplement. Additionally, minor impurities such as $Ni(OH)_2$ and $CaSO_4 \cdot 2H_2O$ do not necessitate further separation, as they can function as soil conditioners, thus achieving the efficient utilization of recovered phosphorus resources.

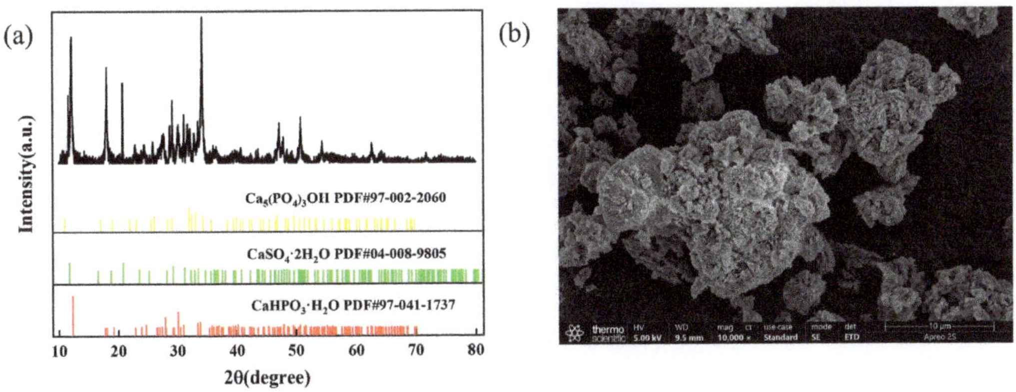

Figure 8. Analysis of precipitation products. ((**a**) XRD image; (**b**) SEM image).

The SEM analysis demonstrated that the precipitate consisted of particles of relatively uniform size. While most particles exhibited good dispersion, noticeable agglomeration into larger clusters was also observed.

4. Conclusions

In this study, an integrated approach combining ED and advanced oxidation processes was developed to achieve the efficient recovery of nickel and phosphorus resources from chemical nickel-plating wastewater. Taking into account the overall cost-effectiveness and operational simplicity, the recommended operational parameters for the ED process include a current density of 20 mA/cm^2, a current speed of 1.0 m^3/h, a sodium sulfate concentration of 20 g/L in the concentrate chamber, and a volume ratio of concentrate-to-dilute water of 1:1. Under these optimized conditions, the maximal migration of phosphite ions was achieved while minimizing the excessive loss of hypophosphite. The regenerated plating bath retained approximately 60% of its original plating activity, and the coating quality exhibited only minimal differences compared to freshly prepared plating solutions, demonstrating considerable potential for practical reuse.

PDS demonstrated outstanding oxidation performance for hypophosphite removal. At an optimal dosage of 0.3 mol/L, nearly the complete oxidation of hypophosphite ions was achieved. Reaction temperatures above 30 °C notably accelerated the oxidation reaction, fully oxidizing hypophosphite within 50 min. Furthermore, efficient hypophosphite oxidation was sustained within an initial pH range of 3–7, while oxidation efficiency was markedly inhibited under strongly acidic or alkaline conditions, indicating the clear pH sensitivity of the PDS oxidation process.

Additionally, phosphorus recovery was mostly effective at a calcium-to-phosphorus (Ca/P) molar ratio of 2.0, achieving removal efficiencies greater than 98% for TP and over 93% for nickel ions. The main components identified in the recovered precipitates included $CaHPO_3 \cdot H_2O$ and hydroxyapatite, along with minor amounts of $CaSO_4 \cdot 2H_2O$ and $Ni(OH)_2$. These recovered precipitates could potentially be utilized directly as phosphorus fertilizers and soil conditioners, thus realizing resource recycling without necessitating further separation processes.

In summary, the innovative combined treatment process developed in this research provides a practical, straightforward, efficient, and economical solution for recovering valuable nickel and phosphorus resources from chemical nickel-plating wastewater. This proposed approach not only facilitates waste recycling but also aligns well with contemporary sustainable development principles and goals.

Author Contributions: Data curation, X.X.; Project administration, H.L. Software, X.X.; Supervision, W.W.; Validation, H.L.; Writing—original Draft, X.X. and K.C. All authors have read and agreed to the published version of the manuscript.

Funding: This work was supported by the Science and Technology Innovation Tackle Plan Project of Anhui Province (Grant No.202423110050041).

Data Availability Statement: The original contributions presented in this study are included in the article. Further inquiries can be directed to the corresponding authors.

Conflicts of Interest: Author Wenming Wu was employed by the company Zhongxin Link Environmental Technology (Anhui) Co., Ltd. The remaining authors declare that the research was conducted in the absence of any commercial or financial relationships that could be construed as a potential conflict of interest.

References

1. Lu, K.; Qin, J.; Hu, M.; Hu, L.; Mao, M.; Li, X.; Lin, Z.; Liu, W. High-efficiency nickel recovery from spent electroless nickel plating solution: Effective degradation of high-concentration nickel complexes to form a nickel ferrite nanomaterial via Fe_3O_4 catalytic oxidation. *Environ. Sci. Nano* **2024**, *11*, 900–910.

2. Li, S.; Dai, M.; Ali, I.; Bian, H.; Peng, C. Recovery of nickel from actual electroplating wastewater by integrated electrodeposition with adsorption pretreatment technique. *Process Saf. Environ. Prot.* **2023**, *172*, 417–424.
3. Wang, S.; Zhang, D.; Wang, W.; Zhong, J.; Feng, K.; Wu, Z.; Du, B.; He, J.; Li, Z.; He, L.; et al. Grave-to-cradle upcycling of Ni from electroplating wastewater to photothermal CO_2 catalysis. *Nat. Commun.* **2022**, *13*, 5305.
4. Yu, X.; Hou, Y.; Ren, X.; Sun, C.; Wang, M. Research progress on the removal, recovery and direct high-value materialization of valuable metal elements in electroplating/electroless plating waste solution. *J. Water Process Eng.* **2022**, *46*, 102577.
5. Huang, X.; Shi, X.; Zeng, H. How far does the Copper/Nickle recovery from the practical application in the electroplating wastewater? *Resour. Conserv. Recycl. Adv.* **2023**, *19*, 200170.
6. Li, S.; Dai, M.; Wu, Y.; Fu, H.; Hou, X.; Peng, C.; Luo, H. Resource utilization of electroplating wastewater: Obstacles and solutions. *Environ. Sci. Water Res. Technol.* **2022**, *8*, 484–509.
7. Liu, Y.; Lian, R.; Wu, X.; Dai, L.; Ding, J.; Ye, X.; Chen, R.; Ding, R.; Liu, J.; Van der Bruggen, B. Nickel recovery from electroplating sludge via bipolar membrane electrodialysis. *J. Colloid Interface Sci.* **2023**, *637*, 431–440.
8. Molaei, M.; Atapour, M. Nickel-based coatings as highly active electrocatalysts for hydrogen evolution reaction: A review on electroless plating cost-effective technique. *Sustain. Mater. Technol.* **2024**, *40*, e00991.
9. Li, L.; Takahashi, N.; Kaneko, K.; Shimizu, T.; Takarada, T. A novel method for nickel recovery and phosphorus removal from spent electroless nickel-plating solution. *Sep. Purif. Technol.* **2015**, *147*, 237–244.
10. Wang, D.; Yang, B.; Ye, Y.; Zhang, W.; Wei, Z. Nickel speciation of spent electroless nickel plating effluent along the typical sequential treatment scheme. *Sci. Total Environ.* **2019**, *654*, 35–42.
11. Bolger, P.T.; Szlag, D.C. Current and emerging technologies for extending the lifetime of electroless nickel plating baths. *Clean Prod. Process.* **2001**, *2*, 209–219. [CrossRef]
12. Stagnaro, S.M.; Mesquida, C.; Zysler, R.; Stábile, F.; Yañez, R.A.; Soldati, A.; Ramos, S. Comparative study of Ni nanoparticles synthetized using electroless Ni plating waste and an analytical Ni reagent. Characterization and possible application in magnetic fluids. *J. Magn. Magn. Mater.* **2024**, *611*, 172594.
13. Li, Z.; Xu, N.; Liu, S.; Wang, Y.; Rajput, V.D.; Minkina, T.; Fan, F.; Gao, W.; Zhao, Y. Electrocatalysis coupled super-stable mineralization for the efficient treatment of phosphorus containing plating wastewater. *Chem. Eng. Sci.* **2025**, *302*, 120910.
14. Shaikh, A.; Singh, B.K.; Purnendu, K.; Kumari, P.; Sankar, P.R.; Mundra, G.; Bohm, S. Utilization of the nickel hydroxide derived from a spent electroless nickel plating bath for energy storage applications. *RSC Sustain.* **2023**, *1*, 294–302. [CrossRef]
15. Tsai, T.H.; Chou, H.W.; Wu, Y.F. Removal of nickel from chemical plating waste solution through precipitation and production of microsized nickel hydroxide particles. *Sep. Purif. Technol.* **2020**, *251*, 117315. [CrossRef]
16. Zhang, H.; Pan, S.; Sheng, J.; Ma, J.; Yan, Y.; Nie, Y.; Yuan, S. High-efficiency phosphorus recovery from hypophosphite and phosphite-containing wastewater via a new electrodialysis enrichment and hydrothermal oxidation coupling process. *Chem. Eng. J.* **2024**, *496*, 154088.
17. Yang, L.; Jiao, Y.; Xu, X.; Pan, Y.; Su, C.; Duan, X.; Sun, H.; Liu, S.; Wang, S.; Shao, Z. Superstructures with atomic-level arranged perovskite and oxide layers for advanced oxidation with an enhanced non-free radical pathway. *ACS Sustain. Chem. Eng.* **2022**, *10*, 1899–1909.
18. Yan, Y.; Shu, W.; Mao, J.; Ma, J.; Zhang, H. Efficient simultaneous removal of nickel, phosphorus and COD from electroless nickel plating wastewater by three dimensional electrodialytic system. *Sep. Purif. Technol.* **2024**, *345*, 127417.
19. Jiang, T.; Guan, W.; Fu, M. Recovery of nickel from electroless nickel plating wastewater based on the synergy of electrocatalytic oxidation and electrodeposition technology. *Water Environ. Res.* **2022**, *94*, e10741.
20. Ding, Y.; Wang, A.; Yin, H.; Xie, X. Adsorption performance of tetratitanate nanowhiskers for Ni^{2+} from aqueous solution and removal of Ni^{2+} from actual electroplating wastewater via successive oxidation and adsorption processes. *J. Water Process Eng.* **2023**, *53*, 103703.
21. Chen, K. Research on Metal Ion Recovery from Chemical Nickel Plating Waste Liquid Based on Membrane Separation Technology. *Forum Res. Innov. Manag.* **2025**, *3*.
22. Fu, X.Z.; Yang, Y.R.; Liu, T.; Guo, Z.Y.; Li, C.X.; Li, H.Y.; Cui, K.P.; Li, W.W. Biological upcycling of nickel and sulfate as electrocatalyst from electroplating wastewater. *Water Res.* **2024**, *250*, 121063. [CrossRef] [PubMed]
23. Liang, H.; Xiao, K.; Wei, L.; Yang, B.; Yu, G.; Deng, S.; Duan, H.; Zhu, C.; Li, J.; Zhang, J. Decomplexation removal of Ni (II)-citrate complexes through heterogeneous Fenton-like process using novel CuO-CeO_2-CoOx composite nanocatalyst. *J. Hazard. Mater.* **2019**, *374*, 167–176. [CrossRef] [PubMed]
24. Scarazzato, T.; Panossian, Z.; Tenório, J.A.S.; Pérez-Herranz, V.; Espinosa, D.C.R. A review of cleaner production in electroplating industries using electrodialysis. *J. Clean. Prod.* **2017**, *168*, 1590–1602. [CrossRef]
25. Song, Q. Study on Electrochemical Regeneration of Electroless Nickel Waste. Master's Thesis, Guangdong University of Technology, Guangzhou, China, 2013.

26. Liu, P.; Li, C.; Liang, X.; Lu, G.; Xu, J.; Dong, X.; Zhang, W.; Ji, F. Recovery of high purity ferric phosphate from a spent electroless nickel plating bath. *Green Chem.* **2014**, *16*, 1217–1224. [CrossRef]
27. Liu, Y.; Sheng, X.; Dong, Y.; Ma, Y. Removal of high-concentration phosphate by calcite: Effect of sulfate and pH. *Desalination* **2012**, *289*, 66–71. [CrossRef]
28. De-Bashan, L.E.; Bashan, Y. Recent advances in removing phosphorus from wastewater and its future use as fertilizer (1997–2003). *Water Res.* **2004**, *38*, 4222–4246. [CrossRef]
29. Dong, Z.; Zhao, Z.; Wang, F.; Wang, F.; Xia, M. Integrating multi-method approaches for the green separation and retrieval of nickel and phosphorus from spent electroless nickel plating solutions. *Green Chem.* **2023**, *25*, 10576–10586. [CrossRef]
30. Li, J.; Liang, Y.; Jin, P.; Zhao, B.; Zhang, Z.; He, X.; Tan, Z.; Wang, L.; Cheng, X. Heterogeneous metal-activated persulfate and electrochemically activated persulfate: A review. *Catalysts* **2022**, *12*, 1024. [CrossRef]
31. Bai, C.; Guo, W.; Liu, Q.; Li, G.; Guo, S.; Chen, R. $Cu_2O/BiVO_4$ heterostructure controllably triggers radical and non-radical persulfate activation via light "on-off" for efficient organic contaminants degradation. *Appl. Catal. B Environ. Energy* **2024**, *344*, 123606. [CrossRef]
32. Zhang, J.; Jing, B.; Tang, Z.; Ao, Z.; Xia, D.; Zhu, M.; Wang, S. Experimental and DFT insights into the visible-light driving metal-free C_3N_5 activated persulfate system for efficient water purification. *Appl. Catal. B Environ.* **2021**, *289*, 120023.
33. Lin, C.C.; Huang, C.Y. Degradation of sulfamethazine in water in thermally activated persulfate process. *J. Taiwan Inst. Chem. Eng.* **2024**, *155*, 105229.
34. Zhang, Y.; Liu, H.; Xin, Y.; Shen, Y.; Wang, J.; Cai, C.; Wang, M. Erythromycin degradation and ERY-resistant gene inactivation in erythromycin mycelial dreg by heat-activated persulfate oxidation. *Chem. Eng. J.* **2019**, *358*, 1446–1453.
35. Huang, M.; Han, Y.; Xiang, W.; Zhong, D.; Wang, C.; Zhou, T.; Wu, X.; Mao, J. In situ-formed phenoxyl radical on the CuO surface triggers efficient persulfate activation for phenol degradation. *Environ. Sci. Technol.* **2021**, *55*, 15361–15370.
36. Xu, C.; Yang, G.; Li, J.; Zhang, S.; Fang, Y.; Peng, F.; Zhang, S.; Qiu, R. Efficient purification of tetracycline wastewater by activated persulfate with heterogeneous Co-V bimetallic oxides. *J. Colloid Interface Sci.* **2022**, *619*, 188–197.
37. Liu, Z.; Ren, X.; Duan, X.; Sarmah, A.K.; Zhao, X. Remediation of environmentally persistent organic pollutants (POPs) by persulfates oxidation system (PS): A review. *Sci. Total Environ.* **2023**, *863*, 160818. [CrossRef]
38. Hou, X.; Dong, H.; Li, Y.; Xiao, J.; Dong, Q.; Xiang, S.; Chu, D. Activation of persulfate by graphene/biochar composites for phenol degradation: Performance and nonradical dominated reaction mechanism. *J. Environ. Chem. Eng.* **2023**, *11*, 109348.
39. Lou, Z.; Song, L.; Liu, W.; Wu, S.; He, F.; Yu, J. Deciphering CaO-induced peroxydisulfate activation for destruction of halogenated organic pollutants in a low energy vibrational mill. *Chem. Eng. J.* **2022**, *431*, 134090.
40. Tang, M.; Wan, J.; Wang, Y.; Ye, G.; Yan, Z.; Ma, Y.; Sun, J. Overlooked role of void-nanoconfined effect in emerging pollutant degradation: Modulating the electronic structure of active sites to accelerate catalytic oxidation. *Water Res.* **2024**, *249*, 120950.
41. Li, Y.T.; Du, W.Y.; Chen, J.L.; Zhang, J.J.; Liu, X.Y.; Liu, H.; Zhang, Y.-L.; Hu, Y.Y. Removal of N-methyldiethanolamine in wastewater by using electro-activated persulfate: Mechanisms and applicability. *J. Environ. Chem. Eng.* **2022**, *10*, 108008.
42. Checa-Fernández, A.; Santos, A.; Romero, A.; Dominguez, C.M. Remediation of real soil polluted with hexachlorocyclohexanes (α-HCH and β-HCH) using combined thermal and alkaline activation of persulfate: Optimization of the operating conditions. *Sep. Purif. Technol.* **2021**, *270*, 118795.
43. Zhu, Y.; Zhang, K.; Hu, Q.; Liu, W.; Qiao, Y.; Cai, D.; Zhu, P.; Wang, D.; Xu, H.; Shu, S.; et al. Accelerated spent coffee grounds humification by heat/base co-activated persulfate and products' fertilization evaluation. *Environ. Technol. Innov.* **2023**, *32*, 103393. [CrossRef]
44. Liu, P.; Li, C.; Liang, X.; Xu, J.; Lu, G.; Ji, F. Advanced oxidation of hypophosphite and phosphite using a UV/H_2O_2 process. *Environ. Technol.* **2013**, *34*, 2231–2239. [CrossRef] [PubMed]
45. *GB/T 11893-1989*; Water Quality—Determination of Total Phosphorus—Ammonium Molybdate Spectrophotometric Method. The State Bureau of Quality and Technical Supervision: Beijing, China, 1989.
46. Benvenuti, T.; Krapf, R.S.; Rodrigues, M.A.S.; Bernardes, A.M.; Zoppas-Ferreira, J. Recovery of nickel and water from nickel electroplating wastewater by electrodialysis. *Sep. Purif. Technol.* **2014**, *129*, 106–112.
47. Liu, C.; Wu, B. Sulfate radical-based oxidation for sludge treatment: A review. *Chem. Eng. J.* **2018**, *335*, 865–875. [CrossRef]
48. Zhou, S.; Yu, Y.; Zhang, W.; Meng, X.; Luo, J.; Deng, L.; Shi, Z.; Crittenden, J. Oxidation of microcystin-LR via activation of peroxymonosulfate using ascorbic acid: Kinetic modeling and toxicity assessment. *Environ. Sci. Technol.* **2018**, *52*, 4305–4312. [CrossRef]
49. Peng, Q.; Ding, Y.; Zhu, L.; Zhang, G.; Tang, H. Fast and complete degradation of norfloxacin by using $Fe/Fe_3C@$ NG as a bifunctional catalyst for activating peroxymonosulfate. *Sep. Purif. Technol.* **2018**, *202*, 307–317.

50. Wang, Y.; Sun, H.; Ang, H.M.; Tadé, M.O.; Wang, S. Facile synthesis of hierarchically structured magnetic $MnO_2/ZnFe_2O_4$ hybrid materials and their performance in heterogeneous activation of peroxymonosulfate. *ACS Appl. Mater. Interfaces* **2014**, *6*, 19914–19923. [CrossRef]
51. Ramasahayam, S.K.; Guzman, L.; Gunawan, G.; Viswanathan, T. A comprehensive review of phosphorus removal technologies and processes. *J. Macromol. Sci. Part A* **2014**, *51*, 538–545. [CrossRef]
52. Qin, Z.; Shober, A.L.; Scheckel, K.G.; Penn, C.J.; Turner, K.C. Mechanisms of phosphorus removal by phosphorus sorbing materials. *J. Environ. Qual.* **2018**, *47*, 1232–1241.

water

MDPI

Review

Problems, Effects, and Methods of Monitoring and Sensing Oil Pollution in Water: A Review

Nur Nazifa Che Samsuria [1], Wan Zakiah Wan Ismail [1,*], Muhammad Nurullah Waliyullah Mohamed Nazli [2], Nor Azlina Ab Aziz [3,*] and Anith Khairunnisa Ghazali [3]

1 Faculty of Engineering and Built Environment, Universiti Sains Islam Malaysia, Nilai 71800, Negeri Sembilan, Malaysia; nncsamsuria@gmail.com
2 Tombak Technology Sdn. Bhd., No.6A, Jalan Emas 1, Bandar Sungai Emas, Banting 42700, Selangor, Malaysia; nurullah.waliyullah95@gmail.com
3 Faculty of Engineering and Technology, Multimedia University, Ayer Keroh 75450, Melaka, Malaysia; anith.ghazali@mmu.edu.my
* Correspondence: drwanzakiah@usim.edu.my (W.Z.W.I.); azlina.aziz@mmu.edu.my (N.A.A.A.)

Abstract: Oil pollution in water bodies is a substantial environmental concern that poses severe risks to human health, aquatic ecosystems, and economic activities. Rising energy consumption and industrial activity have resulted in more oil spills, damaging long-term ecology. The aim of the review is to discuss problems, effects, and methods of monitoring and sensing oil pollution in water. Oil can destroy the aquatic habitat. Once oil gets into aquatic habitats, it changes both physically and chemically, depending on temperature, wind, and wave currents. If not promptly addressed, these processes have severe repercussions on the spread, persistence, and toxicity of oil. Effective monitoring and early identification of oil pollution are vital to limit environmental harm and permit timely reaction and cleanup activities. Three main categories define the three main methodologies of oil spill detection. Remote sensing utilizes satellite imaging and airborne surveillance to monitor large-scale oil spills and trace their migration across aquatic bodies. Accurate real-time detection is made possible by optical sensing, which uses fluorescence and infrared methods to identify and measure oil contamination based on its particular optical characteristics. Using sensor networks and Internet of Things (IoT) technologies, wireless sensing improves early detection and response capacity by the continuous automated monitoring of oil pollution in aquatic settings. In addition, the effectiveness of advanced artificial intelligence (AI) techniques, such as deep learning (DL) and machine learning (ML), in enhancing detection accuracy, predicting leak patterns, and optimizing response strategies, is investigated. This review assesses the advantages and limits of these detection technologies and offers future research directions to advance oil spill monitoring. The results help create more sustainable and efficient plans for controlling oil pollution and safeguarding aquatic habitats.

Keywords: effects of oil pollution; oil pollution sensing method; oily water

1. Introduction

Water pollution is a global issue that threatens human health, wildlife, and ecosystem equilibrium. Contaminated water sources can lead to severe health conditions, including gastrointestinal disease, respiratory disease, and even cancer [1]. The presence of pollutants in water bodies destabilizes delicate ecosystems, reducing biodiversity and threatening marine life. Industrial waste is also one of the primary sources of water pollution, as factories and manufacturing plants tend to release harmful chemicals and pollutants into nearby water bodies, making them hazardous to human and animal consumption [2]. Agricultural activities are also a leading source of water contamination. Runoff of pesticides, fertilizers, and animal waste into lakes and rivers may cause algae blooms and oxygen loss, further deteriorating water quality. Agriculture is also the most water-intensive sector, which is a massive threat to water sustainability and availability [3].

Oil pollution in water bodies has become a critical environmental issue, driven by industrial activities and urbanization. The adverse impacts of oil spills extend beyond immediate ecological harm, affecting biodiversity, water quality, and local communities. Basterrechea et al. (2021) found that industrial oil discharge can severely damage irrigation ditches and nearby ecosystems [4], highlighting the need for efficient monitoring systems to ensure water safety. Understanding the sources and environmental consequences of oil pollution is essential for creating effective management strategies.

Oil spills have a great impact on the environment and diversity of organisms occupying spill sites, raising significant policy and regulatory questions on the transportation of oil, spill clean-up, and site rehabilitation in cases of spills. Increasing global demand for petroleum and petroleum derivatives raises the risks of their extraction, transport, and use. Accidental releases of oil, especially at elevated levels, are extreme environmental risks unless the oil is effectively diluted or degraded. The effects of oil spills are longerterm in the Arctic due to lower temperatures and the presence of ice, which inhibit natural processes of degradation compared to other more temperate habitats. In marine conditions, oil spreads across vast distances easily as opposed to land, where it remains localized in specific locations. In cold climates, oil pollution is regulated by biological and physical mechanisms such as dispersion, hardening, biodegradation, bioaccumulation, and dissolution. Although chemical processes such as photolysis affect oil pollution, they are limited in low-temperature regions [5]. The rapid expansion of oil production, transportation, and processing reflects the continuously expanding growth of the world economy. However, this expansion has led to a sudden increase in the number of crude oil pollution accidents, i.e., fuel oil spills, offshore platform leaks, and tanker discharges [5]. When spilled oil deteriorates, its chemical nature changes with the passage of time. Different parts of the oil evaporate, dissolve, and degrade at varying rates, and thus the relative concentration of the contaminants in the water varies with time and space. The toxicity of the fractions depends on their solubility, molecular structure, mode of action, metabolic pathways, and environmental conditions [6].

1.1. Relationship Between Oil Pollution and Water Pollution

Mass transportation and extraction of oil and gas always lead to oil leakage, which in turn pollutes the soil and groundwater. Alteration of the wettability of the soil due to contamination is one of the severe impacts of the pollution. Studies have proved that in clean soil, capillary water ascends considerably higher over the same time than in oil polluted soil. As the level of contamination increases, the ability of water to rise through the soil is reduced, further decreasing its ability to absorb and transmit water. Pollution by oil also makes it more challenging for water to penetrate the soil, thus adding to the problems of water movement and holding [7].

Oil pollution is a global environmental problem, with an estimated 2.4 million tonnes of oil entering marine waters each year. This pollution results from both natural and anthropogenic processes. Major sources include discharges from storage facilities and refineries, discharge of ballast water from ships, leakage of boreholes, pipeline breaks, carry-over discharges, urban and industrial effluents, and even atmospheric deposition. The ubiquity of oil pollution necessitates the imperative of effective monitoring and mitigation actions to safeguard soil, water resources, and marine ecosystems from its deleterious effects [8].

1.2. The Importance of Monitoring and Managing Oil Pollution in Water

Oil contamination of water bodies is a rapidly emerging worldwide concern with significant human health, aquatic life, and environmental implications. As industrial processes and energy demand continue to escalate, oil spillage has increased, with dire consequences for coastal economies and marine biodiversity. Pollution by oil not only disrupts aquatic life but also affects the human societies that rely upon unpolluted water sources for drinking, fishing, and other simple activities. The moment the oil is released into oceans, it undergoes a sequence of chemical and physical transformations, depending upon environmental conditions such as temperature, wind, wave currents, and salinity. These phenomena dictate the distribution of oil over the water surface and its influence upon marine organisms and eventually the beach. Newly released oil is composed of composite hydrocarbons of varied chemical composition and molecular weight. Components with low boiling points are more soluble and thinner and disperse quickly across the water surface, increasing the spread of pollution. Over time, oil thickens and persists, becoming harder to remove and causing more environmental harm [9].

Since there are severe environmental and economic impacts associated with oil pollution, effective and timely monitoring is required to minimize harm and direct successful response activities. Oil spill monitoring is required for a number of significant reasons: it enables prompt emergency measures, offers the legal framework for environmental surveys and damage claims, and assists in long-term ecological restoration activities. Implementation of efficient monitoring systems lies at the heart of improving safety in transportation and maritime industries, particularly where oil shipping and extraction activities are high [10]. Detection and control of oil spills at an early stage are crucial for minimizing their adverse effects since untreated oil has the potential to spread over a large area, hence rendering containment and clean-up processes laborious and expensive.

Technology advancements have significantly improved the detection and tracking of oil spills. Conventional techniques, including visual observations and satellite imagery, have been complemented by cutting-edge strategies that leverage artificial intelligence (AI) and machine learning (ML), which improve detection precision, accelerate the identification process, and ensure real-time monitoring. AI-driven systems are especially valuable in the management of massive oil spills, as they automatically detect oil slicks and forecast their spreading in relation to environmental conditions [11]. Furthermore, the utilization of deep learning (DL) models enables the investigation of extensive data sets, enhancing the precision of oil spill identification and minimizing false alarms [12]. Aside from making response more efficient, such technologies yield important information for examining environmental effects and establishing preventive measures.

In recent years, cost-effective monitoring methods have also developed as crucial instruments for oil spill management, particularly in developing nations where resources for complex systems may be restricted. For instance, a study in [13] presented a low-cost operational monitoring system that integrates open-source images, citizen science, and affordable multi-platform devices to identify oil spills in real-time. This technology has proven its usefulness during three major oil spills in June and July 2023, presenting a feasible solution for continuous monitoring and timely alarm issuance. Such technologies are especially helpful for countries highly reliant on offshore oil extraction, where continual monitoring is vital to balancing industrial activities with environmental protection.

Oil pollution significantly affects both ecosystems and human health. Demoner et al. (2023) revealed the rapid spread of oil plumes in freshwater systems under various hydrological conditions [14]. Effective response measures must be implemented swiftly, ideally within hours, to prevent severe and long-lasting damage (He et al., 2023) [15]. These findings emphasize the importance of real-time monitoring and sensing technologies to track oil pollution accurately and enable timely interventions.

To address these challenges, various monitoring and detection methods have been developed. For instance, low-cost optical sensors, as proposed by Basterrechea et al. (2021) [4], offer effective solutions for detecting oil in water. Additionally, numerical simulations and qualitative monitoring approaches developed by Demoner et al. (2023) [14,15] provide critical insights into oil dispersion dynamics and emergency response planning. This synthesis of recent research underscores the urgent need for advanced monitoring systems to mitigate oil pollution's effects and ensure the sustainability of water resources.

This work seeks to give a complete assessment of the difficulties, consequences, and methodologies involved with oil pollution monitoring and sensing in aquatic bodies. It analyses the chemical behavior of spilled oil, the environmental and socio-economic repercussions of oil pollution, and the technical developments used in detection and monitoring. Special focus is given to modern sensing techniques, including remote sensing, optical technologies, and AI-based solutions, as well as the limitations of current methodologies. By identifying gaps in existing monitoring systems and researching novel solutions, this review intends to offer insights into enhancing oil spill response and supporting sustainable environmental management practices.

The review consists of five sections. Section 1 discusses the compound structure of oil. Section 2 describes the causes and effects of oil pollution. Section 3 explains the problems of oil pollution in oceans, rivers, and lakes. Section 4 discusses the effects of oil pollution towards the environment. Section 5 describes the methods used to monitor and sense oil pollution. Section 6 discusses the parameters used in the oil pollution sensing method. Section 7 concludes with the findings of this review. By exploring various technological applications and remote sensing, the review aims to enhance the understanding of oil pollution detection and monitoring techniques. The findings contribute valuable insights into the strengths and limitations of different methods, emphasizing innovative oil pollution sensing and monitoring approaches.

2. Compound Structure of Oil

Crude oil, also known as petroleum, is described as a blend of hydrocarbons that originally exist in the liquid state within natural underground reservoirs and retain their liquid form at atmospheric pressure following the passage through surface separating facilities [16]. Alkanes are often referred to as paraffins, cycloalkanes as naphthene, and compounds containing nitrogen, sulphur, and/or oxygen as NSOs [17,18]. These compounds can exist as single molecules or with alkyl side chains attached.

The interaction between oil and water involves intermolecular forces, primarily hydrophobic interaction and Van der Waals forces [19]. Hydrophobic interaction occurs due to the non-polar nature of oil molecules, causing them to repel water molecules because of differences in polarity. On the other hand, Van der Waals forces are attractive forces between molecules that contribute to the cohesion of oil molecules and their interactions with water molecules [20]. These forces play a crucial role in phenomena like oil—water separation by affecting the behavior and solubility of oil in water.

When oil is dropped into water, it resists dispersal due to its non-polar nature and the cohesive forces among its molecules. Oil, being hydrophobic, does not readily dissolve in water, resulting in phase separation or immiscibility. Thus, the mixture of both liquids can produce emulsions. Emulsions tend to divide into an oil and a water phase due to droplet formation, which makes them thermodynamically unstable. Emulsion kinetic stabilization can be achieved by delaying this formation among various molecules present at the oil—water interface [21]. The interaction between oil and water is characterized by limited solubility due to the disparate polarities of the two substances. While small oil droplets may disperse in water, true dissolution does not occur due to lack of molecular compatibility [22]. The separation between oil and water is a physical phenomenon driven by differences in polarity, leading to the formation of distinct layers or droplets. The time required for oil to disperse or separate in water depends on variables such as types of oil, the presence of emulsifying agents, and environmental conditions [23].

3. Oil Pollution Problems in Oceans, Rivers, and Lakes

Water pollution is a detrimental issue. Many studies have been done to monitor water pollution [24–27]. Oil spills are one of the causes of water pollution that normally occur in the ocean due to shipwreck [28]. Ocean water is denser than river water due to salt content. The effects of an oil spill in a river are different than an oil spill in the ocean because river water is less dense than certain types of oils. Oil also can interact with sediment carried by rivers, where the sediment can stick to the oil droplets when the droplets drift into the river. Then, the sediment—oil substance will precipitate to the bottom of the river after the water slows down [29]. River water is often utilized for irrigation to support crop growth in agriculture. Thus, contamination of river resources has resulted in the pollution of irrigation systems.

Anyanwu et al. explored the ecological consequences of extensive oil exploitation and global changes in the African Niger Delta. This research focuses on characterizing seasonal dynamics and pollution, particularly total-petroleum-hydrocarbons (TPHs), heavy metals (HMs), and nutrient loads, in correlation with climate-driven variables. There were increased of nutrient levels, turbidity, salinity, temperature, and sulphate (SO_4^{2-}) which influenced TPHs/HMs variability, notably during the wet season [30]. These findings emphasize the immediate need for enhanced pollution control in the Niger Delta, considering the observed spatio-temporal variations and the exacerbating effects of climate change. The study calls for future assessments to include exposure effects and bioaccumulation in biota, incorporating potential climate change scenarios, especially for communities relying intensively on the system for drinking water, food supply, and livelihood.

4. Environmental Effects of Oil Pollution

Oil pollution has significant and strong impacts on ecosystems, human health, and other environmental elements. Aquatic life can be severely harmed by oil spills. Oil prevents light from reaching underwater life, which disturbs the plant's growth and photosynthesis process. Birds and marine mammals that have oil on their fur and feathers face difficulties staying warm and float around. Oil also can cause direct or indirect harm to fish and other aquatic organisms [31].

The marine environment of the Gulf of Guine in Nigeria, is important for local communities that heavily rely on fish as a primary source of animal protein. Unfortunately, the contribution of fish to food security faces growing challenges due to unsustainable practices and the impacts of climate change. The study sheds light on the oil spillages in the Bonga Oil Field that harm the marine environment. The article states that pollution exacerbates food insecurity in the region by damaging vital ecosystems.

Another study on the effects of climate change on oil pollution concluded that increasing temperatures considerably influence the stability of oil and gas production systems and associated infrastructure. Permafrost melting reduces the amount of ice-based transportation, weakens foundations of buildings constructed on permafrost, and lowers the load-carrying capacity of these structures. In the Arctic (Beaufort Sea), an American oil firm was unable to build transport and extraction infrastructure due to a shortage of sea ice, which stopped its plan to open the first oil drilling facility in Arctic waters. Melting permafrost in Russia accounted for about 23% of technical failures in 2021, disrupting oil and gas production. Similarly, in Texas, USA, harsh cold conditions in 2021 led to a decline in oil refinery output by 2.4 million barrels per day [32].

Individuals exposed to oil spills or living in areas with chronic oil pollution may experience adverse health effects. Inhalation of oil vapors or fumes near affected areas can cause respiratory problems in humans. Hwang et al. [33] discussed the effect of a post-spill incident near Taean, South Korea, which affected people's health over a long period. The study focused on adults who lived close to the accident site. The study shows that being near the oil spill or helping to clean it up can be linked to health issues. These issues include asthma, allergies, and skin problems. It shows that oil spills can affect the health of people for a long time.

In addition, oil spills can cause harmful effects on fishery and mariculture resources, causing physical contamination and toxic impacts on fish and other marine life [34]. Oil pollution can have severe economic consequences for industries such as fisheries. Contaminated water can lead to the decline of fish populations, affecting both commercial and subsistence fishing. Oil spills have the potential to significantly minimize business operations, particularly along the coast. The severity of the impact varies with factors such as the type of oil spilled, the cause of the spill, and the extent of its influence on fishing and agriculture in the sea locally. The research [34] focuses on understanding the consequences of ship-source oil pollution on fishing and mariculture. It offers guidance on response measures and management strategies to help reduce the severity of oil spill impacts. By exploring effective protective measures and cleanup efforts, the article provides insights that may prevent or minimize the damage caused by oil spills in aquatic environments, supporting the health of seafood production and the well-being of coastal communities.

Oil pollution causes serious consequences to the environment, ecosystems, and human well-being. In aquatic ecosystems, oil spills can harm marine life, affect fish, mammals, and birds, and contaminate water sources. Human health is directly affected through skin irritation and other health problems, while economic sectors like fisheries suffer due to declines in fish populations. The long-term environmental damage disrupts ecosystems, affecting food chains and habitats. While some natural processes can break down oil over time, the persistence of pollution and its far-reaching effects highlight the urgent need for preventive measures and effective environmental regulations to safeguard ecosystems and human health.

Ewim et al. [35] conducted a comprehensive analysis of the wastewater issues resulting from oil spills and pollution in Nigeria's Niger Delta region. Utilizing a secondary data collection approach, the study drew on diverse sources to systematically review the existing literature, identifying challenges and proposing mitigation and remediation strategies. Moreover, the study reveals the insufficiency of existing measures for oil spill response, cleanup, compensation, and remediation, perpetuating adverse consequences for local communities. The article suggests innovative technologies, including advanced oil spill detection, bioremediation, and renewable energy to enhance mitigation.

5. Oil Pollution Sensing and Monitoring Methods

There are three common methods to detect oil pollution in oceans, rivers, or lakes, namely remote, optical, and wireless sensing. Table 1 summarizes the methods of sensing and monitoring oil pollution.

Table 1. Various methods of oil pollution sensing.

Categories	Methods/Techniques	Description of the Methods	Areas for Future Improvement
Remote sensing	Camera in visible and infrared spectrum [36]	Some methods include passive observation, optical techniques, visible spectrum, infrared sensors, radar, satellite radar systems, laser fluorosensors, and slick thickness measurements using ultrasonics and chemical analysis	Currently, both visible and infrared cameras are installed on drones. To gain confidence with various oil kinds and conditions, the sensor needs to be used more frequently. The tactical and operational support platform of the future might be the automated aerial drone.
	Underwater acoustic method [37]	The study aims to assess the detection feasibility of Hazardous and Noxious Substances (HNS) using two methods. Firstly, a tank experiment with 200-kHz active sonar utilized castor oil as a chloroform alternative to examine underwater acoustic detection, and secondly, a computer simulation evaluated the potential of side scan sonar for detecting chloroform and tetrachloride randomly distributed on different seabed types.	The side scan sonar may be an effective instrument for monitoring the Hazardous and Noxious Substances (HNS) resting on the sediment interface, according to the results of simulating the sonar image using a sonar equation.
	Multi-temporal approach [38]	Change Vector Analysis (CVA) was employed to examine the present image of the region and compare it to a reference image dataset that had undergone statistical analysis to lessen sea spectral variability across time to develop an automatic detection algorithm.	Further validation and training needed by enriching the oil spill datasets.
Optical sensing	Light Emitting Diode (LED) and Light Dependent Resistor (LDR) [4]	The study proposes an optical sensor system using LEDs to monitor oil concentration in irrigation ditches. The sensor differentiates between the presence of diesel engine oil, identifies the gasoline engine oil source, and estimates the concentration of oil from 0 to 0.20 mL/cm^2 range. The algorithm combines signals from white, blue, and red LEDs for comprehensive oil monitoring in the irrigation system.	Conduct experiments under actual conditions to assess the developed sensor's performance in a dynamic scenario, providing insights into its quantification capacity for oil in irrigation systems.
	Light Detection and Ranging (LIDAR) sensor [39]	A 905nm laser LiDAR system, with 3 channels and 3 Hz scanning, distinguishes seawater from oil. It detects various oil thickness, presenting thick oil as brighter pink on the seawater surface. Key parameters include sensed area, spill thickness, detection Field-of-View (FoV), and time.	Conduct a field test to evaluate the LiDAR system's performance in real-world conditions, considering the impact of waves and tidal currents.
	Laser-induced fluorescence (LIF) sensor & Unmanned Aerial Vehicle (UAV) [40]	The article discusses LIF spectral features for different states of oil products in seawater and thin slicks. It introduces a calibrated LIF method for identifying and measuring ocean pollution from bilge water disposal, along with a small-sized LIF sensor for unmanned aerial vehicles (UAVs) to monitor oil pollution at sea.	Utilize LIF spectroscopy to identify oil products in slicks, with a focus on determining the dependence of LIF spectra parameters on slick thickness.

Table 1. *Cont.*

Categories	Methods/Techniques	Description of the Methods	Areas for Future Improvement
	Oceanic polarization lidar (P-lidar) [41]	Expanding detection dimensions using P-lidar and considering parameters such as polarization degree, polarization angle, and ellipticity.	The research emphasizes the need for a customized P-lidar tailored for oceanic remote sensing by addressing issues related to spatial resolution and radiometric accuracy.
	Integration with FerryBox system and moored SmartBuoy [42]	To evaluate the feasibility of using fluorometric sensors in flow-through systems for real-time detection of oil spills	The impact of measurement depth on oil spill detection
	Optical reflectance [43]	The proposed method effectively estimated the thickness of oil samples in laboratory conditions and accurately determined the volume of thicker oil samples in outdoor settings using RGB images.	Accurately estimate the thickness of oil based on emulsions.
Wireless sensor	Wireless Sensor Networks (WSNs) [44]	The method involves incorporating intelligence at multiple aggregation levels, transforming sensor nodes into active, intelligent observers, thus enhancing the efficiency of the network and contributing to the improvement of oil spill detection in oceanographic research.	Upgrade the intelligence of sensor nodes to enable decision-making regarding oil spills.
	Internet-of-Things (IoT) [45]	The study describes the creation of an autonomous oil—water separator that is Internet of Things (IoT)-based and supports water ecosystems' oil pollution monitoring system. The water-oil separation was carried out by applying semi-permeable membrane nanotechnology.	The potential for further improvement through device design modifications.
	Satellite Aperture Radar (SAR) [46]	This study introduces a novel method for detecting oil spills using satellite synthetic aperture radar (SAR) systems. The technique employs convolutional neural networks (CNNs) trained with an adversarial loss function for image-to-image translation, demonstrating promising results in discriminating between real oil spills and lookalikes in SAR data from Radarsat-2 and Sentinel-1 over the Mediterranean Sea, Atlantic Ocean, and the North Sea.	Further refining the semantic model solution for oil spill detection by exploring modifications to the generator in the Generative Adversarial Network (GAN) architecture, aiming to enhance visual results, accuracy, and Jaccard index scores beyond current state-of-the-art solutions.

5.1. Remote Sensing

Remote sensing refers to the collection of information about an object without direct physical contact [47]. The most popular method of remote sensing for detecting oil spills is mapping the sea surface using passive observation. One of the remote sensing methods involves the usage of cameras in the visible and infrared spectrum [36]. The method is cost-effective to monitor oil spills, especially for operational countermeasures.

Ultraviolet and near infrared methods are less often applied, probably because of the light source and detector's constraint [36]. Satellite and airborne sensors are employed to capture data related to water quality, temperature, and other environmental variables. For seawater, remote sensing helps in monitoring the sea surface temperature, chlorophyll concentration, and ocean color, which indicates phytoplankton abundance and overall water quality [48]. For lakes and rivers, remote sensing can provide information on water turbidity, sediment transport, and shoreline changes [49].

In addition, Babatunde et al. [50] studied acoustic methods, which involve the usage of sound waves to study underwater features and properties. For example, sonar systems can be used to map the bathymetry of the seafloor, assess sediment characteristics, and even detect underwater objects [37]. Doppler devices can measure water velocity and flow patterns [51]. Acoustic methods are particularly useful to study the physical characteristics of water bodies, providing insights into the structure and dynamics of underwater environments.

There are numerous sources of oil waste, and it takes decades to dispose of them. A multi-temporal approach has been used on the problem of oil leak detection. The current image of the area of interest and a dataset of reference images were compared using this implementation [38].

5.2. Optical Sensing

Basterrechea et al. has developed an optical sensor to monitor the discharge of industrial oil in irrigation ditches [4]. A low-cost system was created to detect and measure the presence of oil in irrigation ditches, which could help prevent the contamination of crops and water sources. Five different colors of LEDs were used as light sources, consisting of blue, yellow, red, green, white. Then, two photodetectors were positioned 0° and 180° from the LEDs to detect light that was absorbed by the oil. The amount of oil changed the percentages of light that were absorbed or reflected, leading to various LDR values. The light emitted by the LEDs reached the oil layer after passing through the water column. There was light reflected by the oil layer and returned to the LDR at a different point where it crossed to reach the LDR at a 180° angle [4].

The proposed sensor operated by setting threshold values, detecting oil presence using LEDs and an LDR, and employing a turbidity sensor to prevent false positives caused by water particles. When the oil presence was confirmed and turbidity was below the threshold, an alarm was triggered, and data about the oil's concentration were stored. Otherwise, data were stored in an SD card of the node without triggering an alarm. The system had a limit of 3600 records stored, and once it was reached, it sent the stored data, erased the memory, and waited for a new measurement to start [4]

Jung Hwan Moon et al. [39] discussed the geometric properties of spilled oil on seawater using a LiDAR sensor, employing a near-infrared 905nm wavelength laser. The LiDAR gathered information from the target at each laser point in order to sense the surroundings.

The compact Laser-induced fluorescence (LIF) sensor is presently undergoing testing on Unmanned Aerial Vehicles (UAVs) to acquire on-site comparative data [40]. The utilization of UV radiation presents notable advantages, including the ability to monitor slicks and dissolved oil hydrocarbons during daylight hours, owing to the elevated quantum efficiency of LIF when subjected to UV radiation. This technology holds particular significance in the context of optical sensors designed for environmental monitoring on UAVs, providing valuable insights for research and contributing to the overall body of knowledge in the field.

Xudong Liu et al. [41] discussed the challenges encountered by oceanic polarization lidar (P-lidar) due to strong seawater attenuation, which limited the penetration of the 532 nm laser energy to a depth of 30 m. The research discussed fundamental principles of polarization and P-lidar. The methodology involved expanding detection dimensions using P-lidar and considering parameters such as polarization degree, polarization angle, and ellipticity. Results were presented from a 3D flash P-lidar based on a micro-polarizer camera, showcasing cross-polarization configurations at different visibility distances. The research emphasizes the need for a customized P-lidar tailored for oceanic remote sensing by addressing issues related to spatial resolution and radiometric accuracy.

Moreover, a large proportion of oil spills occur near busy marine fairways. Oil spill monitoring relies on satellite remote sensing, aerial vehicles, and ships. These methods have limitations in detecting and tracking oil spills effectively. Fluorometric sensors were tested for real-time oil detection in flow-through systems. The sensors detected diesel oil for 20 days under laboratory conditions. Detection was impacted by CDOM, turbidity, and algae-derived substances, particularly algae extract. The sensors were integrated into a

FerryBox and a moored SmartBuoy for field testing. Field tests confirmed that interference compounds caused measurement variations. No oil spill peaks were observed during the two-month field study. Both autonomous systems provided reliable real-time data. Improved calibration and advanced sensors are necessary to address natural interferences [42].

Koirala et al. [43] evaluates visible, near-infrared, and shortwave infrared regions for oil spill monitoring. A physical model was developed for oil thickness and volume estimation using optical reflectance. The method accounts for variations in acquisition and illumination conditions. An artificial neural network algorithm was designed for detecting spilled oil. Training samples were generated using the proposed physical model. Experiments were conducted indoors and outdoors on four oil types for validation. Hyperspectral datasets were developed using Agrispec and Imec snapscan cameras under controlled conditions. Oil samples with thicknesses between 500 μm and 5000 μm were analyzed. An RGB drone camera demonstrated outdoor oil monitoring potential using visible wavelengths. The results highlight the method's robustness and applicability in diverse scenarios.

In addition, fluorescence and spectroscopy techniques are employed to study the optical properties of water. Fluorometers measure the fluorescence emitted by substances in the water, such as chlorophyll-a, which can indicate the presence of phytoplankton and algal blooms [52]. Spectroscopy helps in analyzing the absorption and reflection of light at different wavelengths, providing information on water quality parameters like dissolved organic matter [53], turbidity and nutrient concentrations.

5.3. Wireless Sensing

One of the issues that cause oil spill in the ocean is breakage of oil pipes, leakage from ships, and industrial waste. To address the problem, a proposed solution was done by implementing intelligent Wireless Sensor Networks (WSNs) at multiple aggregation levels, creating an inter-network of intelligent nodes to enhance the efficiency of detecting and responding to oil spills without relying on light propagation [44]. To overcome the issue, an Internet of Things (IoT)-based oxygen reduction potential sensor, integrated into wireless sensor networks (WSNs) and underwater robotic submarines, was proposed [54]. Tabella et al. [55] used Wireless Sensor Networks (WSNs) for oil spill detection and localization, where a Fusion Center (FC) aggregated local decisions, enhancing the reliability of global binary decisions. The approach not only provides an estimated position of the leak source but specifically focuses on detecting the leakage source rather than just the presence of oil.

Sai et al. [56] discussed oil spill management system in the seawater by controlling oil pollution. A new water sprinkler system was developed to hasten cloud formation. The Internet of Things (IoT) and remote sensing were employed to expand the capability of the system to collect, store and display data. Various sensors, including barometers, temperature sensors, a 6-axis gyroscope and accelerometers, and ambient light sensors, have been used to ensure the successful operation of the systems. The main objective of the system was to remove or separate the oil spill from the water bodies. The sponges were used to separate crude oil from water by absorbing it. By reducing the amount of energy needed to evaporate a given amount of water from a surface, the sprinkler system assisted in cloud formation. The data obtained through IoT and remote sensing aid in the equipment's correct operation.

The local system aboard the boat was isolated for data management. It was managed by an onboard software stack that was abstracted into multiple levels, from the application server at the top to the hardware firmware at the bottom. Sensory signal processing and storage layers were located between the storage levels of Structured Query Language (SQL) and the Linux kernel. The application servers periodically uploaded the data into the Cloud by using available connectivity options [56].

Meanwhile, Ronci et al. [46] combined the backscattering properties captured in SAR imagery with the capabilities of deep learning models. The process began with the collection of SAR images from satellites or airborne platforms. These images underwent preprocessing steps to enhance their quality and remove noise. A large, labeled dataset was then generated by manually annotating the SAR images, marking oil spill areas as the target class. Data were trained using the labeled dataset to learn relevant features and classify pixels or regions as oil spill or non-oil spill based on Convolutional Neural Network (CNN). The trained model was evaluated using a validation dataset and subsequently applied to new, unseen SAR images for oil spill detection. The method utilized SAR imagery, deep learning algorithms, a labeled dataset for training, and preprocessing techniques to achieve accurate and automated detection of oil spills, facilitating timely response and mitigation efforts. There are limitations in building the system where it is computationally expensive. Other than that, the method is not fully automated when it requires human intervention to select the SAR images to be used for training and testing.

The detection of oil spills in SAR images is essentially a segmentation problem. To address this, researchers have investigated an image-to-image translation method using CNN architectures that have been modified for the task [46]. Other than deep learning-based techniques, recent research has also focused on practical applications of SAR imaging in the detection and management of oil spills in various aquatic environments.

In recent years, oil pollution has emerged as a major concern for water ecosystems worldwide. Oil pollution also occurs in lake and rivers [30,35,57]. Al Bayaty et al. [57] focused on analyzing Synthetic Aperture Radar (SAR) images containing oil spills using a semi-automated processing approach. They found that a combination of visible image interpretation and image processing yielded highly accurate outcomes, allowing for effective monitoring and initial estimation of the disaster state during oil spill events. However, the study does not specifically address the incorporation of Internet of Things (IoT) systems, suggesting a potential research gap in exploring the synergy between oil pollution sensing methods and IoT technologies for enhanced environmental monitoring and response strategies.

In order to enable an oil pollution monitoring system in aquatic ecosystems, Budiman et al. developed an Internet of Things-based automatic oil-water separation device [45]. The constructed apparatus, which was made up of sensors, an actuator, and a microcontroller, could automatically distinguish between water and oil and identify oil pollution.

6. Important Parameters of Oil Pollution Sensing and Monitoring Methods

Oil pollution sensing and monitoring methods are investigated based on many parameters such as delay in time detection, oil compound, types of oil, types of oil and light, and spectral reflectance. Table 2 shows the parameters employed for the detection of oil spills in water.

Table 2. Parameters studied in the oil pollution sensing method.

Methods	Parameters	Future Study
Fluorometric sensors [42]	Detection time, fluorescent compounds	Utilize advanced mathematical protocols or more sophisticated sensors to enhance the differentiation between actual oil pollution and optical interference.
An optical sensor system based on the absorption and dispersion of light [4]	Oil types, light sources, light receptor, water column heights	Test the sensor under dynamic conditions to simulate real-world scenarios where the oil stains may have movement on the water surface due to various factors such as vibration.
Ultrasonic and Image Processing Methods [58]	Oil slick thickness, area of thick oil slicks	Conduct more lab experiments to further validate the technical feasibility and accuracy of the proposed method in the oil spill response facility where real sea conditions can be simulated
Light refraction [59]	Estimation of oil thickness using refraction of light, laser beam wavelength, angle of incident, and angle of refraction in oil	The influence of environmental factors such as temperature, pressure, and humidity on the accuracy and consistency of oil thickness measurements
Spectral reflectance [60]	Reflectivity, wavelength, oil thickness	Impact of different types of oil on sensor performance

6.1. Detection Time/Period

One of the methods to detect oil spills is by using fluorometric sensors that can detect diesel oil for at least 20 days in laboratory conditions. Part et al. [42] evaluates the feasibility of using fluorometric sensors in flow-through systems for real-time detection of oil spills using the integration of FerryBox system and moored SmartBuoy. According to the results, both autonomous systems provide real-time data. It was found that the measured data had systematic gaps, and a maintenance visit was made to reprogram the datalogger and manually clean the sensors.

In the second lab experiment, many detection techniques were compared to study the detection period of the sensors to identify oil molecules in seawater. There were multiple gaps in the Turner C3 data and one gap in the EnviroFlu-HC data due to lagging software. After 165 h, UviLux was moved because there were significant oscillations in the signal following the third addition of water accommodated fraction (WAF), indicating that it was not responding appropriately. Fluorescent compounds derived from diesel oil were found to present for at least 20 days from the beginning of the experiment, according to the responses from all sensors and the conventional HELCOM oil monitoring approach [42].

6.2. Types of Oil

Numerous studies have been conducted to explore and refine the parameters associated with oil spill detection methods utilizing remote sensing technologies. The sensor can distinguish the presence or absence of diesel engine oil regardless of the LED used [4]. The use of Light Dependent Resistors (LDRs) in conjunction with Light Emitting Diodes (LEDs) of various colors, enables the differentiation between the presence and absence of diesel engine oil and the quantification of gasoline engine oil concentrations, showcasing the effectiveness of the sensor in detecting oil contamination in water.

The author examines the variations in the output from various contaminants. The lowest and highest resistance values measured with two oils (diesel and gasoline engines) are shown in Table 3. The difference between maximum and minimum resistance values of both oils is quite similar for green and white light. Meanwhile, larger resistance variations of the two tested oils can be observed for yellow, red, and blue lights [4]

Table 3. Maximum and minimum resistance measured for two oils [4].

Color	Gasoline		Diesel	
	Minimum (kΩ)	Maximum (kΩ)	Minimum (kΩ)	Maximum (kΩ)
Yellow	138.68	160.91	150.01	178.81
Red	239.5	318.83	256.97	359.19
Blue	70.56	80.66	57.09	76.94
Green	50.88	65.57	51.44	63.85
White	10.35	19.29	11.98	19.29

6.3. Volume of Oil

Volume of oil is one of the parameters that have been studied in sensing and monitoring oil pollution. Du et al. discusses the method to assess the volume of oil slicks by measuring the oil thickness using ultrasonic sensor and image processing [58]. The method calculates oil slick thickness by analyzing ultrasonic traveling time through the cross-correlation method, while capturing images of the oil slick using an airborne drone equipped with an optical camera facilitates area calculation through image processing algorithms. The system is implemented on a remotely operated vehicle (ROV) with ultrasonic immersion transducers [58].

The measured thickness and area are directly multiplied to determine the amount of the oil slick. The oil slick volumes are seen to be periodically varying, with minimum and maximum of about 7 litres and 25 litres, respectively. This observation may be the result of the oil slick's nonuniform thickness brought on by surface water flow [58].

The average oil slick volume was determined for each session to reduce the impact of thickness on measurement accuracy. The average amounts of the oil slick measured over each session varied between 9.0 L and 14.0 L and 14.7 L and 17.0 L, respectively, based on the addition of 10 L and 15 L of crude oil to the tests.

The volume of the oil slick was calculated by multiplying its thickness by its area. The volume correlated closely with the variation in thickness when the pump was activated or deactivated, indicating that thickness significantly influenced volume measurement more than area did. The measured volume varied by less than 10% from the additional oil, demonstrating great precision. The technology was successfully evaluated in the laboratory and may assist in monitoring oil spills, assessing their dimensions and viscosity, and improving response strategies to such incidents [58].

6.4. Signal to Noise Ratio

Identifying the system design of LiDAR is important toward the effectiveness of the research. Jing et al. discusses the signal-to-noise ratio (SNR) to obtain spectral information and two-dimensional images [61]. Signal-to-noise ratio (SNR) is an important index parameter used to evaluate the imaging performance of the LiDAR system. It reflects the system ability to detect radiation and helps in assessing the optical system parameters, imaging device selection, and subsequent electronic system design [61].

SNR estimation considers the radiological transmission and photoelectric conversion process of laser radiation passing through the atmosphere, target reflection, and reaching the imaging detector for quantification. The SNR of the visible light camera and hyperspectral detector in the system is calculated to ensure accurate and reliable detection of oil spills over a 24-h period. The proposed system design meets technical requirements for monitoring oil spill pollution effectively and can cover a working band of 400 nm–1100 nm [61].

LiDAR imaging involves the radiological transmission and photoelectric conversion of laser radiation through the atmosphere, reflecting off the target, reentering the atmosphere, and influencing the optical system until it reaches the imaging detector, where it is then detected, analyzed, and measured. LiDAR is impacted by the target's spectral albedo during transmission, in addition to being impacted by other atmospheric components' absorption and scattering of laser light energy.

6.5. Refraction of Light

Kumar et al. has studied parameters used to detect oil in water by employing Snell's law to quantify the refraction of laser beams passing through the air, oil, or water layer [59]. The experimental setup involves recording the position of the refracted laser beam using a camera module controlled by a microcontroller, and the captured images are processed to calculate X and Y coordinates.

To estimate the unknown oil thickness t, triangle ABC of Figure 1 is taken into consideration. The refraction angles θ_1, θ_2, θ_3 in the oil and water media are determined using Snell's law in Equation (1):

$$\frac{\eta_1}{sin\ \theta_1} = \frac{\eta_2}{sin\ \theta_2} = \frac{\eta_3}{sin\ \theta_3} \tag{1}$$

where η_1, η_2, and η_3 are the refractive indices of air, oil, and water, respectively. The thickness of oil t can be calculated from the ratio between refracted beams and the geometry of triangle. Thickness equation t is formulated in Equations (2) and (3), while the total horizontal distance in Equation (4) is as follows:

$$X = t \times tan\ \theta_2 \tag{2}$$

$$D = W \times tan\ \theta_3 \tag{3}$$

$$E = X + D \tag{4}$$

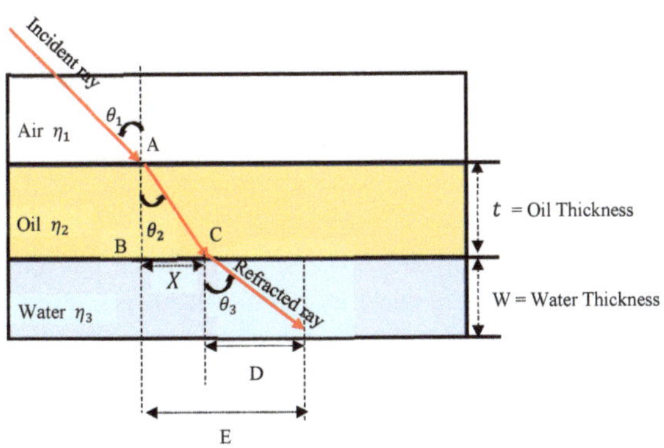

Figure 1. Estimating oil thickness using Snell Law.

Figure 1 depicts the path of a laser beam as it refracts while transitioning between different media specifically from air to oil and later to water. According to Snell's Law, the refraction of a beam is dependent upon the refractive indices of the media traversed. The diagram illustrates the refraction of the laser beam at the boundary of each medium, resulting from variations in optical density, thus clarifying the fundamental principles of light transmission and refraction in layered media.

Kumar et al. [59] investigated the thickness of oil to identify the entire oil runoff using laser refraction, where the brightness of *X* and *Y* coordinates are measured. The findings of the study is shown in Table 4.

Table 4. Varying oil thickness, the resulting Y coordinate, and the percentage error [60].

Oil Thickness	Measured Y Coordinates	Analytic Y Coordinate	Estimated Oil Thickness (mm)	Relative Error in Thickness
1	1340	1340.0729	1.00005440	5.44×10^5
2	1334	1332.8490	1.99827436	8.63×10^4
3	1324	1325.0283	3.00232998	7.77×10^4
4	1316	1316.6108	4.00185653	4.64×10^4
5	1308	1307.5965	4.99845756	3.08×10^4
6	1297	1297.9854	6.00455852	7.60×10^4
7	1290	1287.7775	6.98793992	1.72×10^3
8	1275	1276.9728	8.01237835	1.55×10^3
9	1265	1265.5713	9.00406458	4.52×10^4
10	1257	1253.573	9.97273667	2.73×10^3
11	1239	1240.9779	11.0175600	1.60×10^3
			Percentage error = 0.10246	

Table 4 gives an overview of the measured data and the relative error in oil thickness. The values of the *X* and *Y* coordinates are obtained by image processing, and since oil thickness varies, there is a non-linear relationship between Y coordinates and oil thickness. The Y coordinates change with oil thickness and vice versa. The created quantitative remote sensing model can assess the oil thickness since the estimated values are almost similar.

6.6. Spectral Reflectance

Yin et al. examines the oil thickness detection at two interfaces, namely, water—oil and air—oil [60]. This study introduces a gold-film fiber-optic surface plasmon resonance (FOSPR) sensor for real-time oil slick thickness measurement, fabricated using polydopamine-accelerated wet chemical plating. Salinity and temperature have very little impact on the thickness of an oil slick. The manufactured fiber-optic surface plasmon resonance (FOSPR) sensor can both measure the oil thickness and identify its presence.

The thickness of the oil slick in the atmosphere is measured by the optical fiber sensor [60]. To detect thickness at the air—oil interface, oil was added until a thickness range of 0–5 mm was reached. The sensing region stays in the air while the silver end is submerged in water at the air—oil interface.

7. Conclusions

In conclusion, oil spills present significant challenges to environmental sustainability, particularly due to their adverse effects on aquatic ecosystems. Effective monitoring and detection of oil pollution are essential for mitigating the impacts and safeguarding marine environments, minimizing harm to wildlife and ecosystems. Through the review of current methods and exploration of various oil pollution sensing techniques, including remote, optical, and wireless sensing, it becomes evident that a diverse approach is necessary for comprehensive oil pollution monitoring. Understanding the compounds in oil, the issues caused by oil pollution, the various methods for detecting an oil spill, and the parameters involved in the sensing methods are crucial for developing and implementing effective strategies to address oil spills and protect our marine environments for future generations.

Oil pollution in watery environments is a significant threat for human health, ecological integrity, and the global economy. Efficient monitoring and detection are essential to mitigate the environmental consequences of oil spills and to prevent delays in response activities. Although current technologies, including remote sensing, optical sensing, wireless sensing, and artificial intelligence techniques such as machine learning and deep learning, have significantly enhanced the detection of oil spills and the response time for such incidents, numerous challenges persist. These encompass enhancing the precision and real-time detection capabilities of monitoring systems, identifying cost-effective strategies for large-scale application, and integrating these methods into diverse and dynamic environmental conditions.

Future research must investigate several critical avenues to enhance the efficacy of oil spill detection and response efforts. Initially, there is a necessity for ongoing advancement and refinement of AI-driven algorithms to improve the prediction and detection of oil spills. The incorporation of sophisticated machine learning and deep learning methodologies will significantly improve the real-time identification of oil slicks, facilitating faster localization of spill sites and more accurate estimations of their magnitude. Enhanced predictive models can assist decision-makers in formulating more precise and effective response strategies. In addition to AI technology, there is an urgent need for the creation of affordable, scalable monitoring systems, especially in areas with constrained technical or financial resources. Most locations, particularly in developing nations or those dependent on offshore oil extraction, are limited in their ability to adopt costly, advanced monitoring systems. Investigations into economical and scalable remote sensing platforms, wireless sensor networks, and IoT-based solutions can provide the requisite apparatus for continuous real-time monitoring in these areas. Economical solutions utilizing open-source photography and citizen science serve as sufficient and sustainable alternatives to conventional high-cost monitoring methods.

Moreover, a combination of wireless sensing, optical sensing, and remote sensing technologies can yield more precise and accurate oil spill detecting systems. A multi-sensor system employing the aforementioned technologies could enhance precision and accuracy for real-time detection and identification of oil contamination, especially in challenging or inaccessible situations. It would not only improve detection but also increase efficiency in response activities.

A crucial domain for future investigation is determining the long-term ecological consequences of oil pollution. Although considerable focus is placed on rapid spill response, further research is necessary to evaluate the long-term effects of oil pollution on aquatic ecosystems, biodiversity, and the potential for recovery in affected regions. Such studies would enhance cleanup procedures and improve long-term management and restoration practices for oil-contaminated ecosystems.

Future research should concentrate on enhancing bioremediation and biodegradation technologies, which provide an eco-friendly and sustainable approach to oil spill remediation. Through augmenting the natural degradation of oil in aquatic environments or creating novel bioremediation agents, more efficient and environmentally friendly alternatives could be devised to supplement existing mechanical and chemical remediation techniques. Significant advancements have been achieved in the detection and response to oil spills; yet, considerable potential for innovation and enhancement persist. Through following these guidelines for future research, the industry can advance towards more precise, efficient, and environmentally sustainable methods for mitigating oil pollution and safeguarding vulnerable aquatic ecosystems.

Author Contributions: Conceptualization, N.N.C.S., W.Z.W.I., M.N.W.M.N. and N.A.A.A.; methodology, N.N.C.S., W.Z.W.I. and M.N.W.M.N.; validation, W.Z.W.I., M.N.W.M.N., N.A.A.A. and A.K.G.; investigation, N.N.C.S. and W.Z.W.I.; writing—original draft preparation, N.N.C.S., W.Z.W.I. and M.N.W.M.N.; writing—review and editing, W.Z.W.I., M.N.W.M.N. and A.K.G.; visualization, N.N.C.S. and W.Z.W.I.; supervision, W.Z.W.I. All authors have read and agreed to the published version of the manuscript.

Funding: This research is funded by a grant from Ministry of Higher Education, Malaysia, for Fundamental of Research Grant Scheme (FRGS/1/2024/WAS02/USIM/02/1), and the APC is funded by Multimedia University for (MMUE/210013).

Data Availability Statement: No new data were created or analyzed in this study. Data sharing is not applicable to this article.

Acknowledgments: We would like to acknowledge the support given by the Universiti Sains Islam Malaysia and Multimedia University towards this project.

Conflicts of Interest: Author M.N.W.M.N was employed by the company Tombak Technology Sdn. Bhd. The remaining authors declare that the research was conducted in the absence of any commercial or financial relationships that could be construed as a potential conflict of interest.

Abbreviations

The following abbreviations are used in this manuscript:

IoT	Internet of Things
TPHs	total-petroleum-hydrocarbons
HMs	heavy metals

References

1. Yadav, S.C. Water Pollution: The Problems and Solutions. *Sci. Insights* **2024**, *44*, 1245–1251. [CrossRef]
2. Burkholder, J.; Libra, B.; Weyer, P.; Heathcote, S.; Kolpin, D.; Thorne, P.S.; Wichman, M. Impacts of waste from concentrated animal feeding operations on water quality. *Environ. Health Perspect.* **2007**, *115*, 308–312. [CrossRef] [PubMed]
3. Ingrao, C.; Strippoli, R.; Lagioia, G.; Huisingh, D. Water scarcity in agriculture: An overview of causes, impacts and approaches for reducing the risks. *Heliyon* **2023**, *9*, e18507. [CrossRef]
4. Basterrechea, D.A.; Rocher, J.; Parra, L.; Lloret, J. Low-cost system based on optical sensor to monitor discharge of industrial oil in irrigation ditches. *Sensors* **2021**, *21*, 5449. [CrossRef] [PubMed]
5. Jha, S.; Dahiya, P. Impact analysis of oil pollution on environment, marine, and soil communities. In *Advances in Oil-Water Separation: A Complete Guide for Physical, Chemical, and Biochemical Processes*; Elsevier: Amsterdam, The Netherlands, 2022; pp. 99–113. [CrossRef]
6. French-McCay, D.P.; Parkerton, T.F.; de Jourdan, B. Bridging the lab to field divide: Advancing oil spill biological effects models requires revisiting aquatic toxicity testing. *Aquat. Toxicol.* **2023**, *256*, 106389. [CrossRef] [PubMed]
7. Wei, Y.; Li, G. Effect of Oil Pollution on Water Characteristics of Loessial Soil. *IOP Conf. Ser. Earth Environ. Sci.* **2018**, *170*, 032154. [CrossRef]

8. Effendi, H.; Mursalin, M.; Hariyadi, S. Rapid Water Quality Assessment as a Quick Response of Oil Spill Incident in Coastal Area of Karawang, Indonesia. *Front. Environ. Sci.* **2022**, *10*, 757412. [CrossRef]
9. Asif, Z.; Chen, Z.; An, C.; Dong, J. Environmental Impacts and Challenges Associated with Oil Spills on Shorelines. *J. Mar. Sci. Eng.* **2022**, *10*, 762. [CrossRef]
10. Li, K.; Ouyang, J.; Yu, H.; Xu, Y.; Xu, J. Overview of Research on Monitoring of Marine Oil Spill. *IOP Conf. Ser. Earth Environ. Sci.* **2021**, *787*, 012078. [CrossRef]
11. Alotaibi, E.; Nassif, N. Artificial intelligence in environmental monitoring: In-depth analysis. *Discov. Artif. Intell.* **2024**, *4*, 84. [CrossRef]
12. Singh, U.; Acharya, D.; Mishra, S. Oil Spill Detection & Monitoring with Artificial Intelligence: A Futuristic Approach. *CEUR Workshop Proc.* **2022**, *3314*, 12–22.
13. Uribe-Martínez, A.; Espinoza-Tenorio, A.; Cruz-Pech, J.B.; Cupido-Santamaría, D.G.; Trujillo-Córdova, J.A.; García-Nava, H.; Flores-Vidal, X.; Gudiño-Elizondo, N.; Herguera, J.C.; Appendini, C.M.; et al. An affordable operational oil spill monitoring system in action: A diachronic multiplatform analysis of recent incidents in the southern Gulf of Mexico. *Environ. Monit. Assess* **2024**, *196*, 1069. [CrossRef] [PubMed]
14. Demoner, S.C.; Teixeira, M.R.; de Abreu, C.H.M.; da Cunha, A.C. Numerical Simulation of Oil Spills in the Lower Amazonas River. *Water* **2023**, *15*, 2197. [CrossRef]
15. He, F.; Ma, J.; Lai, Q.; Shui, J.; Li, W. Environmental Impact Assessment of a Wharf Oil Spill Emergency on a River Water Source. *Water* **2023**, *15*, 346. [CrossRef]
16. Nandakumar, V.; Jayanthi, J. Petroleum system and the significance of HCFI study. In *Hydrocarbon Fluid Inclusions in Petroliferous Basins*; Elsevier: Amsterdam, The Netherlands, 2021; pp. 75–106. [CrossRef]
17. Wilkes, H.; Schwarzbauer, J. Hydrocarbons: An Introduction to Structure, Physico-Chemical Properties and Natural Occurrence. In *Handbook of Hydrocarbon and Lipid Microbiology*; Springer: Berlin/Heidelberg, Germany, 2010. [CrossRef]
18. Schobert, H. Composition, classification, and properties of petroleum. In *Chemistry of Fossil Fuels and Biofuels*; Cambridge University Press: Cambridge, UK, 2013; pp. 174–191. [CrossRef]
19. Pullanchery, S.; Kulik, S.; Rehl, B.; Hassanali, A.; Roke, S. Charge transfer across C–H···O hydrogen bonds stabilizes oil droplets in water. *Science* **2021**, *374*, 1366–1370. [CrossRef]
20. Tao, A.R. Intermolecular Forces. In *Chemical Principles of Nanoengineering*; John Wiley & Sons: Hoboken, NJ, USA, 2023; Chapter 1; pp. 7–46.
21. Sjöblom, J.; Stenius, P.; Simon, S.; Grimes, B.A. Emulsion Stabilization. In *Encyclopedia of Colloid and Interface Science*; Springer: Berlin/Heidelberg, Germany, 2013; pp. 415–454. [CrossRef]
22. Costa, C.; Medronho, B.; Filipe, A.; Mira, I.; Lindman, B.; Edlund, H.; Norgren, M. Emulsion formation and stabilization by biomolecules: The leading role of cellulose. *Polymers* **2019**, *11*, 1570. [CrossRef]
23. Saad, M.A.; Kamil, M.; Abdurahman, N.H.; Yunus, R.M.; Awad, O.I. An Overview of Recent Advances in State-of-the-Art. *Processes* **2019**, *7*, 470. [CrossRef]
24. Razman, N.A.; Ismail, W.Z.W.; Razak, M.H.A.; Ismail, I.; Jamaludin, J. Design and analysis of water quality monitoring and filtration system for different types of water in Malaysia. *Int. J. Environ. Sci. Technol.* **2023**, *20*, 3789–3800. [CrossRef]
25. Razman, N.A.; Ismail, W.Z.W.; Kamil, N.A.I.M.; Zainurin, S.N.; Ismail, I.; Jamaludin, J.; Sahrim, M.; Ariffin, K.N.Z.; Balakrishnan, S.R. A Review on Water Quality Monitoring Methods Based on Electronics and Optical Sensing. *J. Adv. Res. Appl. Sci. Eng. Technol.* **2022**, *26*, 1–7. [CrossRef]
26. Tyagi, I.; Singh, P.; Karri, R.R.; Dehghani, M.H.; Goscianska, J.; Tyagi, K.; Kumar, V. Sustainable materials for sensing and remediation of toxic pollutants: An overview. In *Sustainable Materials for Sensing and Remediation of Noxious Pollutants*; Elsevier: Amsterdam, The Netherlands, 2022. [CrossRef]
27. Singh, N.; Poonia, T.; Siwal, S.S.; Srivastav, A.L.; Sharma, H.K.; Mittal, S.K. Challenges of water contamination in urban areas. In *Current Directions in Water Scarcity Research*; Elsevier: Amsterdam, The Netherlands, 2022. [CrossRef]
28. Ashworth, J. World War Shipwrecks are Leaking Pollutants into the World's Oceans. National History Museum, 18 October 2022. Available online: https://www.nhm.ac.uk/discover/news/2022/october/world-war-shipwrecks-leaking-pollutants-into-worlds-oceans.html (accessed on 15 January 2025).
29. Oil Spills in Rivers. National Oceanic and Atmospheric Administration. Available online: https://response.restoration.noaa.gov/oil-and-chemical-spills/oil-spills/resources/oil-spills-rivers.html (accessed on 3 September 2024).
30. Anyanwu, I.N.; Beggel, S.; Sikoki, F.D.; Okuku, E.O.; Unyimadu, J.-P.; Geist, J. Pollution of the Niger Delta with total petroleum hydrocarbons, heavy metals and nutrients in relation to seasonal dynamics. *Sci. Rep.* **2023**, *13*, 14079. [CrossRef]
31. Okafor-Yarwood, I. The effects of oil pollution on the marine environment in the Gulf Of Guinea—The bonga oil field example. *Transnatl. Leg. Theory* **2018**, *9*, 254–271. [CrossRef]

32. Dong, J.; Asif, Z.; Shi, Y.; Zhu, Y.; Chen, Z. Climate Change Impacts on Coastal and Offshore Petroleum Infrastructure and the Associated Oil Spill Risk: A Review. *J. Mar. Sci. Eng.* **2022**, *10*, 849. [CrossRef]
33. Hwang, S.H.; Lee, Y.-J.; Choi, Y.-H.; Huh, D.-A.; Kang, M.-S.; Moon, K.W. Long-term effects of the Hebei Spirit oil spill on the prevalence and incidence of allergic disorders. *Sci. Total Environ.* **2024**, *912*, 168801. [CrossRef] [PubMed]
34. International Tanker Owners Pollution Federation Ltd. Effects of Oil Pollution on Fisheries and Mariculture—ITOPF Technical Information Paper. 2011, pp. 2–11. Available online: https://www.itopf.org/fileadmin/uploads/itopf/data/Documents/TIPS_TAPS_new/TIP_11_Effects_of_Oil_Pollution_on_Fisheries_and_Mariculture.pdf (accessed on 20 January 2025).
35. Ewim, D.R.E.; Orikpete, O.F.; Scott, T.O.; Onyebuchi, C.N.; Onukogu, A.O.; Uzougbo, C.G.; Onunka, C. Survey of wastewater issues due to oil spills and pollution in the Niger Delta area of Nigeria: A secondary data analysis. *Bull. Natl. Res. Cent.* **2023**, *47*, 116. [CrossRef]
36. Fingas, M.; Brown, C.E. A review of oil spill remote sensing. *J. Sens.* **2018**, *18*, 91. [CrossRef]
37. Han, D.G.; Choi, J.W.; Son, S.U. Tank experiment and simulation of sunken hazardous and noxious substances detection using high frequency active sonar. *J. Phys. Conf. Ser.* **2018**, *1075*, 012054. [CrossRef]
38. Roberto, L.; Giovanni, L. Oil Spill Detection Using Otical Sensors: A Multi-Temporal Approach. *Satell. Oceanogr. Meteorol.* **2023**, *3*. [CrossRef]
39. Moon, J.; Jung, M. Geometrical Properties of Spilled Oil on Seawater Detected Using a LiDAR Sensor. *J. Sens.* **2020**, *2020*, 5609168. [CrossRef]
40. Bukin, O.; Proschenko, D.; Alexey, C.; Korovetskiy, D.; Bukin, I.; Yurchik, V.; Sokolova, I.; Nadezhkin, A. New solutions of laser-induced fluorescence for oil pollution monitoring at sea. *Photonics* **2020**, *7*, 36. [CrossRef]
41. Liu, X.; Zhang, L.; Zhai, X.; Li, L.; Zhou, Q.; Chen, X.; Li, X. Polarization Lidar: Principles and Applications. *Photonics* **2023**, *10*, 1118. [CrossRef]
42. Pärt, S.; Kankaanpää, H.; Björkqvist, J.-V.; Uiboupin, R. Oil Spill Detection Using Fluorometric Sensors: Laboratory Validation and Implementation to a FerryBox and a Moored SmartBuoy. *Front. Mar. Sci.* **2021**, *8*, 778136. [CrossRef]
43. Koirala, B.; Mboga, N.; Moelans, R.; Knaeps, E.; Sels, S.; Winters, F.; Samsonova, S.; Vanlanduit, S.; Scheunders, P. Study on the Potential of Oil Spill Monitoring in a Port Environment Using Optical Reflectance. *Remote Sens.* **2023**, *15*, 4950. [CrossRef]
44. Hegde, A.D.; Achari, D.P.; Nithya Sree, K.N.; Chantar, N.S.; Abhijith, H.V. IoT Based Oil Spill Detection System. *Int. J. Adv. Res. Comput. Sci.* **2018**, *9*, 183–185.
45. Budiman, F.; Ismardi, A.; Hardinah, T.; Muhammad, R.; Nurwijayadi; Hartaman, A.; Nurhidayat, A.; Sasto, I.H.; Sutapa, I.D. Strengthening oil pollution monitoring system in aquatic environment through development of IoT-based Oil-Water Separator Device. *Ecohydrol. Hydrobiol.* **2023**, *24*, 617–623. [CrossRef]
46. Ronci, F.; Avolio, C.; di Donna, M.; Zavagli, M.; Piccialli, V.; Costantini, M. Oil Spill Detection from SAR Images by Deep Learning. In Proceedings of the International Geoscience and Remote Sensing Symposium (IGARSS), Brisbane, Australia, 3–8 August 2025; pp. 2225–2228. [CrossRef]
47. Muller-Karger, F.E. Remote Sensing Applications Ocean. In *Encyclopedia of Sustainability Science and Technology*; Springer: New York, NY, USA, 2020; pp. 8919–8939. [CrossRef]
48. Dierssen, H.M. Earth System Monitoring. In *Earth System Monitoring*; Springer: New York, NY, USA, 2013. [CrossRef]
49. Liu, W.; Wang, S.; Yang, R.; Ma, Y.; Shen, M.; You, Y.; Hai, K.; Baqa, M.F. Remote sensing retrieval of turbidity in alpine rivers based on high spatial resolution satellites. *Remote Sens.* **2019**, *11*, 3010. [CrossRef]
50. Babatunde, D.; Pomeroy, S.; Lepper, P.; Clark, B.; Walker, R. Autonomous deployment of underwater acoustic monitoring devices using an unmanned aerial vehicle: The flying hydrophone. *Sensors* **2020**, *20*, 6064. [CrossRef]
51. Fricker, P. Analyzing and Visualizing Flows in Rivers and Lakes with MATLAB. MathWorks. Available online: https://www.mathworks.com/company/technical-articles/analyzing-and-visualizing-flows-in-rivers-and-lakes-with-matlab.html (accessed on 14 April 2025).
52. Marghany, M.; Mansor, S. Genetic algorithm for oil spill automatic detection using synthetic aperture radar. *Glob. Nest J.* **2015**, *17*, 858–869. [CrossRef]
53. Zhang, H.; Yao, B.; Wang, S.; Wang, G. Remote sensing estimation of the concentration and sources of coloured dissolved organic matter based on MODIS: A case study of Erhai lake. *Ecol. Indic.* **2021**, *131*, 108180. [CrossRef]
54. Abhijith, H.V.; Raj, S.D.; Babu, H.S.R. Intelligent Boundary Determination of Oil Spill Detection Using IOT. In Proceedings of the 3rd International Conference on Internet of Things and Connected Technologies (ICIoTCT), Jaipur, India, 26–27 March 2018. [CrossRef]
55. Tabella, G.; Paltrinieri, N.; Cozzani, V.; Rossi, P.S. Wireless Sensor Networks for Detection and Localization of Subsea Oil Leakages. *IEEE Sens. J.* **2021**, *21*, 10890–10904. [CrossRef]

56. Sai, K.R.; Nayak, P.J.; Kumar, K.V.A.; Dutta, A.D. Oil Spill Management System Based on Internet of Things. In Proceedings of the 2020 IEEE-HYDCON International Conference on Engineering in the 4th Industrial Revolution, HYDCON 2020, Hyderabad, India, 11–12 September 2020. [CrossRef]
57. Al Bayaty, H.J.A.; Tarrish, A.; Alsultan, Z. Detection of oil spill pollution on water surface using microwave remote sensing techniques. *IOP Conf. Ser. Mater. Sci. Eng.* **2020**, *737*, 012249. [CrossRef]
58. Du, H.; Fan, H.; Zhang, Q.; Li, S. Measurements of the Thickness and Area of Thick Oil Slicks Using Ultrasonic and Image Processing Methods. *Remote Sens.* **2023**, *15*, 2977. [CrossRef]
59. Kumar, V.; Park, S.; Koh, J. Oil Thickness Measurement Using Laser Refraction. *J. Korea Acad. Ind. Coop. Soc.* **2021**, *22*, 332–339. [CrossRef]
60. Yin, H.; Chen, S.; Huang, R.; Chang, H.; Liu, J.; Qi, W.; He, Z.; Su, R. Real-Time Thickness Measurement of Marine Oil Spill by Fiber-Optic Surface Plasmon Resonance Sensors. *Front. Mar. Sci.* **2022**, *8*, 764970. [CrossRef]
61. Jing, L.; Ying, C.; Shuang, L.; Zhaoxin, W.; Kun, Y. Design of Lidar System Based on Marine Oil Spill Monitoring. *E3S Web Conf.* **2020**, *165*, 03052. [CrossRef]

environments

MDPI

Article

Efficient and Sustainable Removal of Phosphates from Wastewater Using Autoclaved Aerated Concrete and Pumice

Oanamari Daniela Orbuleț, Cristina Modrogan, Magdalena Bosomoiu, Mirela Cișmașu (Enache), Elena Raluca Cîrjilă (Mihalache), Adina-Alexandra Scarlat (Matei), Denisa Nicoleta Airinei, Adriana Miu (Mihail), Mădălina Grinzeanu and Annette Madelene Dăncilă *

Faculty of Chemical Engineering and Biotechnologies, National University of Science and Technology Politehnica Bucharest, Gheorghe Polizu Street, No. 1-7, 011061 Bucharest, Romania; oanamari.orbulet@upb.ro (O.D.O.); cristina.modrogan@upb.ro (C.M.); magdalena.bosomoiu@upb.ro (M.B.); enachemirela19@yahoo.com (M.C.); carjila.raluca@gmail.com (E.R.C.); alexandramatei974@yahoo.com (A.-A.S.); denisa.airinei@yahoo.com (D.N.A.); mihailadriana82@gmail.com (A.M.); madalina.grinzeanu@gmail.com (M.G.)

* Correspondence: madelene.dancila@upb.ro

Abstract

Phosphates are key pollutants involved in the eutrophication of water bodies, creating the need for efficient and low-cost strategies for their removal in order to meet environmental quality standards. This study presents a comparative thermodynamic evaluation of phosphate ion adsorption from aqueous solutions using two sustainable and readily available materials: autoclaved aerated concrete (AAC) and pumice stone (PS). Batch experiments were conducted under acidic (pH 3) and alkaline (pH 9) conditions to determine equilibrium adsorption capacities, and kinetic experiments were carried out for the best-performing adsorbent. Adsorption data were fitted to the Langmuir and the Freundlich isotherm models, while kinetic data were evaluated using pseudo-first-order and pseudo-second-order models. The Freundlich model showed the best correlation ($R^2 = 0.90 - 0.97$), indicating the heterogeneous nature of the adsorbent surfaces, whereas the Langmuir parameters suggested monolayer adsorption, with maximum capacities of 1006.69 mg/kg for PS and 859.20 mg/kg for AAC at pH 3. Kinetic results confirmed a pseudo-second-order behavior, indicating chemisorption as the main mechanism and the rate-limiting step in the adsorption process. To the best of our knowledge, this is the first study to compare the thermodynamic performance of AAC and PS for phosphate removal under identical experimental conditions. The findings demonstrate the potential of both materials as efficient, low-cost, and thermodynamically favorable adsorbents. Furthermore, the use of AAC, an industrial by-product, and PS, a naturally abundant volcanic material, supports resource recovery and waste valorization, aligning with the principles of the circular economy and sustainable water management.

Keywords: phosphate adsorption; autoclaved cellular concrete; pumice stone; wastewater treatment; adsorption isotherms; circular economy

1. Introduction

Phosphate pollution in aquatic environments has become a critical concern worldwide due to its role in accelerating eutrophication. a process leading to excessive algal blooms, oxygen depletion, and overall degradation of water quality [1,2]. The presence of phosphates in wastewater originates primarily from domestic sewage, agricultural runoff, detergents, and industrial discharges. The presence of phosphates in wastewater originates primarily from domestic sewage, agricultural runoff, detergents, and industrial discharges. To prevent ecological imbalances and comply with environmental regulations, the efficient removal of phosphates from wastewater has become a priority in modern water treatment practices.

Under natural conditions, surface waters maintain phosphorus levels in a balanced state, with concentrations meeting but not greatly exceeding the needs of aquatic ecosystems. However, when external phosphorus inputs exceed what aquatic microorganisms and plants can assimilate, the surplus triggers rampant algal growth and eutrophication. This phenomenon not only diminishes water quality but also significantly increases water treatment costs for affected water bodies [3]. Even relatively small increases in phosphate concentration can stimulate nuisance algal blooms; accordingly, stringent effluent standards have been established to limit phosphorus in treated wastewater [4,5].

Phosphorus in wastewater is present in various forms, including organic phosphorus compounds, condensed inorganic phosphates (polyphosphates), and inorganic orthophosphate [4,5]. Organic and polyphosphate species—originating from sources like biological waste, fertilizers, and detergents—can undergo hydrolysis or biodegradation and eventually convert into orthophosphate, the form most readily taken up by algae. Effective treatment processes must therefore capture or transform all these species, ultimately removing the orthophosphate to prevent ecological impact. In addition to environmental protection, there is a resource recovery incentive: phosphorus is a non-renewable element essential for agriculture, so recovering it from waste streams aligns with the growing demand for sustainable mineral resources. This dual concern (eutrophication control and resource sustainability) has heightened interest in technologies for phosphorus removal and recovery from wastewater [6,7].

The recovery of phosphorus from wastewater is becoming an increasingly important issue, both to meet the growing demand for sustainable mineral resources and to prevent the eutrophication of water bodies.

Improving the efficiency of municipal wastewater treatment plants can be achieved by developing high-performance processes aimed at reducing phosphate ion concentrations in the effluent. The discharge of wastewater containing excessive phosphate concentrations into receiving waters leads to the occurrence of undesirable eutrophication phenomena.

Numerous methods have been developed and implemented to remove phosphate from wastewater. Conventional treatments include chemical precipitation using alum or iron salts and enhanced biological phosphorus removal in activated sludge systems. While these methods can effectively reduce phosphate levels, they have well-documented drawbacks: high reagent costs, the generation of large volumes of sludge requiring further handling, and limited opportunities for phosphorus recovery [8]. Achieving ultra-low residual phosphate concentrations with such methods can be challenging and often requires additional treatment steps or polishing units [9,10]. Furthermore, specialized processes for phosphorus recovery, such as controlled crystallization of struvite (magnesium ammonium phosphate) or calcium phosphate, have been explored to reclaim phosphorus as a reusable product [11]. These recovery processes can produce valuable fertilizers, but they necessitate strict operational control and can be costly to implement at scale [11,12].

In contrast, adsorption has emerged as an attractive and sustainable alternative for phosphate removal in water treatment [9]. Adsorption-based processes employ solid sorbent materials to capture phosphate ions from solution, and they offer several advantages over conventional methods. Notably, adsorption is relatively simple to operate and, in many cases, does not require the addition of harsh chemicals. This means that the process produces minimal secondary waste—no extensive sludge production—and does not significantly alter the pH or other quality parameters of the treated water [8]. These benefits translate into lower operational costs and fewer downstream disposal issues. Accordingly, developing high-performance adsorbents for phosphate removal is seen as a promising strategy to meet stringent effluent standards in an economical and environmentally friendly manner [9,13].

According to the Romanian Technical Norm for Water Protection NTPA 001/2002, the permissible limits for phosphorus load in industrial and municipal wastewater discharged into natural receptors apply to all categories of effluents, whether or not originating from treatment plants, as presented in Tables 1 and 2 [4–6].

Table 1. Permissible phosphorus loading limits for industrial and municipal wastewater [5].

Quality Indicator	M.U.	Limit Values	Reference
Total phosphorus	mg/L	1 (2)	SR EN 1189:1999 [14]
Synthetic detergents	mg/L	0, 5	SR ISO 7825/1—1996 [15] SR ISO 7825/2—1996 [16]

Table 2. Permissible phosphorus loading limits for industrial and municipal wastewater discharged into municipal sewer networks [6].

Quality Indicator	M.U.	Limit Values	Reference
Total phosphorus	mg/L	5	STAS 10064—75 [17]

Globally, phosphorus pollution is a significant environmental issue, with municipal and industrial wastewater, as well as agricultural runoff, contributing millions of tons of phosphorus annually to aquatic ecosystems. For example, the European Environment Agency reports that in some EU river basins, average total phosphorus concentrations exceed 0.1 mg/L—the threshold above which eutrophication risks increase substantially. Current removal methods, such as chemical precipitation, can cost between 1–3 EUR/m^3 of treated water, depending on the scale and technology used, which can be prohibitively expensive for small or resource-limited treatment plants [3]. In recent years, numerous low-cost adsorbents—including industrial by-products (e.g., fly ash, steel slag, red mud), agricultural wastes (e.g., rice husk biochar, coconut shell biochar), and natural minerals (e.g., zeolites, limestone, laterite)—have been investigated for phosphate removal, often achieving adsorption capacities between 5 and 50 mg P/g, depending on preparation and operational conditions. Comparative studies indicate that materials rich in calcium, magnesium, and aluminum tend to exhibit higher phosphate affinity through precipitation and complexation mechanisms, making them attractive for cost-effective and sustainable water treatment solutions.

Recent research has focused on identifying low-cost, natural, or waste-derived materials that can serve as effective phosphate adsorbents [10]. The use of such materials can reduce treatment costs and promote a circular economy approach by repurposing industrial by-products or natural minerals for pollution control. Among the promising candidates are autoclaved aerated concrete (AAC) and pumice stone (PS). AAC is a lightweight, highly porous form of concrete produced from silica-rich materials (sand, cement, lime, etc.) and an expanding agent, which results in a material rich in calcium, silica, and alumina. Crushed AAC, often obtained as waste from the construction industry, provides numerous binding sites for phosphates due to its rough, porous texture and alkaline surface chemistry [11]. Pumice stone, on the other hand, is a naturally occurring volcanic rock characterized by its low density and microporous structure. With a high specific surface area and a composition dominated by silica (along with significant alumina content), pumice has been utilized as an adsorbent for various contaminants in water. Both raw and modified pumice have demonstrated the capacity to remove pollutants—for example, natural and iron-coated pumice have been used to adsorb nutrients and even dyes from wastewater [13]. These attributes suggest that ACC and PS could serve as cost-effective sorbents for phosphate, an idea that this study investigates in detail.

The objective of this study is to evaluate the potential of AAC and PS as alternative adsorbent materials for phosphate removal from wastewater. Equilibrium batch adsorption experiments were conducted under varying pH conditions (3 and 9) to determine the phosphate uptake capacity of each material. The resulting adsorption data were analyzed using established isotherm models—Langmuir and Freundlich—as well as kinetic models, namely the pseudo-first-order and pseudo-second-order equations, to characterize the adsorption behavior and estimate key parameters. Through this investigation, we aim to gain a deeper understanding of the mechanisms involved in phosphate adsorption onto AAC and PS and to assess the feasibility of utilizing these readily available materials in sustainable wastewater treatment applications. Unlike previous studies that investigate only one of these materials in isolation or involve costly chemical modifications, this research provides a direct comparison of the two untreated materials under conditions relevant to wastewater treatment, thus contributing to the development of sustainable and applicable solutions within a circular economy context.

The sorption process of phosphorus is complex and involves several steps and retention mechanisms. The apparent adsorption of phosphorus can be viewed as a combination of successive processes, including a rapid (almost instantaneous) reversible sorption process on particle surfaces and various slower processes. These slower processes can be further divided into relatively fast and very slow components. The retention mechanisms include electrostatic interaction, formation of surface complexes, formation of hydrogen bonds, ion exchange, etc. Opinions differ about the degree to which these processes are reversible [18–20].

There are many nonlinear models that are used to describe the experimental isotherms for the adsorption of ions from aqueous phase. These include the Temkin, Freundlich, Langmuir models, the double layer Langmuir model, or the Elovich model. In their simplest form, these equations are defined to assume the instantaneous approach to equilibrium, but they can also be modified to represent a time-dependent approach to equilibrium [21–25].

The mechanisms governing phosphate adsorption onto AAC and PS are presumed to involve electrostatic attraction between negatively charged phosphate ions and positively charged surface sites, ion exchange with calcium or aluminum ions, and surface precipitation on alkaline substrates.

It has been previously reported, in the case of some adsorbents, that phosphorous forms strong bonds with metallic ions in the process of adsorption [26–29]. This increases the adsorption capacity of the material. Some metals may flow back into the current stream, causing secondary pollution [18,30]. For the materials, we tested the risk of releasing dangerous metal ions back into the water, which is minimized by the fact that both materials contain a large amount of SiO_2, which contains very stable or harmless Ca^{2+} and Mg^{2+} ions. The Al^{3+} ions have been reported to be blocked by the adsorbed phosphate ions [12].

While several adsorbents have shown promising results, many rely on chemical modification to enhance performance, which increases cost and limits practical application.

Therefore, there is a need to explore affordable, unmodified materials with sufficient efficiency. AAC and PS offer distinct advantages such as low cost, high porosity, and alkaline composition, as well as the potential—reported in previous studies [31]—for reuse or safe disposal. In this study, we focus on evaluating their adsorption performance and exploring their suitability as sustainable alternatives for tertiary wastewater treatment.

2. Materials and Methods

2.1. Chemical Reagents

All reagents used in this study were of analytical grade and prepared in accordance with the international standard protocol SR ISO 6878:2005 [32]. The chemicals included 63% nitric acid (ρ = 1.41 $g \cdot cm^{-3}$), vanadomolybdate color reagent, and KH_2PO_4 solutions with concentrations ranging from 10 to 100 $mg \cdot L^{-1}$. Sodium hydroxide (2 M) was used for pH adjustment, and the phosphate ion (PO_4^{3-}) was expressed as phosphorus (P) throughout the study.

All reagents were purchased from Sigma-Aldrich (Merck KGaA, Darmstadt, Germany) and used without further purification. Sulfuric acid, NaOH, and nitric acid (63%) were also supplied by Merck. Distilled water was used for the preparation of the adsorbent, standard solutions, and synthetic wastewater.

The raw pumice stone was collected from the volcanic area of the Călimani Mountains, Romania, while the AAC material was obtained from construction waste generated at a local construction site in Iași. Both materials were used in their natural form, without chemical or thermal pre-treatment.

2.2. Pumice Stone Material

Pumice stone material (PS) is an igneous rock formed by the rapid cooling and solidification of volcanic lava at the Earth's surface. Igneous rocks, in general, include both those formed deep underground through the solidification of magma, as well as those resulting from the cooling of lava at the surface. The most common types of magmas are silicate-based (containing more than 30% silicon dioxide—SiO_2), though smaller amounts of carbonate, sulfuric, or oxide magmas may also occur.

Silicate magmas are classified according to their silica content: acidic magmas—containing more than 63% SiO_2; intermediate (neutral) magmas—between 52% and 63% SiO_2; basic magmas—below 52% SiO_2.

Basic magmas contain higher amounts of volatile substances, which gives them greater fluidity compared to acidic magmas. Pumice is a type of volcanic rock known as rhyolite. It is characterized by a high silica content and low levels of iron and magnesium, which result in its low hardness and distinctive light color. Although rhyolite is of volcanic origin, its chemical composition is similar to that of granite. Thus, pumice contains minerals such as feldspar, quartz, and biotite.

The typical chemical composition of pumice is as follows: 70–77% silicon dioxide (SiO_2), 11–14% aluminum oxide (Al_2O_3), 3–5% potassium oxide (K_2O), 3–5% sodium oxide (Na_2O), 1–3% ferrous oxide (FeO), 1–2% ferric oxide (Fe_2O_3), 0.5–1% magnesium oxide (MgO), and approximately 0.03% water [7–13]. This rock is extremely lightweight due to its porous structure—porosity can reach up to 85% of the total volume. The microporous structure of pumice results in a high specific surface area, making it suitable for adsorption applications.

Depending on its composition, pumice can exhibit either basic or acidic characteristics, but generally it has a high silicon content (60–75% SiO_2), which imparts abrasive properties. Due to its high porosity and significant content of silicon and aluminum, pumice is considered a promising material for use as an adsorbent.

Moreover, some studies have shown that pumice represents an ideal support for metal impregnation due to its structure [11]. Therefore, research into the ability of this rock to remove pollutants from wastewater is of particular interest.

In order to characterize the composition of pumice, an analysis was carried out using certified reference materials (CRM) of igneous rocks, and the chemical composition of pumice was determined using X-ray analysis, as shown in Table 3:

Table 3. Chemical composition of PS.

Compound	SiO_2	Al_2O_3	Fe_2O_3	CaO	MgO	Others
Concentration, %	51.26	14.84	3.16	28.08	2.04	0.61

2.3. *Autoclaved Cellular Concrete*

Autoclaved aerated concrete (AAC) is a construction material that belongs to the category of lightweight concrete. It is produced by mixing components such as sand, cement, lime, water, silica, and an expanding agent—typically aluminum powder—which, through a chemical reaction, releases gases, generating its characteristic porous structure.

AAC was first produced in Yxhult, Sweden, by architect and inventor Johan Axel Eriksson. It is a prefabricated, lightweight material widely used in construction due to its excellent thermal insulation, fire resistance, mold resistance, and load-bearing capacity. Additionally, it contributes to structural rigidity while maintaining a much lower weight compared to conventional concrete—up to five times lighter.

The internal structure of Autoclaved Aerated Concrete (AAC) is characterized by uniformly distributed, small, closed spherical pores, typically ranging from 0.1 to 1 mm in diameter. This porous microstructure results from the chemical reaction between calcium hydroxide ($Ca(OH)_2$), produced during binder hydration, and aluminum powder. The reaction generates hydrogen gas, which causes the mixture to expand, forming the characteristic cellular architecture during the autoclaving process.

The basic mixture for producing AAC includes fine sand (with a grain size comparable to that of cement), cement, lime, gypsum, water, and the expanding agent. After the mixture is prepared and the porous structure develops, the material undergoes thermal treatment in an autoclave, where high pressure and temperature finalize the hardening process and stabilize the structure [18,21,22].

Due to its porous structure and large specific surface area, AAC is not only an excellent construction material but also a potential efficient adsorbent, suitable for use in adsorption processes aimed at removing pollutants from aqueous or gaseous environments.

The chemical composition of the ACC sample as determined by XRF analysis is given in Table 4. Like in the case of PS, the main oxide is SiO_2, followed by CaO and Al_2O_3.

Table 4. Chemical composition for ACC.

Compound	SiO_2	Al_2O_3	Fe_2O_3	CaO	MgO	Others
Concentration, %	61.01	13.05	2.93	21.73	1.01	0.26

2.4. *Characterization of PS and AAC Materials*

The surface morphology of the samples (initial and final PS/AAC with phosphate) was analyzed using a Scanning Electron Microscope (SEM) Quanta Inspect F50 operating at an acceleration voltage of 30 kV, with magnifications ranging from 1000× to 5000×.

A secondary electron detector (ETD) was used in SE mode for topographical contrast. Elemental composition of selected areas was analyzed using Energy Dispersive X-ray Spectroscopy (EDX) having 133 eV resolution at Mn Kα, coupled to the SEM (all from FEI Company, Eindhoven, The Netherlands,).

The crystallographic structure of the samples was analyzed by X-ray diffraction (XRD) using a Rigaku MiniFlex II diffractometer (Tokyo, Japan) equipped with a Cu-Kα radiation source (λ = 1.5406 Å) and calibrated with an Al_2O_3 standard (SRM 676A) [33]. The measurements were performed in the 2θ range of 10–70°, with a scan step of 0.02° and a counting time of 100 s per step.

Specific surface area, pore volume, and pore size distribution of the adsorbents were determined by nitrogen adsorption–desorption isotherms using the BET method. Measurements were carried out with an instrument capable of performing multi-point BET analysis and BJH pore analysis, using adsorption at liquid nitrogen temperature (77 K). The analysis has been performed using a Micromeritics TriStar II Plus (Norcross, GA, USA) surface area and porosity analyzer.

The elemental composition of pumice stone (PS) and autoclaved aerated concrete (AAC) was determined using an energy-dispersive X-ray fluorescence spectrometer (MiniPal PW4025, PANalytical, Almelo, The Netherlands). The instrument is equipped with a rhodium anode X-ray tube (operating voltage: 5–30 kV; current: up to 1 mA) and a high-resolution silicon drift detector (SDD), allowing qualitative and quantitative determination of elements from sodium (Na, Z = 11) to uranium (U, Z = 92) in solid samples. Prior to analysis, samples were oven-dried at 105 °C for 24 h, finely ground to pass a 100 μm sieve, and pressed into pellets (32 mm diameter) under a load of 20 t using a hydraulic press. Calibration was performed using certified reference materials, and measurements were carried out under vacuum to improve detection sensitivity for light elements. Results are expressed as weight percentages (wt%) of oxides, as calculated by the instrument's software package.

2.5. Adsorption Experiments

The batch adsorption studies were performed by mixing the solution containing phosphate ions and the adsorbent, using a magnetic stirrer. The size of the adsorbent material granules was between 0.063 and 2 mm.

The retention of phosphate ions from the solution was calculated as the difference between the initial concentration and the concentration at different contact times. The amount removed per unit mass of adsorbent (a_t, mg/kg) at time "t" results from:

$$a_t = (C_0 - C_t)\ \frac{V}{m}. \tag{1}$$

The removed percentage (RP) of phosphate ions was calculated from:

$$RP = \frac{C_0 - C_t}{C_0} 100, \tag{2}$$

where a_t, mg/kg is the amount of phosphate ions removed per unit mass of the adsorbent at a given time, t; C_0, mg/L is the initial concentration of phosphate ions in the aqueous solution, C_t, mg/L is the concentration of phosphate ions at time, t; V, L is the volume of the solution, m, g, is adsorbent mass.

2.5.1. Adsorption Equilibrium Study

The equilibrium study aimed to obtain adsorption isotherms in order to determine the degree of phosphorus retention on the granules of the adsorbent materials (PS and AAC). To characterize the adsorption process, Langmuir and Freundlich isotherms were used, at a mass ratio of solid to liquid phases of 1:10, and pH values of 3 and 9.

In the thermodynamic study, the adsorbent samples were kept in contact with the standard phosphorus solution (KH_2PO_4) of different concentrations: 10, 20, 40, 60, 80, and 100 mg/L, for one hour to achieve equilibrium. Afterwards the suspension was filtered on a filter paper with pore size of 0.45 µm. The resulting aqueous extract was analyzed for phosphorus content with an X-ray spectrometer, as described in Section 2.4.

The description of the equilibrium state of the adsorbed-adsorbent system was made through the Langmuir and Freundlich models presented below. These models represent the most used method to describe retention of ions from liquid solutions onto the surface of porous solids.

- The Langmuir equation

The Langmuir model assumes that the adsorbent surface has active centers that can hold the adsorbate ions by physisorption (by Vand der Waals forces) or chemisorption. He applies the hypothesis that physical and chemical bonds are formed between the surface of the adsorbent and the molecules of the adsorbed component, and there are no interactions between the molecules of the adsorbed component. The adsorption is monolayer and there are no interactions among the adsorbate species.

The Langmuir isotherm model is described by the following formula:

$$a_e = \frac{K \cdot C \cdot a_{\max}}{1 + K \cdot C}, \tag{3}$$

which gives by linearization of

$$\frac{1}{a_e} = \frac{1}{K \cdot C \cdot a_{\max}} + \frac{1}{a_{\max}}. \tag{4}$$

Performing the variable changes $y_e = \frac{1}{a_e}$ and $x = \frac{1}{C}$ the equation of a straight line is obtained,

$$y_e = A \cdot x + B \; A = \frac{1}{K \cdot a_{\max}}; \qquad B = \frac{1}{a_{\max}},$$

where a_e, a_{max} are the phosphorus content at equilibrium and the maximum adsorption capacity (at saturation, mg P/kg dry adsorbent; C is the concentration of adsorbate (phosphate) solution after adsorption equilibrium is reached, mg/L; K is the adsorption coefficient, L/mg [27,28].

- The Freundlich equation

The Freundlich model starts from the hypothesis of reaching chemical equilibrium when there is a dynamic exchange between adsorbed molecules and those remaining in solution. The adsorbent surface is heterogeneous, and the adsorbate species are distributed in multilayer at the adsorbent surface. This implies that all the adsorption sites have different affinities. Freundlich isotherm is expressed by the equation:

$$a_e = K \cdot C^{1/n}. \tag{5}$$

Equation (5) is linearized by logarithmization obtaining:

$$\ln a_e = \ln K + \frac{1}{n} \ln C. \quad (6)$$

Replacing $y = \ln a_e$ and x = ln C, the equation of a straight line is obtained $y_c = A \cdot x + B; A = \frac{1}{n}; \ B = \ln K$.

2.5.2. Description of Models for Kinetic Study

The kinetic experimental determinations were carried out on the PS (pumice stone) material at two pH values, 3 and 9, maintaining a solid-to-liquid phase mass ratio of 1:10. The pH was adjusted using 2 M NaOH and H_2SO_4 solutions. The samples were brought into contact with KH_2PO_4 solutions of varying concentrations (10, 20, 40, 60, 80, and 100 mg/L) for different contact times: 2, 5, 10, 15, 20, 30, 40, and 60 min, under stirring at 200 rpm. After contact, the suspensions were vacuum filtered, and the resulting aqueous extract was analyzed according to the SR ISO 6878:2005 standard [32]. To understand the adsorption mechanism of phosphorus ions on the adsorbent, a kinetic study was conducted based on the pseudo-first-order and pseudo-second-order models, analyzing efficiency as a function of pH.

(a) The pseudo-second-order adsorption kinetic model is mathematically described according to Equation (7) [23]:

$$\frac{da_t}{dt} = k''(a_{eq} - a_t)^2, \quad (7)$$

where: k'' = pseudo-second-order rate constant of the adsorption process (g/min·mg); a_{eq} = phosphate ion loading of the adsorbent at equilibrium (mg/g); a_t = adsorbent loading at time t (mg/g).

Applying the boundary condition: t = 0, t = t şi $a_t = 0$, $a_t = a_t$ by integrating Equation (7) we obtained:

$$\int_0^{a_t} \frac{da_t}{(a_{eq} - a_t)^2} = k'' \int_0^t dt \Rightarrow \frac{1}{a_{eq} - a_t} - \frac{1}{a_{eq}} = k'' \cdot t. \quad (8)$$

(b) The pseudo-first-order adsorption kinetic model is mathematically described according to Equation (8) [23]:

$$\frac{da_t}{dt} = k'(a_{eq} - a_t). \quad (9)$$

Applying the boundary condition: t = 0, t = t şi $a_t = 0$, $a_t = a_t$ by integrating Equation (9), we obtained: $\log(a_{eq} - a_t) = \log(a_{eq}) - \frac{k'}{2303}t$, where k' and a_{eq} can be determined from the slope of the line.

The model used to determine the evolution of the load over time is given by the balance Equation (10)

$$a_1 = \frac{V_{aq}(C_0 - C)}{m_{PS}} + a_0, \quad (10)$$

where $a_1 = a + a_0$, a_t and C—have the same meaning as in Equation (7), mg/g; C_0 = initial concentration of phosphate in the aqueous phase, mg/L; V_{aq} = volume of the aqueous phase, mL; a_0 = initial phosphate content in the pumice stone, mg/g; m_{PS} = mass of the solid (pumice stone), g.

3. Results and Discussion

3.1. Characterization of PS and AAC Materials

3.1.1. X-Ray Diffraction (XRD) Analysis for PS

The XRD patterns for initial PS sample and for PS after being kept at pH = 3 are given in Figure 1 and confirm the crystalline structure of the material presenting sharp peaks.

The XRD spectra of initial PS sample shows three main components (xCaO SiO_2 zH_2O, SiO_2 and $K_2Ca_6Si_4O_{15}$). The sample that was kept at pH = 3 (Figure 1b) in the presence of phosphate shows same phase rearrangements and the presence of five components (xCaO SiO_2 zH_2O, SiO_2, Mn_2O_3, $K_3(MnH_4)H$, $Mg_2P_2O_7$), as shown in Figure 1.

3.1.2. X-Ray Diffraction (XRD) Analysis for AAC

The XRD patterns for initial ACC sample and for AAC after being kept at pH = 3 are given in Figure 2. The sharp peaks in the XRD patterns confirm the crystalline structure of the AAC material. The XRD spectra of initial ACC sample (Figure 2a) shows four components (Ca_2SiO_4, SiO_2, $Mg_{0.06}Ca_{0.94}CO_3$, and Al_2O_3). The sample that was kept at pH = 3 (Figure 2b) in the presence of phosphate ions shows same phase rearrangements and the presence of five crystalline components (Ca_2SiO_4, SiO_2, $Mg_{0.06}Ca_{0.94}CO_3$, $MgHPO_4$ H_2O, and $Ca_{0.05}Al_{0.1}Si_{1.9}O_4$).

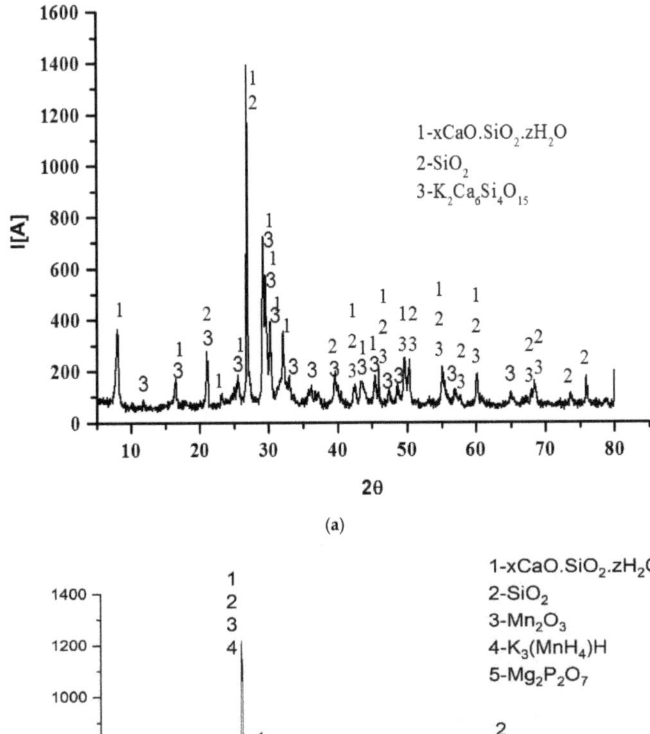

(a)

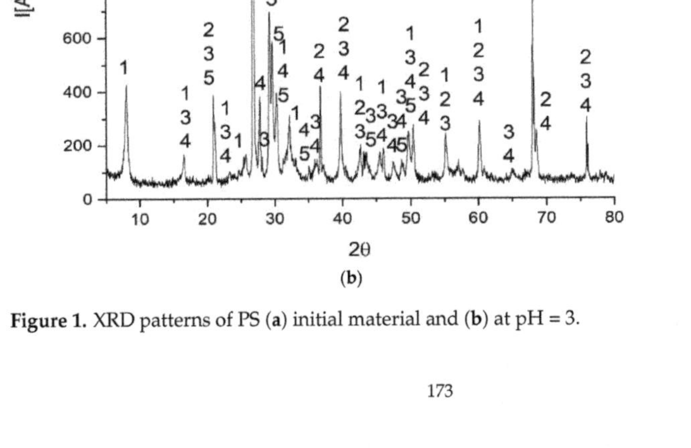

(b)

Figure 1. XRD patterns of PS (**a**) initial material and (**b**) at pH = 3.

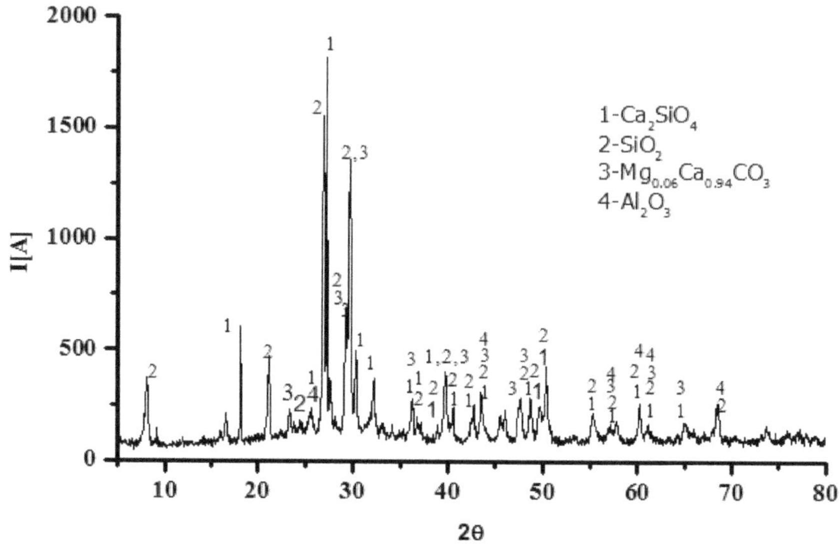

(a)

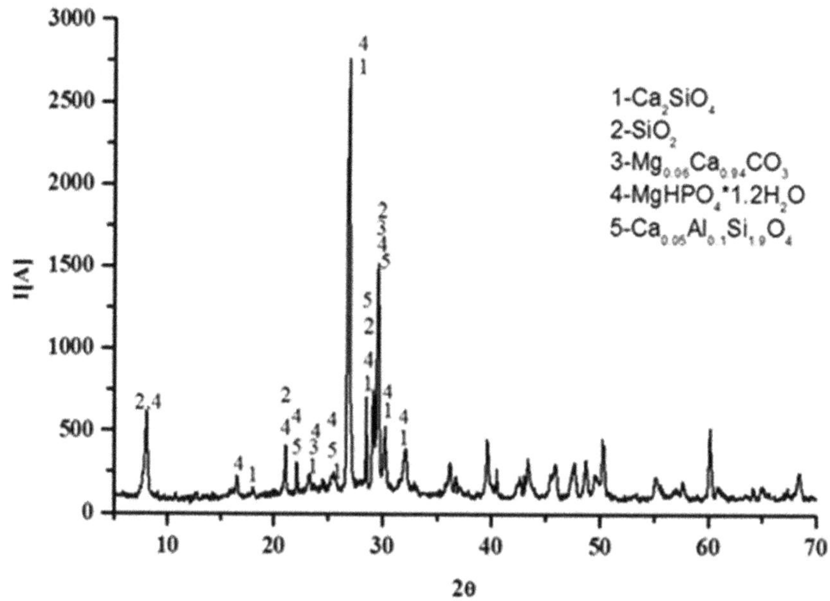

(b)

Figure 2. XRD patterns of AAC (**a**) initial material and (**b**) at pH = 3.

The specific surface of initial ACC sample used in our experiments is 30.99 m^2/g and average pore size 9.2 nm, which is in agreement with previously reported values [19,20,30]. The value of the sample at acidic pH (3) is slightly lower than the initial sample, 23.86 m^2/g and average pore size 9.6 nm, indicating a slight reduction of number of pores at lower pH.

3.1.3. SEM Analysis for PS

The SEM analysis of the control sample highlights the heterogeneous nature of the pumice stone, with particles of varying sizes and morphologies. The bright areas in the image indicate the presence of elements with higher atomic numbers, particularly calcium, while the darker areas likely correspond to compounds rich in silicon and oxygen. To identify the local chemical composition, two EDX analyses were performed in distinct regions of the sample (Figure 3) [13,23].

Initial Material PS

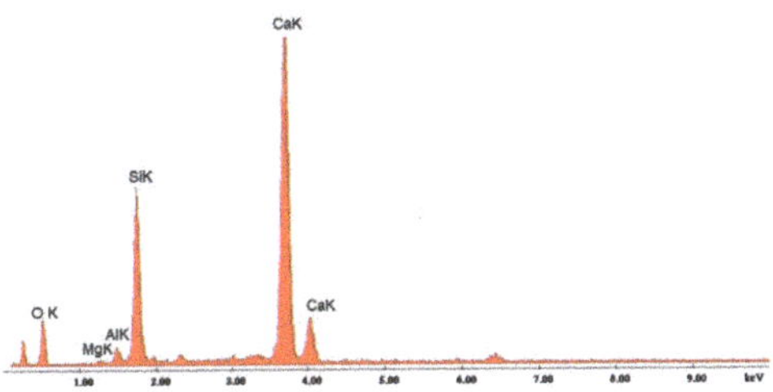

EDX 1

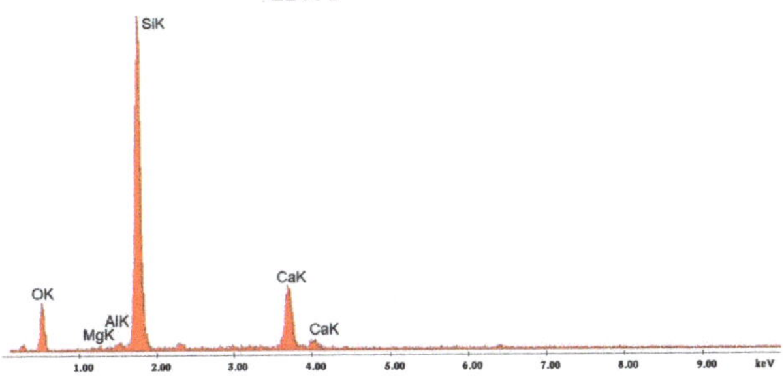

EDX 2

Figure 3. SEM and EDX images for initial material PS.

In the EDX1 area (Figure 3), a high concentration of calcium was observed, accompanied by silicon and small amounts of aluminum and magnesium (at least on the surface), suggesting the presence of calcareous phases. In contrast, the EDX2 spectrum shows a predominance of silicon with a lower calcium content, probably associated with silicates or quartz. Correlating these results with the overall chemical analysis indicates that the sample consists mainly of SiO_2 (51.26%) and CaO (28.08%), along with Al_2O_3 (14.84%) and small amounts of Fe_2O_3, MgO, and other oxides. Overall, the pumice stone exhibits a heterogeneous mineralogical structure, consisting of alternating silicate-rich and calcium-rich zones, which is characteristic of volcanic-origin materials [34].

The SEM analysis of pumice stone (PS) after contact with phosphate solution at pH 3 (Figure 4) reveals a granular morphology with a rough surface and visible porosity, characteristic of aluminosilicate volcanic materials. The analyzed areas (EDX1, EDX2, EDX3) display fine deposits on the particle surfaces, suggesting the adsorption or precipitation of phosphate species. The porous structure remains apparent, indicating that the acidic treatment did not cause extensive dissolution, although partial leaching of certain mineral components is likely to have occurred, leading to surface modification.

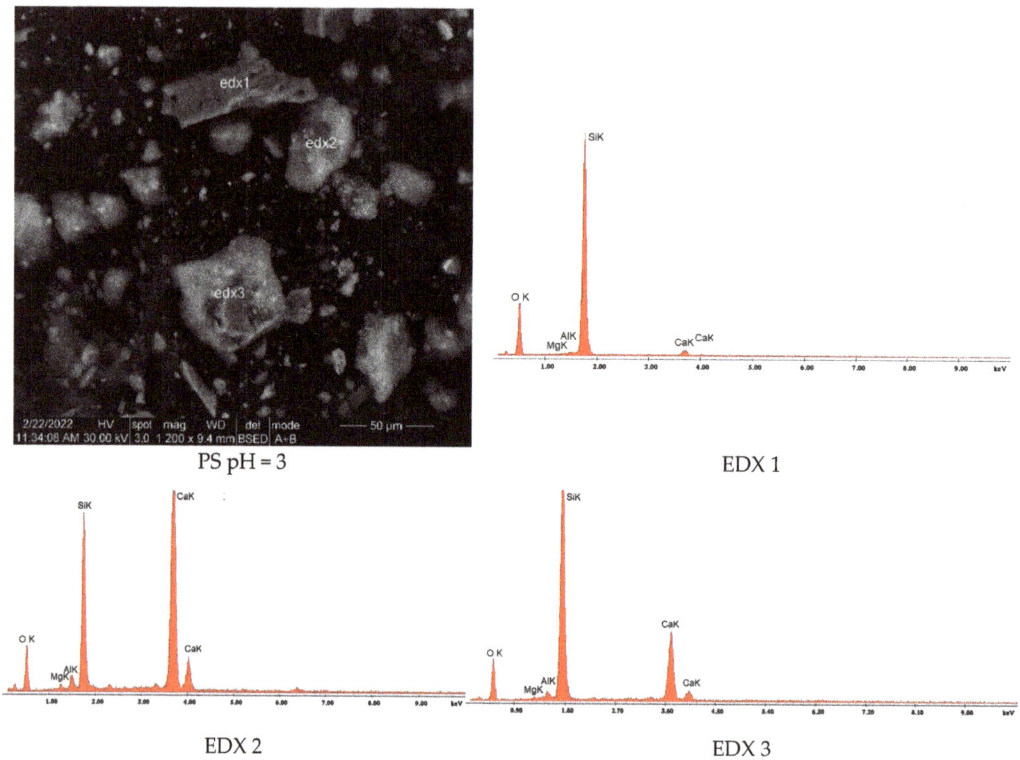

Figure 4. SEM and EDX images for PS at pH = 3.

EDX analysis confirms that silicon (Si) and oxygen (O) are the major elements, corresponding to the aluminosilicate matrix, with aluminum (Al) consistently present (Figure 4). Potassium (K), calcium (Ca), and magnesium (Mg) are also detected in smaller proportions, typical of the pumice mineral composition. The presence of phosphorus (P) in all analyzed points indicates the retention of phosphate from solution, either via adsorption onto the mineral surface or through the formation of calcium or aluminum phosphate precipitates. Variations in phosphorus intensity between points suggest a heterogeneous distribution, likely related to local mineralogical composition and surface reactivity [35].

Overall, the morphological and compositional data indicate that acidic conditions at pH 3 promote partial leaching of exchangeable cations (Ca, Mg, K), generating active sites for phosphate retention while preserving the inherent porosity of the pumice stone.

The SEM images of the initial pumice stone (PS) show a highly porous (Figure 5), rough surface with irregularly shaped particles, typical of aluminosilicate volcanic material. After contact with the phosphate solution at pH 3, the surface morphology exhibits visible deposits partially covering the pores, suggesting phosphate adsorption or precipitation. Despite these surface changes, the porous structure is still evident, indicating that the acidic conditions did not cause severe dissolution of the matrix.

Figure 5. SEM and EDX analysis of PS before and after at pH 3.

EDX analysis of the initial PS reveals silicon (Si) and oxygen (O) as the predominant elements, with aluminum (Al), potassium (K), calcium (Ca), and magnesium (Mg) also present. After treatment at pH 3, phosphorus (P) is clearly detected, confirming phosphate retention. Additionally, a slight reduction in Ca and Mg signals suggests partial leaching under acidic conditions, potentially generating new active sites for phosphate binding.

3.1.4. SEM Analysis for AAC

The initial material AAC sample (Figure 6a) was analyzed morphologically using scanning electron microscopy (SEM), with images captured at magnifications of 1000× and 5000×. The micrographs revealed a non-homogeneous morphology, characterized by randomly oriented acicular lamellar formations. Elemental composition analysis was performed by energy-dispersive X-ray spectroscopy (EDX) at multiple points. At EDX points 1 and 4, high concentrations of calcium (Ca) and silicon (Si) were recorded, along with traces of magnesium (Mg) and aluminum (Al). At EDX point 2, significant amounts of Si, Al, and Ca were identified, while Mg was present in lower proportions. EDX point 3 was dominated by silicon, most likely in the form of silicon dioxide (SiO_2), with other elements being weakly represented [19–21].

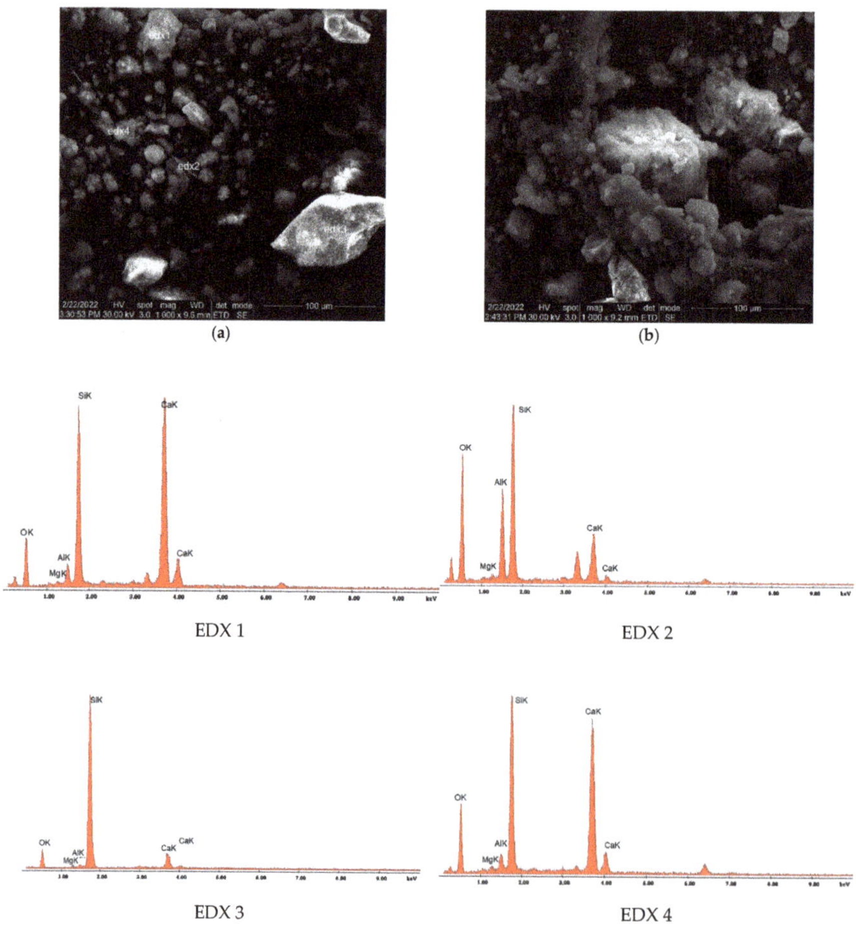

Figure 6. SEM and EDX images for (**a**) initial material AAC, (**b**) AAC at pH = 3.

In the case of the AAC sample treated at pH 3 (Figure 6b), SEM images were taken at magnifications of 1000× and 2000×. Compared to the reference sample, significant surface modifications were observed: over the initial lamellar formations, compact, angular deposits resembling granules appeared. The EDX analysis (Figure 7) indicated, at points 1 and 3, a predominant presence of Ca and Si, distributed in the form of submicron-sized lamellar compounds. At EDX point 2, silicon was dominant, associated with the compact granular formations also observed in the SEM images.

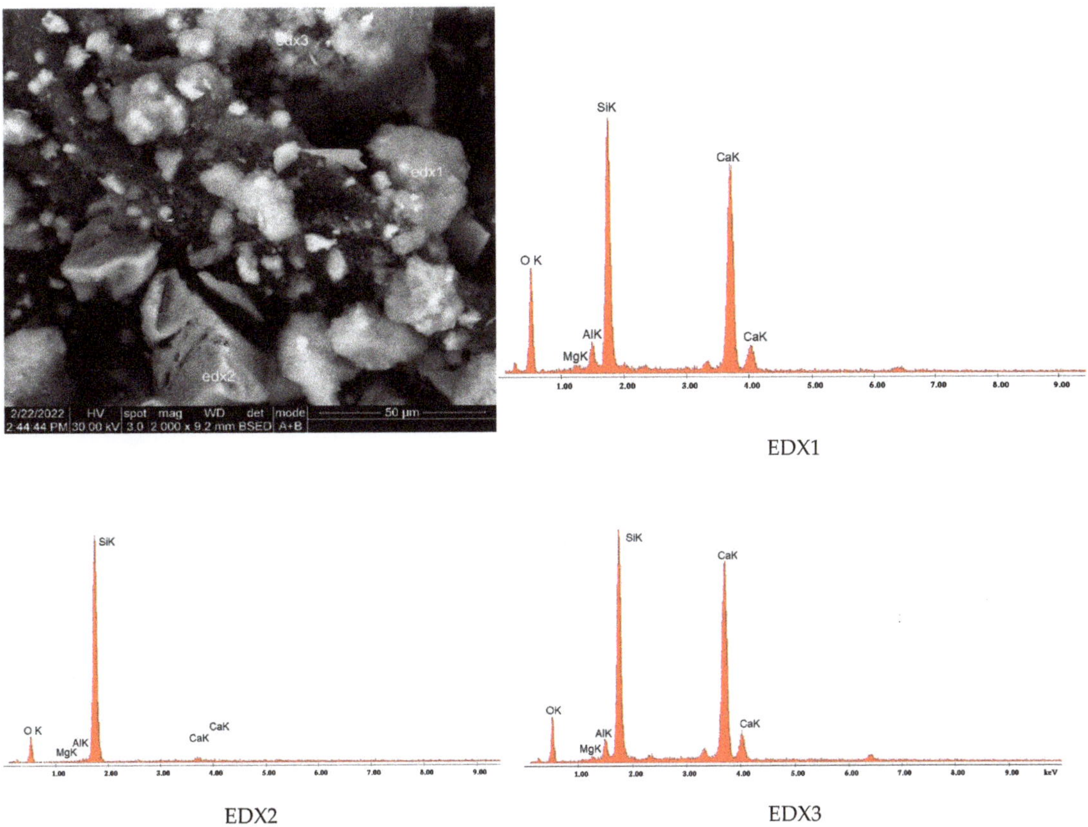

Figure 7. SEM and EDX images for AAC at pH = 3.

By comparing the two samples, it can be concluded that treating the AAC with an acidic solution (pH 3) induces visible morphological changes, suggesting chemical rearrangements and/or the formation of new phases. Although the distribution of the dominant elements, Ca and Si, remains relatively constant, the mode of aggregation and surface structure differ substantially. These changes may directly influence the adsorption capacity of the analyzed material [36–38].

3.1.5. BET Method for PS and AAC

The specific surface of the initial PS sample used in our experiments is 16.38 m^2/g, which is slightly higher than previous values reported in the literature [13,23,30]. The value of the sample at acidic pH (3) is slightly higher than the initial sample, 20.96 m^2/g, indicating some pore structure modifications at lower pH, as shown in Table 5.

Table 5. Specific surface area and pore volume.

Material	BET (m^2/g)	Langmuir (m^2/g)	Adsorbed Pore Volume (cm^3/g)	Observations
initial material PS	16.37	24.67	0.04315	Low specific surface area
PS pH = 3	20.97	31.02	0.04519	Significant increase at acidic pH
initial material AAC	30.99	45.57	0.06761	Highest BET surface area
AAC pH = 3	23.86	35.61	0.05407	Decrease at acidic pH

Pumice exhibits a lower specific surface area than the reference AAC but increases significantly under acidic pH, suggesting pore activation. In contrast, AAC initially has the highest specific surface area, but this decreases at pH 3, likely due to pore clogging or structural changes.

Pore volume follows the same trend: higher in reference AAC and lower in pumice. pH variations influence the porous structure of both adsorbents.

3.2. Adsorption Equilibrium Study

The experimental results establish the equilibrium curves in order to determine the degree of retention of phosphorus on two types of adsorbent materials (PS and ACC) using the Langmuir and the Freundlich isotherms, at a mass ratio of the adsorbent material: water phases of 1:10 at two pH values (3 and 9) for one hour. The experiments were performed in triplicate. The results showed standard deviations lower than 15% of the average measurements. The experiments were conducted at pH 3 and 9, corresponding to conditions where phosphate species differ significantly ($H_2PO_4^-$ in acidic and HPO_4^{2-}/PO_4^{3-} in alkaline media). Neutral pH (around 7) was avoided to prevent overlapping of multiple species, which could obscure the interpretation of adsorption mechanisms. Langmuir and the Freundlich model parameters were calculated by linear regression.

The relevant parameters obtained using the Langmuir and the Freundlich models are presented in Tables 6 and 7.

Table 6. Values of thermodynamic parameters for pumice stone (PS).

pH	Langmuir					Freundlich				
	K (L/mg)	a_{max} (mg/kg)	R^2	RMSE	χ^2	K	1/n	R^2	RMSE	χ^2
3	0.134	1006.69	0.9174	0.169	0.131	141.939	0.597	0.9755	0.071	0.032
9	0.042	924.14	0.9204	0.119	0.060	67.82	0.594	0.9540	0.087	0.034

Table 7. Values of the thermodynamic parameters for ACC.

pH	Langmuir					Freundlich				
	K (L/mg)	a_{max} (mg/kg)	R^2	RMSE	χ^2	K	1/n	R^2	RMSE	χ^2
3	0.087	859.20	0.8271	0.225	0.224	88.66	0.6188	0.9304	0.127	0.088
9	0.058	630.50	0.8209	0.176	0.136	68.369	0.5153	0.9047	0.122	0.072

The Langmuir isothermis characterised by a dimensionless constant called the separation factor, R_L, or the equilibrium parameter. This is used to predict whether an adsorption system is favorable or not [22]. Based on R_L value, the represented isotherm can indicate favorable adsorption, when R_L is between zero and one; unfavorable when R_L is higher than one; in the linear model for adsorption, R_L equals one; irreversible adsorption, $R_L = 0$. R_L is calculated using Equation (9):

$$R_L = 1/(1 + K\,C_0), \tag{11}$$

where K in L/mol is the Langmuir constant and C_0 in mol/L is the highest initial phosphate concentration. The calculated values for PS are 0.069 (pH = 3) and 0.19 (pH = 9), while for ACC are 0.103 (pH = 3) and 0.147 (pH = 9). These values indicate that phosphate adsorption on PS or ACC is favorable.

The Freundlich model provides the best approximation of phosphorus retention isotherms ($R^2 = 0.90 \div 0.97$) for all adsorbent material types compared to the Langmuir model ($R^2 = 0.82 \div 0.92$).

To quantify the quality of fit of the Langmuir and Freundlich adsorption models to the experimental data, two error indicators were calculated: RMSE (Root Mean Square Error) and χ^2 (chi-squared). RMSE represents the average squared deviation of the estimated values from the experimental values (the square root of the mean of squared differences); the lower the RMSE, the better the model approximates the experimental data. The χ^2 statistic was determined using a formula commonly applied in adsorption studies, namely the sum of squared errors normalized by the predicted values, following common practice in adsorption studies.

It can be observed from the table that the Freundlich model yields significantly lower errors than the Langmuir model in all analyzed cases. For each material and pH, both RMSE and χ^2 values are lower for Freundlich compared to Langmuir, suggesting a superior fit of the Freundlich model to the experimental data. This indicates that the adsorbent surface exhibits a heterogeneous character, which favors the applicability of the empirical Freundlich model (which assumes the presence of multiple types of binding sites and possibly multilayer adsorption), whereas the Langmuir model (based on homogeneous surfaces and monolayer adsorption) cannot describe the studied systems as accurately.

In the case of the Freundlich model, the influence of the pH value is also felt on the intensity of the anion exchange. Thus, the slope A = 1/n gives us indications on the intensity of adsorption, and B = lnK gives indications about the adsorption capacity. The values of A and B calculated based on the Freundlich model are presented in Table 8.

Table 8. Parameters of the interpolation line A and B for the Freundlich model.

pH	PS		ACC	
	A (Slope)	**B (Ordinate)**	**A (Slope)**	**B (Ordinate)**
3	0.597	4.9554	0.6188	4.4848
9	0.594	4.2169	0.5153	4.2249

The phosphate anions in the solution are attracted to the positively charged areas on the surface of the adsorbent material particles, according to the anion exchange capacity.

Using the determined parameters, the Langmuir and the Freundlich isotherms for phosphates sorption were calculated and are presented in Figures 8–11. The calculated adsorption isotherm data shown in these figures were fitted to experimental data using the least squares method ($R^2 = 0.82 \div 0.97$). The graph of the isotherms, for the Langmuir and the Freundlich model, represents the loading a (amount of adsorbed phosphate) depending on the concentration of phosphates in the solution, C_e. Generally, the Langmuir isotherm presents three regions: a steep one given by the intense adsorption process, a flat third region characterized by a low intensity of the adsorption due to the saturation of the adsorbent with adsorbate, and the intermediate region. The shape of the Langmuir isotherm is characterized in this case by a steep, practically linear initial region and a curved middle portion; the lack of a horizontal segment demonstrates the high availability of the adsorbent material for the adsorption process. This allure is specific to the phenomena of chemosorption and physical adsorption, where the adsorption is carried out in a monomolecular layer [24–29,39–42].

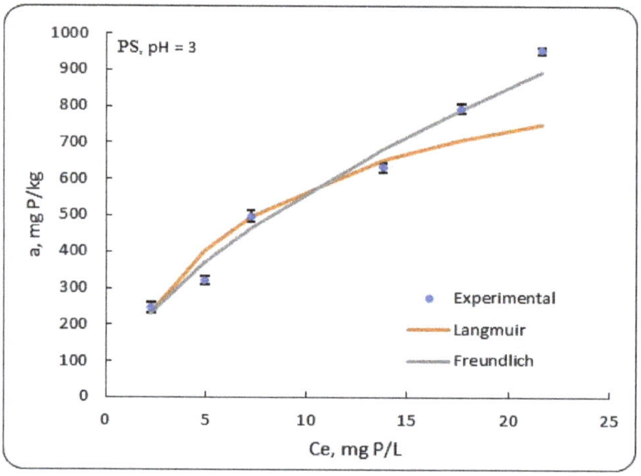

Figure 8. Isotherms for phosphorus adsorption on PS for pH = 3. Note: The points represent experimental data, and the line represents the calculated data according to the mentioned model. Concentrations and adsorption capacities are expressed as phosphorus (P); 1 mg P/L = 3.23×10^{-5} mol/L; 1 mg P/kg = 3.23×10^{-5} mol/kg.

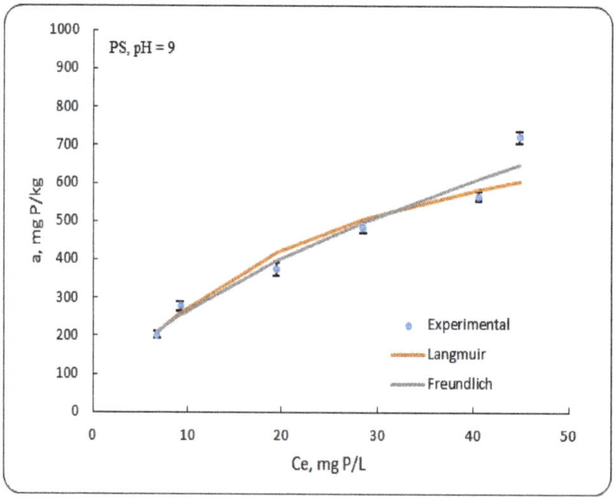

Figure 9. Isotherms for phosphorus adsorption on PS for pH = 9. Note: The points represent experimental data, and the line represents the calculated data according to the mentioned model. Concentrations and adsorption capacities are expressed as phosphorus (P); 1 mg P/L = 3.23×10^{-5} mol/L; 1 mg P/kg = 3.23×10^{-5} mol/kg.

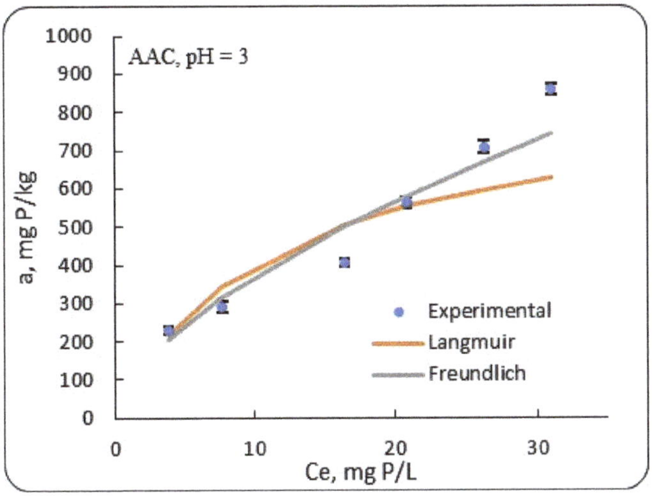

Figure 10. Isotherms for phosphorus adsorption on ACC for pH = 3. Note: The points represent experimental data, and the line represents the calculated data according to the mentioned model. Concentrations and adsorption capacities are expressed as phosphorus (P); 1 mg P/L = 3.23×10^{-5} mol/L; 1 mg P/kg = 3.23×10^{-5} mol/kg.

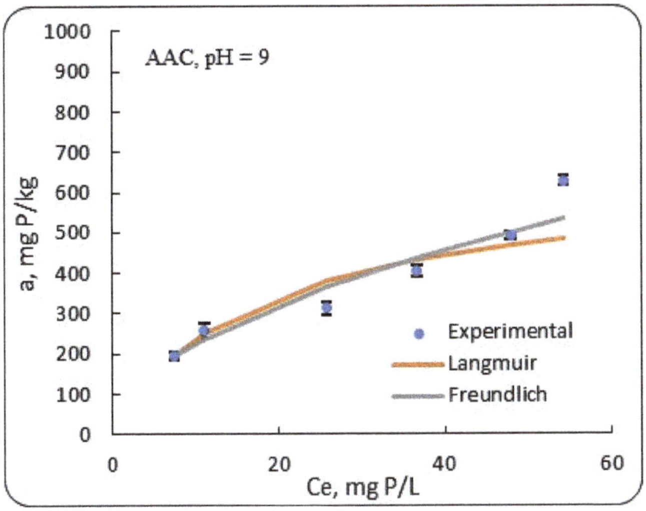

Figure 11. Isotherms for phosphorus adsorption on ACC for pH = 9. Note: The points represent experimental data, and the line represents the calculated data according to the mentioned model. Concentrations and adsorption capacities are expressed as phosphorus (P); 1 mg P/L = 3.23×10^{-5} mol/L; 1 mg P/kg = 3.23×10^{-5} mol/kg.

The extent of phosphorus removal (percentage removed) as a function of aqueous solute concentration was also studied. In Figure 12, the percentage of phosphorus removed is shown; the removal efficiency increases from 5% (pH = 9) to 90% (pH = 3) with increasing solute concentration. It is also observed that at an acidic pH, the studied adsorbent material presents a higher selectivity and adsorption capacity.

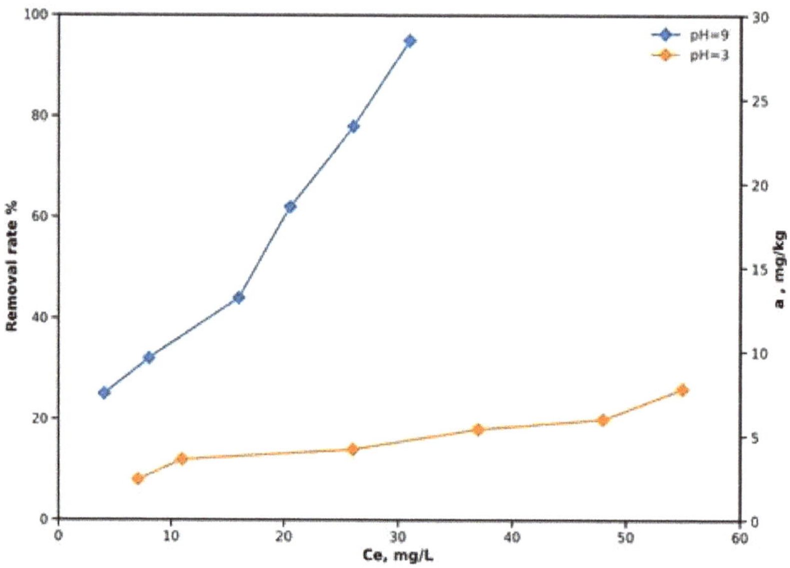

Figure 12. Phosphorus removal rate on PS as a function of solution concentration.

Phosphorus adsorption on the surface of adsorbent materials is determined by the solid surface charge and the protonation state of phosphorus in the water solution [18]. The results of this study show that AAC and pumice are effective adsorbents for the removal of phosphate ions from aqueous solutions. Laboratory studies carried out on PS and ACC have shown that phosphorus sorption varies with the pH value. Based on previous studies [36,42], phosphate adsorbed onto AAC could potentially be recovered and reused in agriculture as a phosphate-based fertilizer.

Although PZC (point of zero charge) or IEP (isoelectric point) measurements were not conducted experimentally in this study, literature values were included to support the interpretation of pH influence on adsorption. For pumice stone, PZC values are generally reported in the range of 6.0–7.5, depending on the mineralogical composition and origin of the material [31,34]. For autoclaved aerated concrete (AAC), PZC values are reported around 10.0–10.5, due to its high calcium content and alkaline nature [38,43]. These values are consistent with the pH conditions we selected (3 and 9) and support the observed behavior that phosphate retention is favored under acidic pH for both materials.

Although the bulk chemical composition of AAC and PS is broadly similar (both being rich in SiO_2, Al_2O_3, and CaO), the XRD analysis revealed distinct mineralogical phases, which directly influence their adsorption mechanisms. Pumice contains species such as $Mg_2P_2O_7$ and Mn_2O_3, while AAC exhibits phases like $Mg_{0.06}Ca_{0.94}CO_3$, Ca_2SiO_4, and $MgHPO_4{\cdot}H_2O$ after phosphate adsorption at acidic pH. These differences suggest that specific crystalline structures, especially those containing Ca, Mg, or Al, may serve as active sites for phosphate immobilization.

The phosphate adsorption efficiency is strongly influenced by the pH-dependent speciation of phosphate ions. At acidic pH (≈3), the dominant species in solution is $H_2PO_4^-$, while at alkaline pH (≈9), HPO_4^{2-} and PO_4^{3-} prevail. Simultaneously, the surface charge of the adsorbent materials changes with pH due to protonation/deprotonation reactions, which impacts the electrostatic interaction between phosphate anions and the adsorbent surface.

This behavior can be better understood by considering the point of zero charge (PZC) of each material. Thus, at pH 3, both materials are positively charged, enhancing electrostatic attraction for phosphate anions (especially $H_2PO_4^-$), which explains the higher removal efficiency observed in acidic conditions. Conversely, at pH 9, both materials are at or beyond their PZC and may carry a negative surface charge, leading to weaker attraction or even repulsion of phosphate species like HPO_4^{2-} and PO_4^{3-}.

Additionally, the presence of Mg, Ca, and Al-based species on the surface promotes inner-sphere complexation or precipitation mechanisms, especially under acidic pH, which increases phosphate fixation. For example, Ca^{2+} and Mg^{2+} can form insoluble phosphates (e.g., $MgHPO_4$, $Ca_3(PO_4)_2$), and Al^{3+} may participate in ligand exchange reactions with phosphate, forming strong Al–O–P bonds.

Therefore, the observed differences in adsorption performance between PS and AAC, despite similar oxide compositions, are attributed to their distinct mineralogical phases and surface charge behavior at varying pH levels, which modulate their affinity for different phosphate species.

In addition to the crystalline phases identified, it is important to consider the contribution of amorphous structures, which are characteristic of volcanic materials such as pumice. These may play a significant role in phosphate adsorption, as discussed below.

The chemical analysis of the PS (pumice stone) material revealed a significant concentration of Al_2O_3 (14.84%); however, this oxide was not detected as a distinct crystalline phase in the XRD pattern. This discrepancy suggests that aluminum is predominantly present in amorphous phases or incorporated into aluminosilicate structures, a characteristic feature of volcanic rocks such as pumice [34,44].

Amorphous phases play an important role in adsorption processes due to their high porosity and the presence of unsaturated reactive groups (e.g., –OH, structural defects). These features provide an extended active surface that facilitates interactions with anions such as phosphate, either through ion exchange mechanisms or by complexation with available cations (e.g., Al^{3+}, Ca^{2+}, Fe^{3+}) present in the disordered network [27,31].

Furthermore, aluminum and calcium ions in amorphous phases are often more chemically available than those in well-defined crystalline phases, which favors the formation of low-solubility compounds such as $AlPO_4$ or $Ca_3(PO_4)_2$ [38,45]. These interactions contribute to the efficient retention of phosphorus from solution, even in the absence of crystalline phases detectable by XRD.

Therefore, the experimental results obtained in this study, which indicate a high phosphate adsorption capacity for the PS material, can be explained, at least in part, by the presence and activity of these amorphous structures.

3.3. Kinetic Study

The kinetics of the phosphate adsorption process were quantified through laboratory assessments conducted using batch experiments.

Following the mathematical analysis of the experimental data using pseudo-first-order and pseudo-second-order kinetic models, the rate constants describing the retention of phosphate ions on the surface of the adsorbent material (PS) were determined. The variation in the amount of phosphate adsorbed as a function of contact time was studied under controlled conditions: the pH of the solution was set at 3 and 9, the initial phosphate ion concentration ranged from 10 to 100 mg/L (10, 40 și 100 mg/L), and the stirring speed was maintained at 200 rpm. The results showed that pH had a significant impact on the adsorption efficiency; as the pH increased from 3 to 9, the average amount of phosphate ions adsorbed decreased. The kinetic parameters calculated by pseudo-first-order and pseudo-second-order kinetic models are presented in Table 9.

Table 9. Kinetic parameters for PS.

pH	C_0 (mg/L)	Pseudo First Order		Pseudo Second Order			
		k' (min^{-1})	R^2	k'' (kg/(min·mg)	v_0 (mg/(g·min))	a_{eq} (mg/kg)	R^2
3	10	0.0271	0.9839	$6.639 \cdots 10^{-4}$	5.8644	93.9815	0.9778
	40	0.0284	0.9730	$1.855 \cdots 10^{-4}$	26.4275	377.4078	0.9869
	100	0.045	0.9344	$2.536 \cdots 10^{-4}$	139.0584	740.4479	0.9909
9	10	0.0240	0.9531	$9.796 \cdots 10^{-4}$	2.9603	54.9713	0.9949
	40	0.0259	0.9722	$2.361 \cdots 10^{-4}$	13.9092	242.7118	0.9848
	100	0.036	0.9254	$2.070 \cdots 10^{-4}$	82.5609	631.5199	0.9974

The data in Table 6 show that the pseudo-second-order model describes the adsorption process with higher accuracy, having R^2 values closer to 1 compared to the first-order model. Furthermore, the equilibrium adsorption capacity (a_{eq}) increases proportionally with the initial concentration and is significantly higher at pH 3 than at pH 9. This behavior can be explained by the stronger electrostatic attraction between phosphate ions and the protonated surface of the adsorbent in an acidic medium.

The graphs in Figures 13–16 illustrate the kinetic behavior of phosphate ion adsorption onto PS (pumice stone) at two distinct pH values (3 and 9) and for three initial concentrations (10, 40, and 100 mg/L).

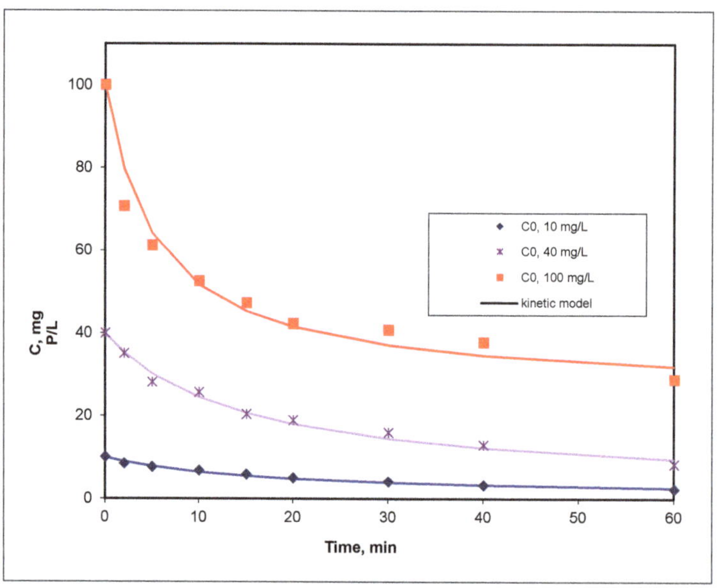

Figure 13. Variation of phosphorus concentration over time for PS at pH = 3. Note: The points represent experimental data, and the line represents the calculated data according to the mentioned model. Concentrations and adsorption capacities are expressed as phosphorus (P); 1 mg P/L = 3.23×10^{-5} mol/L; 1 mg P/kg = 3.23×10^{-5} mol/kg.

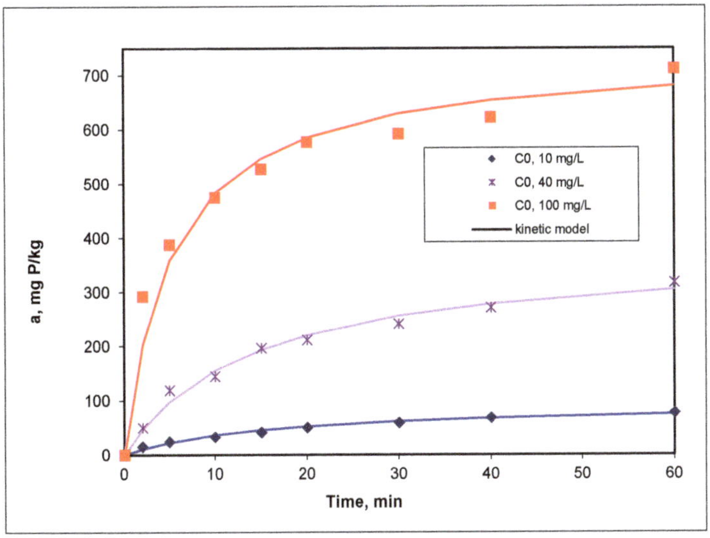

Figure 14. Variation of phosphorus loading over time for PS at pH = 3. Note: The points represent experimental data, and the line represents the calculated data according to the mentioned model. Concentrations and adsorption capacities are expressed as phosphorus (P); 1 mg P/L = 3.23×10^{-5} mol/L; 1 mg P/kg = 3.23×10^{-5} mol/kg.

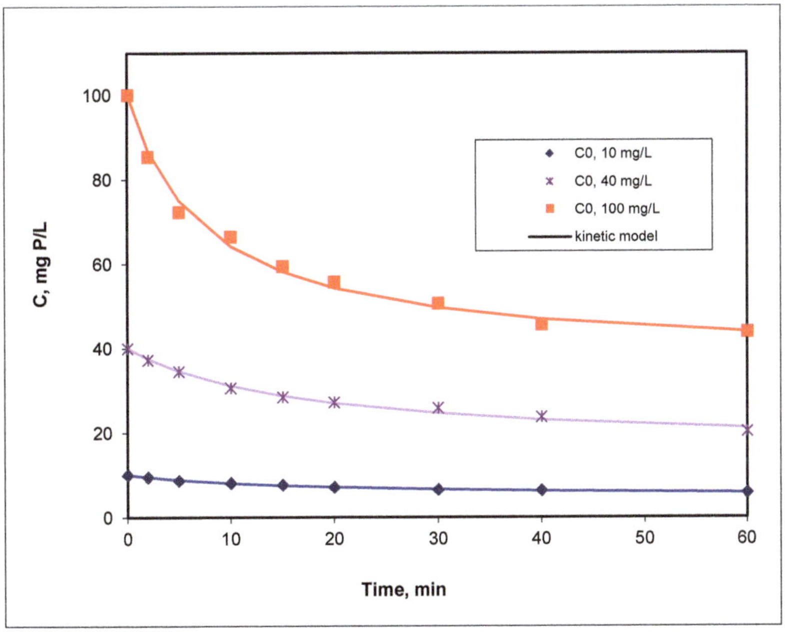

Figure 15. Variation of phosphorus concentration over time for PS at pH = 9. Note: The points represent experimental data, and the line represents the calculated data according to the mentioned model. Concentrations and adsorption capacities are expressed as phosphorus (P); 1 mg P/L = 3.23×10^{-5} mol/L; 1 mg P/kg = 3.23×10^{-5} mol/kg.

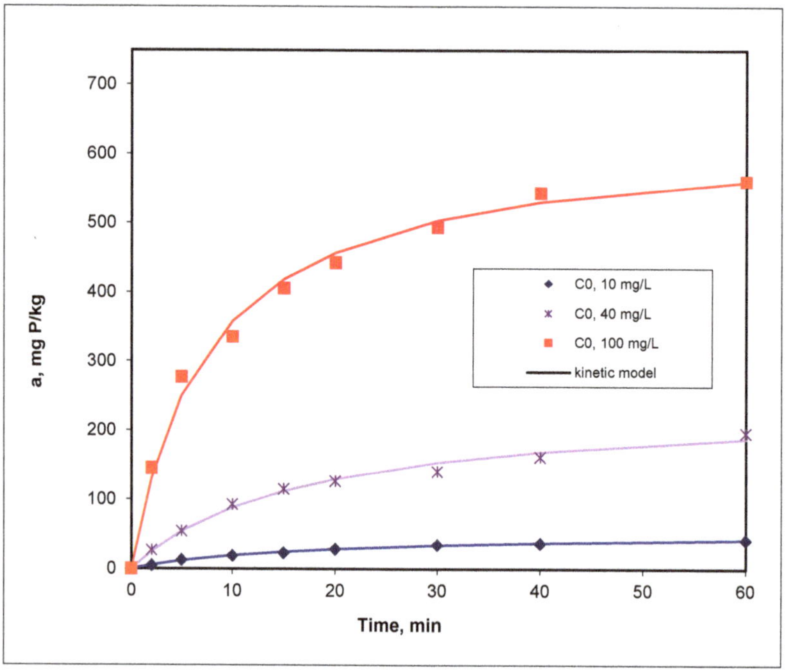

Figure 16. Variation of phosphorus loading over time for PS at pH = 9. Note: The points represent experimental data, and the line represents the calculated data according to the mentioned model. Concentrations and adsorption capacities are expressed as phosphorus (P); 1 mg P/L = 3.23×10^{-5} mol/L; 1 mg P/kg = 3.23×10^{-5} mol/kg.

In Figures 13 and 15, a rapid decrease in phosphorus concentration in solution is observed during the first 10–15 min, followed by a tendency toward stabilization, suggesting that the adsorption process reaches equilibrium relatively quickly. This behavior is typical of a process controlled by mass transfer and surface chemical reactions. The efficiency is noticeably higher at pH 3, where the residual phosphate concentration decreases much more sharply compared to pH 9.

Figures 14 and 16 show the evolution of the adsorbent loading (a, mg/kg) over time. The experimental data (points) are very well fitted by the lines corresponding to the pseudo-second-order model, supporting the hypothesis of a chemisorption-type kinetic (the adsorption phenomenon is controlled by chemical reactions between phosphate ions and the active surface of PS). It is noteworthy that at the initial concentration of 100 mg/L, a_{eq} reaches values above 700 mg/kg at pH 3 and above 600 mg/kg at pH 9—consistent with the values presented in Table 9, confirming the high efficiency of pumice stone in retaining phosphates.

This difference between pH values suggests that an acidic environment favors adsorption, probably due to the positive surface charge of PS, which attracts phosphate anions.

In addition, the shape of the curves in both graphs (a versus time) indicates rapid initial adsorption, followed by a gradual attainment of equilibrium—typical of processes governed by surface reactions rather than solely by diffusion.

The analysis of the correlation between the experimental data and the kinetic models showed that the pseudo-second-order model describes the adsorption process with superior accuracy, indicating that chemical interactions play a dominant role in the mechanism of phosphate retention on PS. The high determination coefficient values ($R^2 > 0.98$) for this model confirm the hypothesis of a higher-order kinetic, influenced by the availability of active sites on the adsorbent surface and the electrostatic nature of the interactions between the solid phase and phosphate ions.

These results are consistent with observations from the literature: Fetene et al. (2020) demonstrated that phosphate adsorption onto pumice reaches equilibrium in approximately 60 min, and the data fit best to the pseudo-second-order model, suggesting a chemisorption mechanism [28]. Furthermore, their study reported an efficiency of up to 94% at pH 7—a value comparable to the performance observed in our study at pH 3, where adsorption was even more efficient, suggesting an increased potential of PS under acidic conditions [28].

Another earlier study, Onar and Ozturk (1993) [46], reported an adsorption efficiency above 80%, with a reversible mechanism combining first- and second-order reactions, as well as good conformity to the Freundlich model at various pH values (3, 5, 7, and 9). Our observations regarding the strong influence of pH, with lower adsorption in basic media, are thus also confirmed by previous findings [35].

Table 10 provides a comparison of the adsorption capacities of several materials.

Table 10. Comparison of phosphate adsorption capacities of pumice and AAC materials.

Material	Adsorption Capacity (mg/g)	Reference
Pumice (natural)	7.4–12.5	[28]
Pumice (natural)	1.114	[46]
Pumice (Fe-coated)	18.9–29.6	[34]
Crushed AAC (construction waste)	7.3–15.8	[8,43]
Modified AAC (calcined)	16.4	[10]
Pumice (this study)	1.006 (pH 3)	This work
AAC (this study)	0.859 (pH 3)	This work

Although the adsorption capacities obtained in this study are lower compared to those reported for chemically or thermally modified materials, the use of unmodified, low-cost, and readily available adsorbents such as natural pumice and construction waste-derived AAC offers significant economic and environmental advantages. Furthermore, the study provides a comprehensive analysis of adsorption behavior and mechanisms, which is less commonly addressed for raw materials. These findings support the potential practical applicability of AAC and PS in cost-effective and sustainable wastewater treatment systems.

Thus, pumice stone demonstrates promising potential as an adsorbent material for phosphate removal from aqueous solutions, particularly under acidic conditions where its efficiency is maximized. The kinetic study provides a solid basis for understanding the adsorption mechanism and for optimizing future processes for the treatment of phosphate-contaminated waters.

Regeneration studies involving at least five adsorption–desorption cycles will be carried out in future work to provide a comprehensive evaluation of the adsorbents' potential for practical applications.

4. Conclusions

This study investigated the efficiency of two low-cost adsorbent materials—pumice stone (PS) and autoclaved aerated concrete (AAC)—for phosphate removal from wastewater, as a function of solution pH and contact time.

PS demonstrated a superior phosphate adsorption capacity (up to ~1000 mg/kg at pH 3), due to the presence of amorphous aluminosilicate phases and reactive Al^{3+}, Ca^{2+}, and Mg^{2+} species, as identified by XRD and XRF analyses.

The removal efficiency was strongly influenced by the solution pH, with optimal performance in acidic medium, where phosphate species (especially $H_2PO_4^-$) interact favorably with the positively charged surfaces of the adsorbents. This behavior was supported by the differences in the point of zero charge (PZC) of the materials.

Isotherm studies showed that the Freundlich model best describes the adsorption process. The presence and role of amorphous phases were confirmed as key factors, especially in the case of PS, significantly contributing to phosphate retention through complexation and precipitation mechanisms (e.g., formation of $MgHPO_4$, $AlPO_4$, $Ca_3(PO_4)_2$).

Kinetic modeling indicated a pseudo-second-order mechanism, confirming that the dominant adsorption mechanism is chemisorption, accurately described by the pseudo-second-order model. The observed performances, including the increased efficiency at lower pH, are consistent with the literature, reinforcing the validity of the method and the potential of pumice stone as an efficient adsorbent for phosphate removal from aqueous solutions.

From an environmental perspective, these results support the sustainable use of natural materials and construction waste for phosphorus removal from domestic and industrial effluents, reducing the risk of eutrophication and enabling nutrient recycling in agriculture.

Although the adsorption capacities determined for AAC and pumice are relatively modest compared to those reported for chemically or synthetically modified materials, these materials offer significant advantages in terms of practical applicability. Their low cost (a synthetic material with a capacity of 10 mg/g can be up to 10 times more expensive than a raw material with 1 mg/g capacity), wide availability, and environmental compatibility make them ideal candidates for use in decentralized wastewater treatment systems or tertiary treatment stages, where phosphate concentrations are low, but effluent quality standards are strict. Moreover, the absence of additional activation or chemical modification steps reduces process complexity and operational costs. Therefore, the use of unmodified materials such as AAC and PS can represent an efficient and sustainable solution within the framework of the circular economy and sustainable phosphorus management [47,48].

In conclusion, this work makes a relevant contribution to the current knowledge, highlighting the importance of mineralogical composition and surface charge behavior in controlling phosphate adsorption, and emphasizing the potential of PS and AAC materials as efficient, reusable, and low-cost solutions for wastewater treatment.

Author Contributions: Conceptualization, O.D.O. and C.M.; methodology, O.D.O.; software, E.R.C., M.B. and A.M.D.; validation, C.M., A.-A.S. and M.C. and D.N.A.; formal analysis, M.G. and O.D.O. investigation, M.C.; resources, A.M.; data curation, A.M.D. and O.D.O.; writing—original draft preparation, O.D.O. and C.M.; writing—review and editing, E.R.C., M.B., C.M. and A.-A.S.; visualization, O.D.O. and D.N.A.; supervision, O.D.O. and C.M. All authors have read and agreed to the published version of the manuscript.

Funding: This research received no external funding.

Data Availability Statement: The original contributions presented in this study are included in the article. Further inquiries can be directed to the corresponding author.

Conflicts of Interest: The authors declare no conflicts of interest.

References

1. Ding, Y.; Yi, Q.; Jia, Q.; Zhang, J.; Zhou, Z.; Liu, X. Quantifying phosphorus levels in water columns equilibrated with sediment particles in shallow lakes: From algae/cyanobacteria-available phosphorus pools to pH response. *Sci. Total Environ.* **2023**, *666*, 39–45. [CrossRef] [PubMed]
2. Bai, X.; Sun, J.; Zhou, Y.; Gu, L.; Zhao, H.; Wang, J. Variations of different dissolved and particulate phosphorus classes during an algae bloom in a eutrophic lake by ^{31}P NMR spectroscopy. *Chemosphere* **2017**, *169*, 577–585. [CrossRef]
3. Simoni, G.; Cheali, P.; Roslev, P.; Haasler, S.; Reitzel, K.; Smith, A.M.; Haferbier, M.H.S.; Christensen, M.L. Flocculating and dewatering of lake sediment: An in-situ pilot study comparing synthetic polymers and biopolymers for restoring lake water quality and reusing phosphorus. *Sci. Total Environ.* **2024**, *913*, 169597. [CrossRef]
4. Council Directive 91/271/EEC of 21 May 1991 Concerning Urban Waste-Water Treatment. Available online: https://eur-lex.europa.eu/eli/dir/1991/271/oj (accessed on 11 February 2024).
5. Council Directive of 21 May 1991 Concerning Urban Waste Water Treatment (91/271/EEC). Available online: https://eur-lex.europa.eu/legal-content/EN/TXT/HTML/?uri=CELEX:31991L0271 (accessed on 11 February 2024).
6. Kumar, P.S.; Korving, L.; van Loosdrecht, M.C.M.; Witkamp, G.J. Adsorption as a technology to achieve ultra-low concentrations of phosphate: Research gaps and economic analysis. *Water Res. X* **2019**, *4*, 100029. [CrossRef]
7. Zhou, H.; Margenot, A.J.; Li, Y.; Si, B.; Wang, T.; Zhang, Y.; Li, S.; Bhattarai, R. Phosphorus pollution control using waste-based adsorbents: Material synthesis, modification, and sustainability. *Crit. Rev. Environ. Sci. Technol.* **2022**, *52*, 2023–2059. [CrossRef]
8. Castellar, J.A.d.C.; Formosa, J.; Chimenos, J.M.; Canals, J.; Bosch, M.; Rosell, J.R.; Silva, H.P.d.; Morató, J.; Brix, H.; Arias, C.A. Crushed autoclaved aerated concrete (CAAC) as a reactive filter medium for enhancing phosphorus removal in nature-based solutions—Preliminary batch studies. *Water* **2019**, *11*, 1442. [CrossRef]
9. Egemose, S.; Sønderup, M.J.; Beinthin, M.V.; Reitzel, K.; Hoffmann, C.C.; Flindt, M.R. Crushed concrete as a phosphate binding material: A potential new management tool. *J. Environ. Qual.* **2012**, *41*, 647–653. [CrossRef]
10. Deng, L.; Dhar, B.R. Phosphorus recovery from wastewater via calcium phosphate precipitation: A critical review of methods, progress, and insights. *Chemosphere* **2023**, *330*, 138685. [CrossRef]
11. Mavhungu, A.; Foteinis, S.; Mbaya, R.; Masindi, V.; Kortidis, I.; Mpenyana-Monyatsi, L.; Chatzisymeon, E. Environmental sustainability of municipal wastewater treatment through struvite precipitation: Influence of operational parameters. *J. Clean. Prod.* **2021**, *285*, 124856. [CrossRef]
12. Oguz, E.; Gurses, A.; Yalcin, M. Removal of phosphate from waste waters by adsorption. *Water Air Soil Pollut.* **2003**, *148*, 279–287. [CrossRef]
13. Heibati, B.; Rodriguez-Couto, S.; Turan, N.G.; Ozgonenel, O.; Albadarin, A.B.; Asif, M.; Tyagi, I.; Agarwal, S.; Gupta, V.K. Removal of noxious dye—Acid Orange 7 from aqueous solution using natural pumice and Fe-coated pumice stone. *J. Ind. Eng. Chem.* **2015**, *31*, 124–131. [CrossRef]
14. *SR EN 1189:1999*; Water Quality. Determination of Phosphorus Content. Spectrometric Method with Ammonium Molybdate. ASRO: Bucharest, Romania, 1999.
15. *SR ISO 7875/1—1996*; Water Quality—Determination of Surfactants—Part 1: Determination of Anionic Surfactants by the Methylene Blue Spec-trometric Method. ASRO: Bucharest, Romania, 1996.
16. *SR ISO 7875/2—1996*; Water Quality—Determination of Surfactants—Part 2: Determination of Anionic Surfactants by the Methylene Blue Spec-trometric Method. ASRO: Bucharest, Romania, 1996.
17. *STAS 10064—75*; Water Quality. Determination of Total Phosphorus Content. ASRO: Bucharest, Romania, 1975.
18. Du, M.; Zhang, Y.; Wang, Z.; Lv, M.; Tang, A.; Yu, Y.; Qu, X.; Chen, Z.; Wen, Q.; Li, A. Insight into the synthesis and adsorption mechanism of adsorbents for efficient phosphate removal: Exploration from synthesis to modification. *Chem. Eng. J.* **2022**, *442*, 136147. [CrossRef]
19. Abou-Elela, S.I.; Abo-El-Enein, S.A.; Hellal, M.S. Utilization of autoclaved aerated concrete solid waste as a biocarrier in an immobilized bioreactor for municipal wastewater treatment. *Desalin. Water Treat.* **2019**, *168*, 108–116. [CrossRef]
20. Bao, T.; Yu, Z.M.; Damtie, M.M.; Wu, K.; Jin, J.; Zhang, Y.; Wei, X.L.; Frost, R.L. Use of autoclaved aerated concrete particles for simultaneous removal of nitrogen and phosphorus as filter media from domestic wastewater. *Environ. Technol.* **2020**, *41*, 3032–3042. [CrossRef]
21. Martemianov, D.; Xie, B.-B.; Yurmazova, T.; Khaskelberg, M.; Wang, F.; Wei, C.-H.; Preis, S. Cellular concrete-supported cost-effective adsorbents for aqueous arsenic and heavy metals abatement. *J. Environ. Chem. Eng.* **2017**, *5*, 3930–3941. [CrossRef]

22. Sayğılı, G.A.; Güzel, F. Chemical modification of a cellulose-based material to improve its adsorption capacity for anionic dyes. *J. Dispers. Sci. Technol.* **2017**, *38*, 381–392. [CrossRef]
23. Kasraee, M.; Dehghani, M.H.; Mahvi, A.H.; Nabizadeh, R.; Arjmand, M.M.; Salari, M.; Hamidi, F.; Mubarak, N.M.; Tyagi, I. Adsorptive removal of humic substances using cationic surfactant-modified nano pumice from water environment: Optimization, isotherm, kinetic and thermodynamic studies. *Chemosphere* **2022**, *307*, 135983. [CrossRef] [PubMed]
24. Asgari, G.; Roshani, B.; Ghanizadeh, G. The investigation of kinetic and isotherm of fluoride adsorption onto functionalize pumice stone. *J. Hazard. Mater.* **2012**, *217–218*, 123–132. [CrossRef]
25. Elkhlifi, Z.; Lahori, A.H.; Shahib, I.I.; Iftikhar, J.; Wang, S. Comparative assessment of phosphate adsorption properties and mechanisms on Mg/Al-engineered sewage sludge biochar in aqueous solution. *J. Water Proc. Eng.* **2023**, *56*, 104443. [CrossRef]
26. Karaca, S.; Gürses, A.; Ejder, M.; Açıkyıldız, M. Kinetic modeling of liquid-phase adsorption of phosphate on dolomite. *J. Colloid Interface Sci.* **2004**, *277*, 257–263. [CrossRef]
27. Kudeyarova, A.Y. Chemisorption of phosphate ions and destruction of organomineral sorbents in acid soils. *Soil Chem.* **2010**, *43*, 635–650. [CrossRef]
28. Fetene, Y.; Addis, T. Adsorptive removal of phosphate from wastewater using Ethiopian Rift pumice: Batch experiment. *Air Soil Water Res.* **2020**, *13*, 1178622120969658. [CrossRef]
29. Li, W.; Zeng, L.; Kang, Y.; Zhang, Q.; Luo, J.; Guo, X. A solid waste, crashed autoclaved aerated concrete, as a crystalline nucleus for the removal of low concentration of phosphate. *Desalin. Water Treat.* **2016**, *57*, 14169–14177. [CrossRef]
30. Kuhn, D.C.; Bottini, R.C.R.; Rambaldi, J.G.; de Luna, F.M.; Perretto, F.; Bottini, R.C.; Lied, E.B.; Mymrin, V.; Nagalli, A.; Passig, F.H.; et al. White cement as a nontoxic and eco-friendly adsorbent for phosphate adsorption/desorption: Characterization, optimization, and modeling studies. *ACS EST Water* **2024**, *4*, 2731–2745. [CrossRef]
31. Lu, S.G.; Bai, S.Q.; Zhu, L.; Shan, H.D. Removal mechanism of phosphate from aqueous solution by fly ash. *J. Hazard. Mater.* **2009**, *161*, 95–101. [CrossRef]
32. *SR EN ISO 6878:2005*; Water Quality. Determination of Phosphorus. Spectrophotometric Method with Ammonium Molybdate. ASRO: Bucharest, Romania, 2005.
33. *SRM 676 A*; Alumina Internal Standard for Quantitative Analysis by X-Ray Powder Diffraction. NIST: Gaithersburg, MD, USA, 2005.
34. Samarghandi, M.R.; Zarrabi, M.; Sepehr, M.N.; Amrane, A.; Safari, G.H.; Bashiri, S. Application of acidic treated pumice as an adsorbent for the removal of azo dye from aqueous solutions: Kinetic, equilibrium and thermodynamic studies. *J. Environ. Health Sci. Engineer.* **2012**, *9*, 9. [CrossRef]
35. Alemayehu, E.; Lennartz, B. Virgin volcanic rocks: Kinetics and equilibrium studies for the adsorption of cadmium from water. *J. Hazard. Mater.* **2009**, *169*, 395–401. [CrossRef] [PubMed]
36. Buriánek, J.; Václavíková, M.; Havelcová, M.; Klementová, M.; Novák, F. Phosphates adsorption from water by recycled concrete. *GeoSci. Eng.* **2014**, *60*, 1–8. [CrossRef]
37. Tran, H.S.; Tran, N.T.V.; Duong, T.H.; Nguyen, L.H.; Kawamoto, K. Autoclaved aerated concrete grains as alternative absorbent and filter media for phosphorus recovery from municipal wastewater: A case study in Hanoi, Vietnam. *Environ. Technol. Innov.* **2023**, *31*, 103175. [CrossRef]
38. Gong, C.; Huo, L.; Wang, Z.; Sun, C.; Wang, Y.; Cheng, X. Phosphorus removal by porous aggregate from aerated concrete waste residues: Simulating porous structure, performance optimisation, adsorption mechanisms. *J. Environ. Chem. Eng.* **2023**, *5*, 111075. [CrossRef]
39. Wang, J. Comparison of linearization methods for modeling the Langmuir adsorption isotherms. *J. Mol. Liq.* **2019**, *296*, 111850. [CrossRef]
40. Mekonnen, D.T.; Alemayehu, E.; Lennartz, B. Adsorptive removal of phosphate from aqueous solutions using low-cost volcanic rocks: Kinetics and equilibrium approaches. *Materials* **2021**, *14*, 1312. [CrossRef] [PubMed]
41. Nechifor, G.; Pascu, D.-E.; Pascu, M. Comparative study of Temkin and Flory-Huggins isotherms for adsorption of phosphate anion on membranes. *UPB Sci. Bull. Ser. B Chem. Mater. Sci.* **2017**, *79*, 129–136.
42. Foo, K.Y.; Hameed, B.H. Insights into the modeling of adsorption isotherm systems. *Chem. Eng. J.* **2010**, *156*, 2–10. [CrossRef]
43. Egemose, S.; Reitzel, K.; Andersen, F.Ø.; Flindt, M.R. Chemical Lake Restoration Products: Sediment Stability and Phosphorus Dynamics. *Environ. Sci. Technol.* **2010**, *44*, 985–991. [CrossRef]
44. Loganathan, P.; Vigneswaran, S.; Kandasamy, J.; Bolan, N.S. Removal and recovery of phosphate from water using sorption. *Crit. Rev. Environ. Sci. Technol.* **2014**, *44*, 847–907. [CrossRef]
45. Zhang, M.; Gao, B.; Yao, Y.; Xue, Y.; Inyang, M. Synthesis of porous MgO-biochar nanocomposites for removal of phosphate and nitrate from aqueous solutions. *Chem. Eng. J.* **2012**, *210*, 26–32. [CrossRef]
46. Onar, A.N.; Öztürk, N. Adsorption of phosphate onto pumice powder. *Environ. Technol.* **1993**, *14*, 605–612. [CrossRef]

47. Chrispim, M.C.; Nolasco, M.A.; von Sperling, M. Phosphorus recovery from municipal wastewater treatment: Critical review of challenges and opportunities for developing countries. *J. Environ. Manag.* **2019**, *248*, 109268. [CrossRef]
48. Egle, L.; Rechberger, H.; Krampe, J.; Zessner, M. Phosphorus recovery from municipal wastewater—An integrated comparative technological, environmental and economic assessment of P recovery processes. *Crit. Rev. Environ. Sci. Technol.* **2016**, *46*, 908–971. [CrossRef]

processes

MDPI

Article

Process Route for Electric Arc Furnace Dust (EAFD) Rinse Wastewater Desalination

Hedviga Horváthová [1], Eduardo Henrique Rotta [2], Tatiane Benvenuti [3], Andréa Moura Bernardes [2], Andrea Miskufova [1,*] and Zita Takáčová [1]

[1] Institute of Recycling and Environmental Technologies, Faculty of Materials, Metallurgy and Recycling, Technical University of Kosice, Letná 1/9, 04200 Kosice, Slovakia; hedviga.horvathova@tuke.sk (H.H.); zita.takacova@tuke.sk (Z.T.)

[2] Programa de Pós-Graduação em Engenharia de Minas, Metalúrgica e de Materiais, Universidade Federal do Rio Grande do Sul, Av. Bento Gonçalves 9500, Setor 4, Prédio 43426, Campus do Vale, Porto Alegre 91509-900, RS, Brazil; eh.rotta@gmail.com (E.H.R.); amb@ufrgs.br (A.M.B.)

[3] Ministério de Ciência, Tecnologia e Inovação, Esplanada dos Ministérios, B E, 382, Brasília CEP 70067-900, DF, Brazil; benvenutitatiane@gmail.com

* Correspondence: andrea.miskufova@tuke.sk; Tel.: +421-556022400

Abstract

This study introduces a two-step treatment method for synthetic and real electric arc furnace dust (EAFD) wastewater, integrating sorption with Mg–Al layered double hydroxides (LDHs) and electrodialysis (ED). The hydrotalcite (LDH), mainly $Mg_6Al_2(CO_3)OH_{16}\cdot 4H_2O$ (hydrotalcite-2H), was characterized by XRD, FTIR, SEM, and EDX, confirming its layered structure and ion-exchange capacity. Calcination at 550 °C was identified as optimal, enhancing sorption efficiency while retaining rehydration potential. Sorption tests demonstrated high effectiveness in removing multivalent ions, achieving over 99% elimination of Ca^{2+}, SO_4^{2-}, and Pb^{2+} ions and Cr from both synthetic and real wastewater. In contrast, monovalent ions such as Na^+ and K^+ were not effectively removed, except for partial removal of Cl^-. To overcome this limitation, electrodialysis was applied in the second step, successfully targeting the remaining monovalent ions and achieving more than 95% conductivity reduction. A key challenge of ED, salt precipitation caused by calcium and sulphate in the concentrate, was effectively mitigated by the prior LDH treatment. The combined process minimized scaling risks, improved overall ion removal (above 97% for Na^+ and K^+), and produced low-salinity effluents (0.84 mS cm^{-1}), suitable for reuse in hydrometallurgical operations. These findings demonstrate that coupling LDH sorption with electrodialysis provides a sustainable and efficient strategy for treating high-salinity industrial wastewaters, particularly those originating from EAFD processes.

Keywords: EAFD wastewater; sorption; electrodialysis; hydrotalcite; desalination

1. Introduction

Crude steel production generates multiple waste streams, including dust, sludge, slag, wastewater, and exhaust gases, which require treatment to prevent the release of hazardous substances and to minimize material losses. Among these, electric arc furnace dust (EAFD) is particularly significant, accounting for approximately 1–2% of the furnace charge and containing up to 40 wt.% Zn and 50 wt.% Fe [1–3]. At the same time, EAFD contains toxic metals such as Pb, Cr, and Cd, and is therefore frequently classified as hazardous waste.

Hydrometallurgy has emerged as a promising approach for EAFD valorization, targeting the recovery of zinc and iron. However, the presence of mineral impurities such as KCl, NaCl, SiO_2, and $CaCO_3$ reduces zinc extraction efficiency and should be removed prior to further processing. Water leaching as a pre-treatment step has been shown to effectively remove water-soluble halides [1,4]. Nevertheless, this process generates large volumes of rinse wastewater with high alkalinity, elevated concentrations of chlorides, calcium, sodium, potassium, and sulphates, and potentially trace amounts of toxic metals such as chromium and lead.

Several technologies are available for the treatment of such saline wastewater, including membrane-based processes (reverse osmosis (RO), electrodialysis (ED)), electrochemical methods (capacitive deionization), sorption using ion exchangers or layered double hydroxides (LDHs), and precipitation techniques [5–16]. LDHs, featuring $[M_{1-x}{}^{2+}M_x{}^{3+}(OH)_2](A^{n-})x/n{\cdot}yH_2O$ (M^{2+} contains Mg^{2+}, Ca^{2+}, Zn^{2+}, etc.; M^{3+} contains Al^{3+}, Fe^{3+}, Cr^{3+} etc.; A^{n-} represents interlayer anions such as $CO_3{}^{2-}$, $NO_3{}^{-}$, $PO_4{}^{3-}$, etc.; x typically ranges from 0.20 to 0.33), have shown high potential for the removal of both anions (Cl^-, $SO_4{}^{2-}$, $CO_3{}^{2-}$, $PO_4{}^{3-}$, $CrO_4{}^{2-}$) and cations (Ca^{2+}, Pb^{2+}, etc.) [17]. These lamellar inorganic solids feature brucite-like structures and generate positively charged hydroxide layers balanced by interlayer anions and water molecules. Their highly specific surface area and tunable interlayer composition provide abundant active sites for selective ion exchange, including chloride and sulphate removal. Interlayer anions are readily exchangeable, with affinity generally ranked as $CO_3{}^{2-} > SO_4{}^{2-} > OH^- > F^- > Cl^- > Br^- > NO_3{}^-$ [18,19]. Their interactions with coexisting ions are selective: nitrate minimally affects Cl^- uptake, whereas sulphate can hinder adsorption of chlorides by stabilizing monosulphate phases. Cations such as Ca^{2+} and Mg^{2+} further influence hydration and anion incorporation, modulating intermediate hydroxide formation. LDHs also interact with polyvalent oxyanions through combined anion and ligand exchange. These properties, alongside recyclability and structural flexibility, render hydrotalcites as highly effective sorbents for halogen anions, oxyanions, and anionic dyes, highlighting their broad potential in water treatment and pollutant removal. LDHs demonstrate reversible anion exchange, high interlayer adsorption capacity, and selective interactions with coexisting ions, making them versatile sorbents for environmental remediation [18]. ED is a membrane-based separation process that utilizes an applied electric field to selectively transport ions through anion and cation exchange membranes, thereby reducing water salinity. Unlike pressure-driven methods such as RO, ED offers distinct advantages in terms of energy efficiency for low to moderate salinities, selective ion removal, and operational flexibility. Recent advancements—particularly in the development of monovalent- or divalent-selective membranes and hybrid system configurations—have significantly enhanced ED's applicability in brackish water desalination and resource recovery. Recent studies have highlighted the following key developments: Comparative assessments show that ED is economically favorable for feed salinities up to 3 g L^{-1}, but at higher salinities, rising membrane and electricity costs increase the levelized cost of water (LCOW), giving reverse osmosis (RO) a clear advantage. Strategies such as reducing ion exchange membrane costs and enhancing membrane properties could help close this economic gap, though at salinities ≥ 5 g L^{-1}, cost reductions alone are insufficient to make ED competitive with RO [20]. Innovative two-stage hydraulic ED systems with optimized stack architectures—featuring resin-filled electrode cells, asymmetric cell pair designs, and counterflow arrangements—demonstrated energy savings of up to ~40% compared to conventional ED systems during brackish water desalination to potable standards [20]. Integrating ED with RO (ED–RO hybrid) in a single operational unit has shown mutual performance enhancement: increased RO permeate flux and improved ion rejection, along with 9–19% energy consumption reduction for RO and 2–3% for ED at feed concentrations of 5–10 g L^{-1} [17]. Commercial ion exchange membranes (IEMs) were modified via interfacial polymerization (IP) to form dense polyamide layers, aiming to enhance monovalent ion selectivity for electrodialysis applications. These IP-modified membranes were characterized and evaluated for their ability to concentrate sodium chloride (NaCl) from seawater reverse osmosis (SWRO) brine, demonstrating improved divalent ion rejection and potential for direct use in the chlor-alkali industry [20]. These advancements collectively underscore ED's potential as a sustainable and versatile desalination technology. Ongoing research is focused on optimizing membrane performance, stack design, energy integration (e.g., coupling with renewable energy sources), and hybrid configurations to overcome remaining limitations and bring ED closer to widespread practical deployment [17,21,22]. Despite these advances, comprehensive studies on integrated treatment of complex industrial wastewater combining sorption and ED for simultaneous removal of chlorides, sulphates, calcium, sodium, potassium, and toxic metals remain limited. Most prior research has focused on individual contaminants. This study addresses this gap by investigating ion removal from synthetic and real EAFD wastewater (RW) via sorption and electrodialysis, both independently and sequentially. The objective is to optimize conditions for reducing salinity and eliminating primary impurities (Cl^-, SO_4^{2-}, Ca^{2+}, Pb^{2+}, Cr, Zn^{2+}, Na^+, K^+), contributing to the development of an effective and integrated wastewater purification system suitable for hydrometallurgical EAFD recycling processes.

2. Materials and Methods

2.1. Wastewater for Experiments

Samples of synthetic and real wastewater were used for sorption and electrodialysis studies. The synthetic solutions, which mimic real EAFD wastewater, were prepared by dissolving precise amounts of the following chemicals in deionized water: $PbSO_4$, $ZnSO_4$, $FeSO_4{\cdot}7H_2O$, Na_2SO_4, K_2SO_4, Na_2CO_3, $CrCl_3{\cdot}6H_2O$, $CaCl_2$, NaOH, and KOH. All chemicals used were of analytical grade and were purchased from Merck Life Science spol.s r.o., Bratislava, Slovakia. The real wastewater was obtained by leaching a sample of EAFD in deionized water with a liquid-to-solid ratio (L:S) of 10. After 60 min of stirring, the EAFD sample was filtered, and the resulting solution was directly used for experiments: (i) sorption using laboratory-prepared commercial Mg-Al LDH sorbent for both RW and synthetic solution (S1), prepared to match RW solution composition; (ii) ED for both RW and S1; and (iii) sorption followed by ED using the synthetic solution (S2) that imitated the composition of RW after sorption. A fresh solution was prepared for each experiment to prevent calcium carbonate precipitation. The composition and characteristics of the RW, S1, and S2 are summarized in Table 1. The values for potassium and sulphates fluctuated in the EAFD dust samples and, consequently, in the real wastewater obtained from the treatment of these samples. This explains why the values for K and SO_4^{2-} in RW differ from those in S1.

The concentrations of Ca^{2+}, K^+, Na^+, Cr, and Pb ions in samples were measured using Atomic Absorption Spectrometry (AAS) (SpectrAA 20 PLUS, Varian, Prague, Czech Republic). Chloride and sulphate ions were quantified through the Mohr titration method and with a multiparameter photometer (Hanna Instrument HI 83 200, Woonsocket, RI, USA). Solution pH and electrical conductivity were monitored using a multimeter (WTW inoLab Multi 9420 IDS, Fisher Scientific, Leicestershire, UK). Chemical analyses of the solutions were conducted in triplicate to ensure accuracy and reproducibility.

Table 1. Composition and characteristics of the input solutions synthetic (S1), EAFD (RW), and synthetic simulation of RW after sorption (S2) used in the experiments.

Parameter	Unit	S1	RW	S2
pH	-	11.75 ± 0.44	12.15 ± 0.48	12.10 ± 0.37
Electrical Conductivity (k)	mS cm^{-1}	8.54 ± 0.05	11.88 ± 0.11	8.95 ± 0.16
Calcium	mg L^{-1}	684 ± 0.09	664 ± 1.17	*
Chloride	mg L^{-1}	1800 ± 1.91	1276 ± 1.61	*
Sulphate	mg L^{-1}	1400 ± 1.98	800 ± 2.22	1250 ± 0.97
Sodium	mg L^{-1}	825 ± 5.29	739 ± 5.77	730 ± 6.1
Potassium	mg L^{-1}	2951 ± 6.03	1379 ± 5.58	1320 ± 6.44
Lead	mg L^{-1}	8.95 ± 0.21	22.3 ± 0.22	2.52 ± 0.17
Chromium	mg L^{-1}	0.17 ± 0.23	0.76 ± 0.34	*
Zinc	mg L^{-1}	-	3.48 ± 0.12	0.145 ± 0.14
Carbonates	mg L^{-1}	-	240 ± 0.14	360 ± 0.7

RW = real EAFD wastewater; S1 = synthetic solution imitating the composition of RW; S2 = synthetic solution imitating the composition of RW after sorption, *—not relevant.

2.2. Sorption Experiments

The sorbent used for sorption experiments was commercial LDH obtained from Sigma-Aldrich (Merck KGaA, Darmstadt, Germany). To identify its chemical and phase composition, atomic absorption spectroscopy using AAS (SpectrAA 20 PLUS, Varian, Prague, Czech Republic) and X-ray diffraction using the PANalytical X'Pert PRO with Co-Kα radiation (XRD, Malvern Panalytical Ltd., Almelo, The Netherlands) were employed.

The microstructure, morphology, and grain size of the LDH precursor sample were observed using the Scanning Electron Microscope (SEM), MIRA FE-SEM, TESCAN (Brno, Czech Republic). Finally, the LDH sample was analyzed using Fourier Transform Infrared Spectroscopy with Attenuated Total Reflectance (FTIR-ATR), Nicolet iS50 FTIR Spectrometer, Thermo Scientific (Waltham, MA, USA), to examine functional groups.

Uncalcined LDH was modified in an electric oven at different temperatures (350 °C, 500 °C, 550 °C, and 600 °C) to obtain its calcined products, which could rehydrate and reconstitute during sorption. To determine the optimal temperature for sorbent calcination, where LDH achieves the highest sorption efficiency and reconstruction ability, calcined products were analyzed by XRD. Additionally, the LDH sample calcined at 550 °C was analyzed by electron microscopy combined with energy-dispersive X-ray spectroscopy (SEM-EDX) (MIRA FE-SEM, TESCAN (Brno, Czech Republic)) to obtain information about the material's chemical composition, grain size, and morphology for use in further experiments.

To determine the rate of sorption and select the appropriate dosage of material, experiments were carried out with S1. Samples of LDH calcined at the selected temperature of 550 °C for 24 h, with different dosages (50, 75, 100, and 150 g L^{-1}), were added to beakers containing 50 mL of S1 solution. After adding the precise amount of LDH to the solution, the beakers were stirred on a magnetic stirrer at 500 rpm. Once the necessary sorption time was achieved, the sample was drawn off and filtered using a vacuum filter. The collected samples, along with the primary S1 solution, were examined for Cl^-, SO_4^{2-}, Ca, K, Na, Cr, Pb, pH value, and conductivity.

Sorption experiments with RW were conducted using 100 and 150 g L^{-1} of LDH calcined at 550 °C for 24 h. The remaining sorption steps were analogous to those for the S1 solution.

The rehydrated and reconstructed sorption material obtained from the experiments with both the S1 solution and RW, using HT dosage of 100 g L^{-1}, was dried at room temperature and analyzed by XRD and SEM-EDX. The content of Ca, K, Na, Cr, and Pb in the liquid/solid samples was determined by Atomic Absorption Spectrometry (AAS, Varian 240). Chlorides and sulphates were determined by the Mohr method and a multiparameter photometer, Hanna Instrument HI 83 200. The pH value and conductivity in the solutions were measured using a multimeter (WTW inoLab Multi 9420 IDS). All sorption experiments and chemical analyses of solutions after sorption were repeated three times. The efficiency of sorption was calculated using the following formula:

$$\%\text{removal} = \left[\frac{100 \times (C_0 - C_e)}{C_0}\right] \quad (1)$$

where C_0 and C_e are the initial and equilibrium concentration of metals ions (mg L^{-1}), respectively.

2.3. Electrodialysis Experiments

A laboratory-scale electrodialysis cell with five compartments was used in the experiments. The compartments were separated by two pairs of heterogeneous cation exchange membranes (CEMs) and anion exchange membranes (AEMs) with 16 cm^2 of effective area, respectively, named as HDX100 (also known as IONSEP-HC-C) and HDX200 (also labelled as IONSEP-HC-A), both provided by Hidrodex® (Bassersdorf, Switzerland). Commercial $Ti/70TiO_230RuO_2$ electrodes were employed as the anode and cathode. The experimental ED setup is shown in Figure 1. More details about the ED setup can be found elsewhere [23]. The concentrated and diluted reservoirs of the cell were filled with 0.5 L of the working solution (S1, S2 or RW), whereas the electrodes' reservoir was filled with 0.5 L of a 12 g L^{-1} Na_sSO_4 solution, in order to maintain the electrical conductivity of the ED setup.

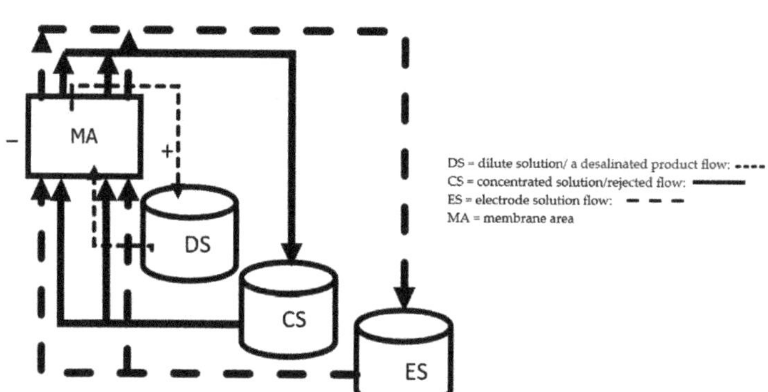

Figure 1. Electrodialysis system used for the experiments (DS = Dilute solution, CS = concentrated solution, ES = electrode solution).

The ED experiments were carried out in triplicate, at room temperature and in galvanostatic mode. The applied current density corresponded to 80% of the value of limiting current density (i_{lim}), determined by current–voltage curves (CVCs) [24,25]. These experiments were conducted until the time necessary to achieve a conductivity value lower than

0.2 μS cm^{-1} (usually reported by the local water supply) or until reaching the maximum potential value of the power supply (64 V).

The pH and electrical conductivity of the concentrated and diluted solution were reported hourly. Additionally, aliquots at pre-established times were collected and the concentrations of calcium, chloride, sodium, sulphate, potassium, and calcium ions were obtained by Ion Chromatography (Dionex®, ICS 3000, Waltham, MA, USA), zinc and lead ions by simultaneous ICP-OES (Perkin Elmer®, Optima 8300 DV, Waltham, MA, USA), and carbonates ions by IC methodology in a TOC analyzer (Shimadzu®, TOC-LCPH, Kyoto, Japan)—all equipment was available at the Federal University of Rio Grande do Sul, Brazil. For the RW, prior to these analyses, the samples were subjected to acidification with 2 M H_2SO_4 (to avoid compounds precipitation) and filtration with an analytical 45 μm filter. The SO_4^{2-} amount added by H_2SO_4 acidification was deducted from the concentration value reported by the analytical method.

The energy consumptions, EC (kWh m^{-3}), for the ED experiments of individual solutions (S1, RW, and S2) were determined using the following Equation (2), adapted from [26]. In this study, the applied current was constant throughout the experiment and, therefore, removed from the integral variation. Moreover, since a multicomponent solution was evaluated, it is not technically feasible to determine the EC for each ionic species; thus, a global EC to treat 1 m^3 of solution was calculated.

$$\mathrm{EC}\ (\mathrm{kWh\ m^{-3}}) = \left(\frac{\mathrm{I}\int_{t=0}^{t}\mathrm{U(t)dt}}{\mathrm{V}}\right) \tag{2}$$

where:

I = current applied during the experiments.
U(t) = electrodialysis cell voltage (V).
t = time (h).
V = volume of treated solution (m^3), diluted compartment.

3. Results and Discussion

3.1. Characterization of Uncalcined LDH

AAS analysis has shown that uncalcined LDH contains 22.94% of Mg and 11.8% of Al. By XRD (Figure 2), dominant phase $Mg_6Al_2(CO_3)OH_{16}{\cdot}4H_2O$ (hydrotalcite-2H, JCPDS: 00-054-1030) was confirmed. Moreover, different polytype variations, which are the result of the metal oxide's gradual transformation at a relatively low temperature (30 °C) [27], were detected, and were mainly $((Mg_4Al_2)(OH)_{12}CO_3(H_2O)_3)_{0.5}$ (JCPDS: 01-070-2151), meixnerite (JCPDS 00-038-0478, 00-050-1684), quintinite-2H $Mg_4Al_2(OH)_{12}(CO_3)(H_2O)_3$ (JCPDS 01-087-1138), and $Mg_2Al(OH)_7$ (JCPDS 35-1275). They are a kind of hydrotalcite-like layered Mg/Al double hydroxide compound with a hexagonal or rhombohedral crystal structure derived from brucite $Mg(OH)_2$, where Al^{3+} cations replace some of the Mg^{2+} cations. In addition, the sample of LDH contained metals like Cu, Ca, and S as a by-product of synthesis-produced or thermal oxidation processes. A small portion of the thermal decomposition products of Cu from synthesis were detected in phases such as Cu_2SO_4 (JCPDS: 00-028-0401), CuO (JCPDS: 01-089-2530), and $Cu_{1.8}S$ (JCPDS: 00-026-0476). The calcium in the LDH sample was in the form of calcite (JCPDS: 01-081-2027). In the structure of LDH were observed boehmite crystals (JCPDS 1-774) with octahedral plates. γ-AlO(OH) as boehmite was formed from gibbsite at approximately 250–300 °C. In the structure was detected γ-Al_2O_3 (JCPDS 10-425) formed by thermal dehydroxylation during the hydrothermal synthesis process (180–250 °C) of LDH production from boehmite. In addition to this, ε-Al_2O_3 (JCPDS 21-10) was detected in the structure and its existence is assumed to relate to the transformation of pseudoboehmite (gelatinous boehmite) to γ-Al_2O_3 and further to ε-Al_2O_3. XRD patterns also indicate the possible presence of metastable χ-Al_2O_3 (JCPDS 4-880). According to [28], gibbsite is the only aluminum hydroxide that produces chi-alumina crystals by thermal dehydroxylation.

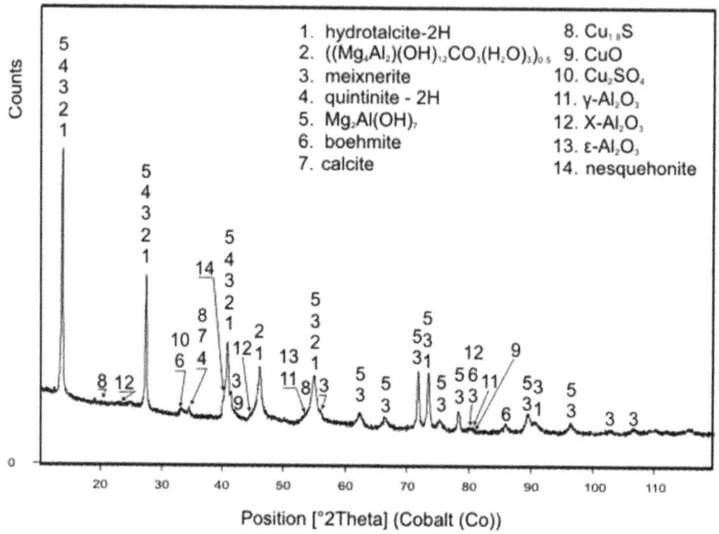

Figure 2. XRD patterns of the uncalcined hydrotalcite-2H (LDH).

FTIR analysis of uncalcined LDH (Figure 3) provided information about the main functional groups present in the structure of the sample. Region 3600–3200 cm^{-1} presents a characteristic broad and robust peak centered at 3408 cm^{-1}, with an O-H stretching vibration at the surface and water molecules in the interlayer space [29]. This vibration peak could be attributed to either boehmite or $Mg_2Al(OH)_7$ [30]. Several signals were detected between 1800 and 1200 cm^{-1}, such as the bending vibration of water around 1600–1700 cm^{-1}, an infrared (IR) vibration associated with the Mg-Al band at 1571 cm^{-1}, and an intense absorption ν3 vibration of interlayer carbonate at 1362 cm^{-1}—with the lattice vibration of the main layers from 500 to 1000 cm^{-1} [31]. The IR bands at 768 cm^{-1} and 560.8 cm^{-1} are assigned to quintinite-2H [32]. The absorption peak at 1560 cm^{-1} is ascribed to κ-(Al_2O_3) [33]. The IR absorption bands from 1000 to 500 cm^{-1} cover several overlapping bands, which may not be fully interpreted considering that the exact material procedure preparation is unknown. However, absorption peaks at 934 cm^{-1} and 768 cm^{-1} could be assigned to the Al-OH bending and stretching vibration, respectively [34].

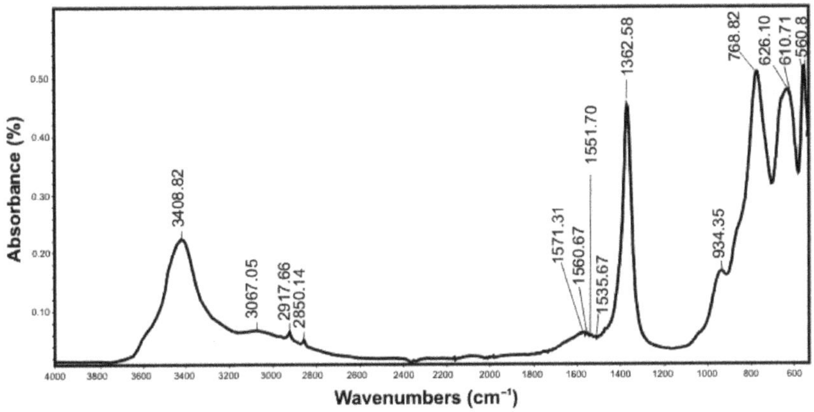

Figure 3. FTIR spectra of the uncalcined hydrotalcite-2H (LDH).

SEM images of the uncalcined LDH sample (Figure 4) showed the crystals of Mg-Al hydroxide with regular mature shapes with an average grain size of 200 to 800 nm.

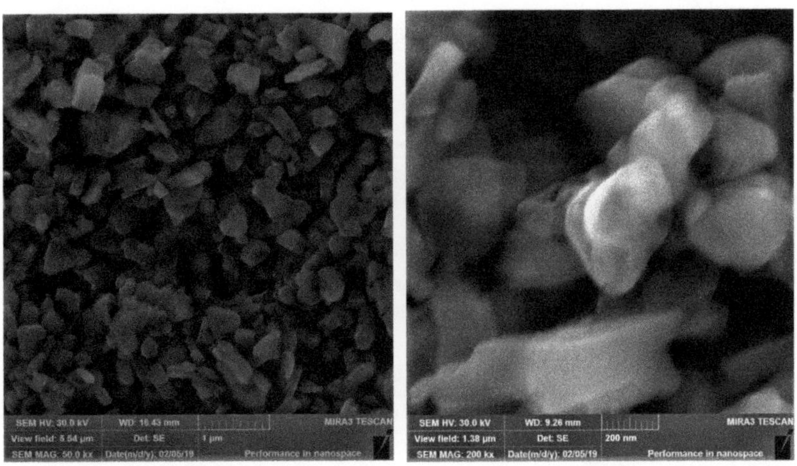

(**a**) (**b**)

Figure 4. SEM images of hydrotalcite-2H (LDH) precursor (**a**) ×50,000, (**b**) ×200,000.

3.2. Characterization of Calcined Products of LDH

The uncalcined LDH sample was subjected to calcination at temperatures of 350 °C, 500 °C, 550 °C, and 600 °C to analyze the changes in phase composition during thermal treatment. This analysis is crucial for identifying the temperature at which the LDH structure irreversibly transforms into a stable spinel-type phase. At this point, it loses its ability to rehydrate and restore its original layered structure—an essential property for ion exchange applications.

The calcined products were characterized using XRD for qualitative phase analysis, and the results were compared to those of the uncalcined LDH and a reference MgO sample (Figure 5). The XRD patterns of the samples calcined at increasing temperatures reveal a progressive transformation of the original LDH structure into a mixed oxide phase with the composition $Mg_{0.8}Al_{0.2}O_{1.1}$. This phase exhibits diffraction patterns similar to periclase (MgO), indicating the formation of a defective cubic oxide structure characteristic of thermally decomposed LDH materials containing both Mg and Al.

From Figure 5, a gradual decomposition and transformation of the well-crystallized LDH phases in the original uncalcined sample is observed as the temperature increases from 350 °C to 600 °C. This transformation leads to the formation of more amorphous phases with smaller grain sizes and crystallites. Upon heating to 350 °C, the basal reflection (003) of the hydrotalcite 2H phase diminishes at the peak position 13.49° (2θ) [28]. The phases from the original LDH gradually fade, including hydrotalcite-2H, meixnerite, and magnesium aluminum carbonate hydroxide. A new phase, quintinite-2H, with the chemical formula $Mg_4Al_2(OH)_{12}(CO_3)\cdot 3H_2O$ (JCPDS: 01-087-1138), appears instead.

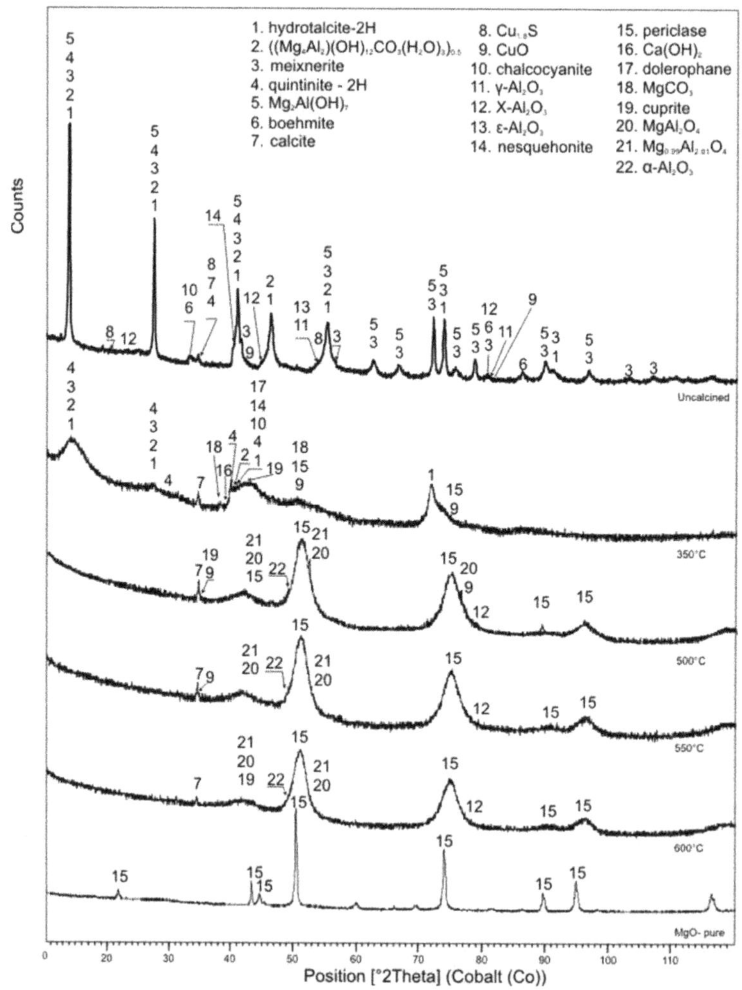

Figure 5. XRD patterns of original input sample of hydrotalcite-2H (LDH) as precursor and calcined products at 350 °C, 500 °C, 550 °C, 600 °C compared with pure MgO.

At the same time, the compound $Mg_2Al(OH)_7$ begins to decompose, leading to the formation of MgO (JCPDS: 01-075-0447), which shows a distinct peak at 50.16° (2θ) in Figure 5. The phase qualitative analysis also reveals minor phases such as nesquehonite

($MgCO_3{\cdot}3H_2O$, JCPDS: 01-070-1433) and its dehydration product, $MgCO_3$ (JCPDS: 01-086-2348). As reported by [35], nesquehonite loses its water below 300–350 °C, while CO_2 is released above 350 °C, typically between 500 and 550 °C, though the exact thermal decomposition mechanism is influenced by experimental conditions.

Additionally, the $Cu_{1.8}S$ phase decreases due to its thermal decomposition into CuO and Cu_2SO_4 [36]. The compound γ-Al_2O_3 (JCPDS: 10-425) is also identified as an amorphous phase in the sample.

Analyzing the structure of the LDH sample calcined at 500 °C, it can be observed that the original diffraction peaks from hydrotalcite-2H, quintinite, and nesquehonite are no longer present. By-products that decomposed below 450 °C led to the formation of α-Al_2O_3 and MgO, while those that decomposed above 500 °C formed $MgAl_2O_4$, accompanied by a sharp increase in MgO dispersity. Between 350 °C and 500 °C, crystallization of MgO (periclase, JCPDS 96-900-6817) was observed. This behavior is likely attributed to the dehydroxylation of brucite-like layers and the removal of interlayer anions (CO_3^{2-}). The abundant hydrotalcite phases were transformed into crystalline, non-stoichiometric MgO (JCPDS: 96-900-6817). The compound $Mg_2Al(OH)_7$ underwent dehydroxylation to form smaller portions of spinels, such as $Mg_{0.99}Al_{2.01}O_4$ (JCPDS: 96-900-1373) and $MgAl_2O_4$. The presence of CuO and $CaCO_3$ is still evident, while new phases of Cu_2O (JCPDS: 01-077-0199) and Cu_2SO_4 (JCPDS: 28-0401) form as a result of CuS decomposition. Amorphous χ-Al_2O_3 and $CaCO_3$ also remain present in the structure [37].

At 550 °C, the calcined LDH sample contains dominant unstable MgO, along with small proportions of stoichiometric and non-stoichiometric spinels. Other identified phases include α-Al_2O_3 (JCPDS 96-210-1053), χ-Al_2O_3, CuO, Cu_2O, and $CaCO_3$.

By 600 °C, dehydroxylation and decarbonization of the LDH are essentially complete. The majority of phases in the LDH sample calcined at 600 °C are periclase MgO (JCPDS: 96-900-6804) and spinels. Minor phases such as α-Al_2O_3 and χ-Al_2O_3, along with impurities like Cu_2O and $CaCO_3$, remain. Based on these observations, it can be concluded that temperatures above 550 °C are not recommended for LDH calcination, as at 600 °C the LDH structure transforms into stable spinel phases, thus losing its ability to rehydrate and reconstruct its structure when in contact with water.

The sample of LDH calcinated at 550 °C was subjected to EDX analysis to obtain its chemical composition, Figure 6.

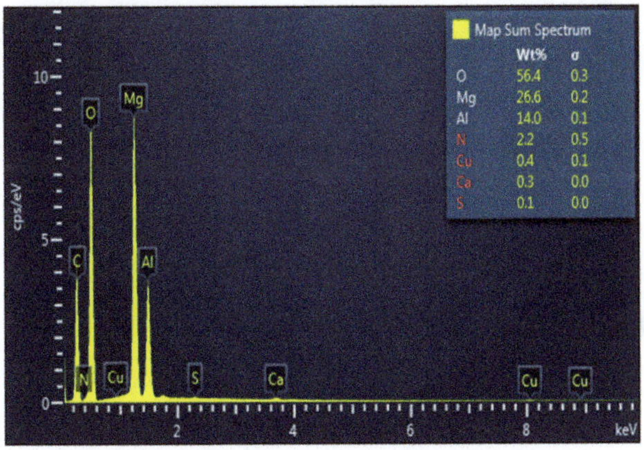

Figure 6. EDX pattern of hydrotalcite-2H(LDH) calcined at 550 °C.

EDX analysis revealed the of presence of the main elements (marked with white colour in the Figure 6) Mg (26.6%) and Al (14%) in the LDH sample. Additionally, the sample contains minor or trace amounts (marked with red colour in the Figure 6) of other elements such as N (2.2%), Cu (0.4%), Ca (0.3%), and S (0.1%). The presence of these elements is likely due to unwashed impurities or ions incorporated into the structure during the coprecipitation and synthesis of the LDH. As mentioned previously, the exact procedure used to prepare the LDH is unknown. Element C was excluded from the spectrum map in Figure 6 because carbon was added to improve the sample's electrical conductivity. Despite the presence of nitrogen in the EDX analysis, XRD did not detect any inorganic phase containing nitrogen. Therefore, it is highly probable that N originated from organic substances added during the coprecipitation process.

After calcination at 550 °C, the sample experienced a weight loss of ~39%, which is consistent with the decomposition and removal of water and other volatile components.

The grain size in the LDH sample after calcination at 550 °C (Figure 7) decreases to a range of 100–400 nm, compared to the original uncalcined LDH sample (200–800 nm, Figure 4). This reduction in grain size is attributed to the processes of dehydration, dehydroxylation, and decarbonization that occur during calcination. Despite this size reduction, the plate-like morphology of the LDH sorbent is largely preserved, with the plates maintaining an almost regular shape even after calcination.

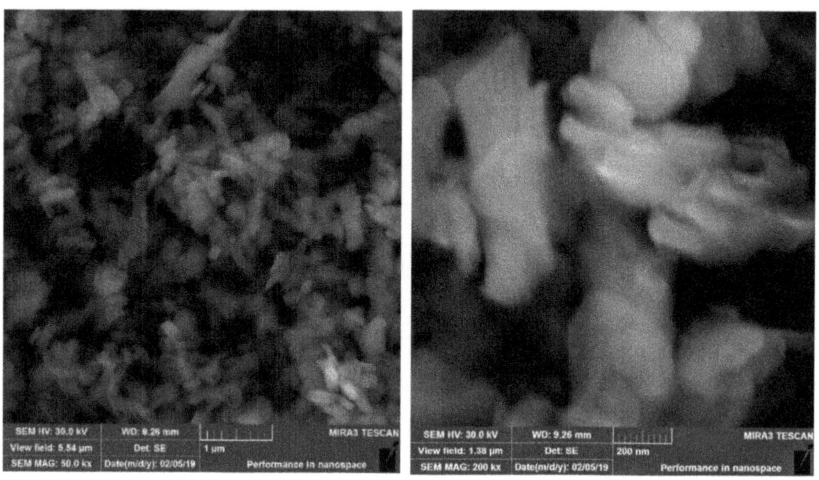

(a) (b)

Figure 7. SEM images of hydrotalcite-2H (LDH) calcined at 550 °C (**a**) ×50,000, (**b**) ×200,000.

3.3. Sorption Tests

Sorption tests were conducted using the S1 solution, a synthetic solution that mimics EAFD wastewater. The primary goal was to optimize the dosage of sorbents and the duration of the sorption process to maximize ion removal efficiency. Based on previously determined optimal calcination at 550 °C, LDH dosages of 50, 75, 100, and 150 g L^{-1} were tested. Concentrations of Cl^{-}, SO_4^{2-}, and Ca^{2+} ions in the S1 solution during the sorption process were measured at specific intervals alongside pH and conductivity changes. The concentrations of other ions, such as K^{+}, Na^{+}, Cr, and Pb, were analyzed after 24 h of sorption. Furthermore, the impact of LDH dosage on removing chloride, sulphates, and calcium ions over time was investigated.

The time-dependent sorption of chlorides by LDH at various dosages from the S1 solution is shown in Figure 8.

The results indicate chloride removal efficiency increases with higher LDH dosages and longer sorption times. Equilibrium sorption of chlorides is achieved after 24 h at both 50 g L^{-1} and 150 g L^{-1} LDH dosages. The maximum removal efficiency of Cl^- (100%) is attained with a 150 g L^{-1} dosage, while 92% of chlorides are removed with 100 g L^{-1} after the same 24 h period. A significant increase in the rate of chloride sorption was observed after 16 h at an LDH dosage higher than 150 g L^{-1}. Lower dosages, specifically 50 and 75 g L^{-1}, achieved only 17% and 54% removal efficiency after 24 h, and at dosages of 75 and 100 g L^{-1}, sorption equilibrium was not reached within 24 h. This indicates that these amounts are insufficient for complete chloride removal, and a longer sorption time would be required to achieve equilibrium. At the 150 g L^{-1} dosage, the solution density increased, initially slowing the sorption rate for approximately 1000 min. During this period, partial settling of the sorbent at the bottom of the reactor was observed; however, once the sorbent became effectively dispersed in the solution, mass transfer was facilitated, resulting in a sudden increase in chloride removal efficiency.

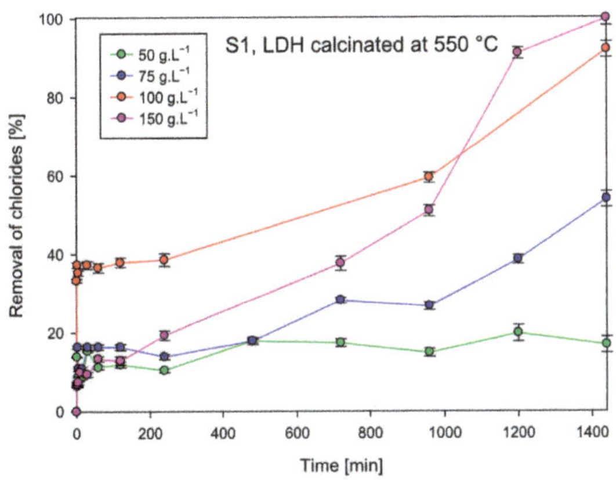

Figure 8. Effect of the dosage of hydrotalcite-2H (LDH) on the removal efficiency of chlorides from synthetic solution S1 with LDH dosage 50, 75, 100, 150 g L^{-1}.

The sorption of sulphates was also investigated using the S1 solution, as shown in Figure 9. The data reveal that the rate of sulphate sorption increases as the LDH dosage increases. Sorption equilibrium is reached within 3 h, with 100% sulphate removal achieved at 100 g L^{-1} and 150 g L^{-1} dosages. To achieve complete sulphate removal with 50 g L^{-1} and 75 g L^{-1} of LDH, it takes 12 h, although more than 95% of SO_4^{2-} ions are removed within the first 2 h.

The sorption of calcium ions from the S1 solution is remarkably swift, as shown in Figure 10. Sorption equilibrium is reached in just 1 min when the LDH concentration is 150 g L^{-1}. In contrast, a lower dosage of LDH takes up to 120 min to reach equilibrium. Interestingly, at 50 g L^{-1} and 75 g L^{-1} dosages, calcium release begins after 6 h of sorption. This behavior may result from the solution becoming enriched with CO_3^{2-} ions, which could be captured from the air during rehydration, leading to ion exchange with SO_4^{2-} or an increase in OH^- groups. Once equilibrium for chloride sorption is reached, calcium readily attaches to the LDH, illustrating the dynamic interplay between these ions.

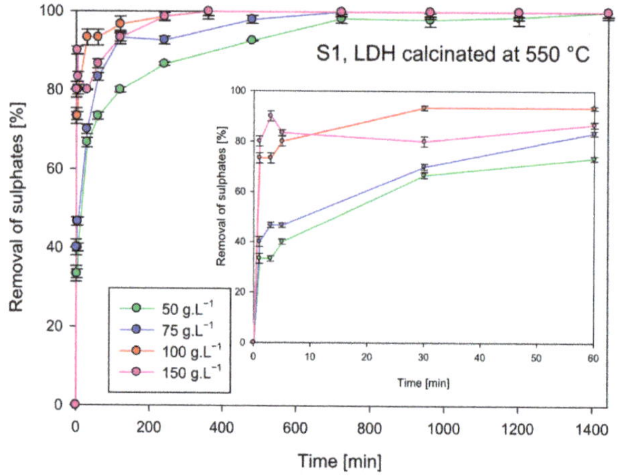

Figure 9. Effect of the dosage of hydrotalcite-2H (LDH) on the removal efficiency of sulphates from synthetic solution S1 with LDH dosage 50, 75, 100, 150 g L^{-1}.

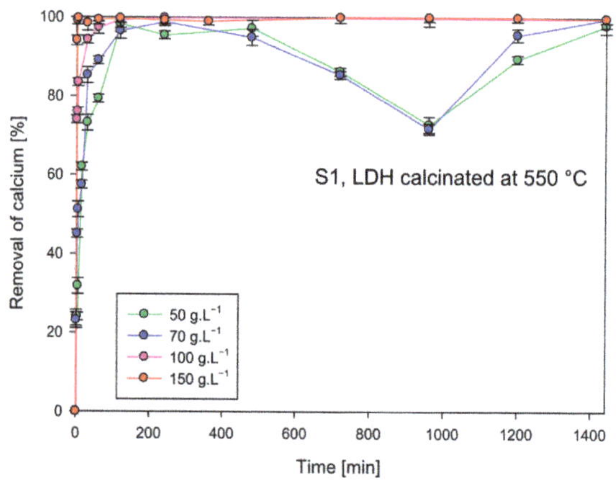

Figure 10. Effect of the dosage of hydrotalcite-2H (LDH) to removal efficiency of calcium from synthetic solution S1 with LDH dosage 50, 75, 100, 150 g L^{-1}.

The sorption efficiency for Cr and Pb ions exceeded 99.9% across all dosages tested. However, the removal of sodium (Na^+) and potassium (K^+) was not detected, as hydrotalcite exhibits a low affinity for monovalent ions. Therefore, no removal of Na^+ or K^+ was observed in the synthetic solutions. The initial concentrations of chromium (0.17 mg L^{-1}) and lead (8.95 mg L^{-1}) decreased below the detection limits at every LDH dosage. In contrast, the removal of sodium (Na^+) and potassium (K^+) was negligible, as hydrotalcite exhibits low affinity for monovalent ions. No significant decrease in Na and K concentrations was observed within 24 h. For instance, at an LDH dosage of 150 g L^{-1}, only 5.3% of Na and 9.3% of K were removed.

Changes in pH and conductivity during the sorption process in the S1 solution were closely monitored, as shown in Figure 11a,b.

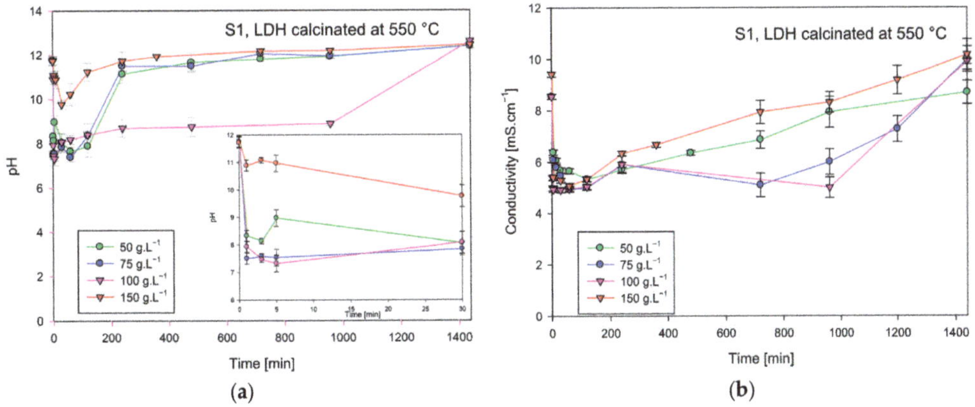

Figure 11. (**a**) pH and (**b**) conductivity changes during sorption with hydrotalcite-2H from S1 with different sorbent concentrations (50, 75, 100, 150 g L^{-1}).

As shown in Figure 11a,b, the dynamics of pH and conductivity were meticulously observed during the sorption process in solution S1. Initially, the pH fluctuated between 9 and 12. At the onset of the process, the pH value slightly decreased before gradually increasing again, eventually reaching equilibrium. The fundamental nature of LDH plays a key role here, as it releases OH^- ions during sorption. This release occurs almost immediately after a significant amount of calcium is removed from the solution, which typically leads to a reduction in OH^- concentration.

Figure 11b presents the changes in conductivity during the same sorption process. A consistent trend was observed across all four LDH dosages tested. Within the first minute after LDH was introduced into the S1 solution, conductivity dropped by approximately 30%, likely due to the rapid removal of calcium ions. After this initial drop, conductivity gradually increased again. This increase is mainly attributed to the release of OH^- ions into the solution, resulting from the rehydration of Mg-Al oxides. The following reaction can describe this transformation:

$$Mg_{0.80}Al_{0.20}O_{1.10} + 1.10\ H_2O \rightarrow Mg_{0.80}Al_{0.20}OH_2(OH)_{0.20} \qquad (3)$$

During this process, rehydrated Mg-Al oxides interact with anions present in the solution, which is accompanied by the release of OH^- ions [38].

3.4. Sorption from EAFD Wastewater

The sorption rates of chlorides (Cl^-), sulphates (SO_4^{2-}), calcium (Ca^{2+}), chromium (Cr), lead (Pb), sodium (Na^+), and potassium (K^+) were tested using RW with an LDH dosage of 150 g.L^{-1} over a period of 24 h. Figure 12 illustrates the sorption dynamics for Ca^{2+}, Cl^-, and SO_4^{2-} over the first 240 min of the process.

The sorption experiments revealed distinct behaviors for different ions. Calcium (Ca^{2+}) and sulphate (SO_4^{2-}) were removed very rapidly, with more than 99% of these ions eliminated within the first five minutes, indicating a highly efficient ion-exchange process with the layered double hydroxide. In contrast, chloride (Cl^-) ions were removed more slowly, with only 36% sorption after five minutes and equilibrium reached after 240 min. When all three anions were present, calcium and sulphate were preferentially and rapidly removed, followed by the slower sorption of chloride. This behavior can be explained by the rehydration and reconstruction of Mg-Al oxides in LDH, which releases hydroxide ions

(OH^-) while enabling the incorporation of anions into the interlayer structure, as described with Reaction (3) in Section 3.3 [38].

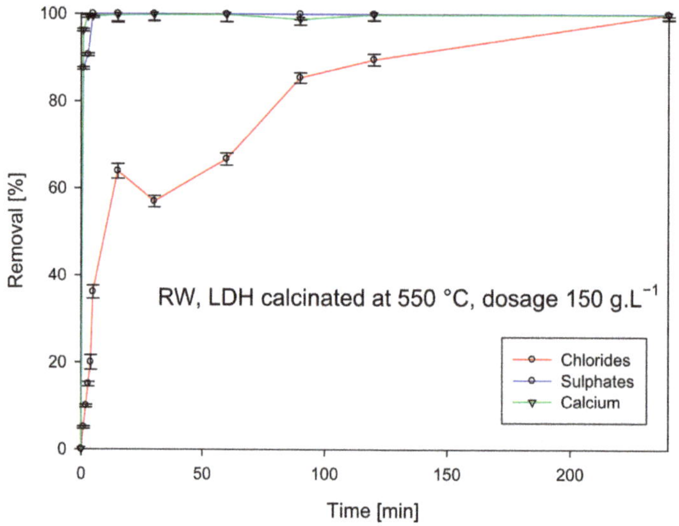

Figure 12. The kinetics of Cl^-, SO_4^{2-}, and Ca^{2+} ions' sorption with calcinated hydrotalcite-2H (LDH) from RW during 240 min.

During sorption in real EAFD wastewater, visible turbidity was observed, likely due to partial dissolution of Al^{3+} ions from the hydrotalcite, which subsequently precipitated with OH^-. This phenomenon contributes to the maintenance of a relatively stable OH^- concentration in the solution, in accordance with the general reaction (4) of anion incorporation into LDH, as reported in previous studies [39]:

$$Mg_{1-1}Al_xO_{1+x/2} + \frac{x}{n}A^{n-} + (1+\tfrac{x}{2})H_2O \rightarrow Mg_{1-1}Al_x(OH)_2A_{x/n} + xOH^- \quad (4)$$

where x is a molar ratio Al/(Mg + Al) and An is an n-valent anion. This phenomenon of the precipitation of Al with OH^- helps to more or less maintain a constant concentration of OH^- groups in solution [7].

The sorption study [40] reports the precipitation of Ca^{2+} ions with hydrotalcite from a $CaCl_2$ solution, forming $Ca(OH)_2$ through OH^- ions released from hydrotalcite. The authors suggest that Cl^- ions are sorbed only when calcium precipitates with the released OH^- ions as $Ca(OH)_2$. Furthermore, chloride sorption is more likely in systems where the solution contains an excess of OH^- groups.

In contrast, in real EAFD wastewater, where carbonates are present, $Ca(OH)_2$ formation is unlikely. Partially, the OH^- groups released from hydrotalcite appear to inhibit chloride sorption because they have a higher affinity for the LDH than chloride ions. Consequently, chloride ions begin to adsorb only after Al^{3+} ions dissolve from hydrotalcite to form $Al(OH)_4^-$, and only when sulphate ions are absent from the solution. Moreover, divalent anions such as SO_4^{2-} or CO_3^{2-} exhibit higher selectivity than monovalent ions. The selectivity of Mg-Al oxides increases with increased anion charge and decreased anion size. Therefore, hydrotalcite does not readily combine with Cl^- ions, as it preferentially reacts with OH^- formed during Reaction (4) [11]. Sorption of heavy metals was highly efficient, with chromium (Cr) and lead (Pb) completely removed within the first minute, largely unaffected by the presence of other anions. Conversely, monovalent cations such as sodium (Na^+) and potassium (K^+) were poorly removed, with less than 10% sorption achieved after 240 min. This indicates a low affinity of LDH for monovalent cations, which are primarily sorbed onto the surface, whereas anions are incorporated into the interlayer space.

During the LDH sorption process, the pH of the RW solution decreased from the initial value of 12.15. Initially, the pH dropped to 8.56 within the first minute of sorption. After this, the pH gradually increased, reaching 12.0 by the end of the process. This increase in pH is attributed to the release of OH^- ions during the sorption process as the LDH interacts with the ions in the wastewater. A similar trend, but with a more intensive change, was observed in the conductivity of the solution during sorption. In the first three minutes, conductivity dropped sharply from 11.88 mS cm^{-1} to a minimum of 2.8 mS cm^{-1}. This decrease corresponds to the removal of almost all calcium and sulphate ions from the solution. After 240 min, conductivity increased to 4.23 mS cm^{-1}, reflecting the rehydration and ion exchange activities occurring within the system. Over the next 24 h of sorption, the conductivity rapidly increased to 9.00 mS cm^{-1}. These changes in pH and conductivity during the sorption process closely mirror the behavior observed in the S1 solution, indicating similar sorption dynamics for both the real wastewater and the synthetic solution.

3.5. SEM Observation and Structural Characterization of the LHD After Sorption

After the sorption experiments were conducted with a dosage of 100 g L^{-1} for both the S1 and RW solutions, the suspension was filtered, and the solid residue was air-dried for further analysis. The dried samples were subjected to SEM, EDX, and XRD analyses to investigate the structural changes in the adsorbent.

The XRD analysis observed that the adsorbent largely retained its original hydrotalcite structure (Figure 13). The amount of adsorbed ions (Cl^-, SO_4^{2-}) was insufficient to form any new, visible crystalline phases. Notably, the phase of meixnerite disappeared in both the S1 and RW samples.

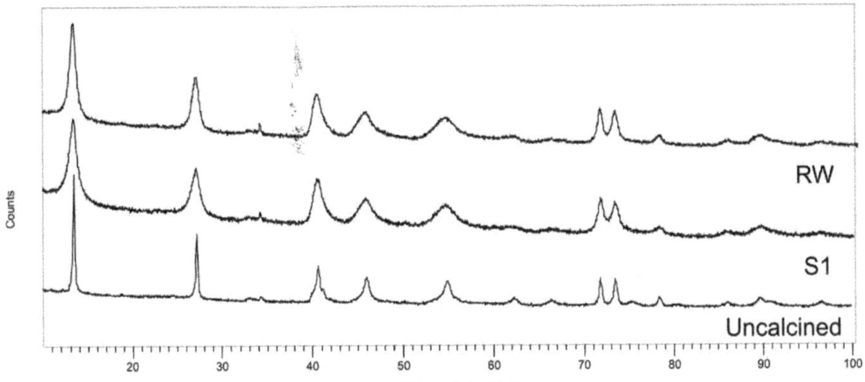

Figure 13. XRD patterns of LDH (100 g L^{-1}) samples after sorption, 24 h from S1 and RW compared with uncalcined LDH.

Further examination of the XRD patterns from the S1 and RW samples revealed a decrease in the intensity of the hydrotalcite peaks. This reduction in peak intensity suggests ion exchange between the Cl^- and SO_4^{2-} ions and the hydrotalcite structure. A noticeable increase in the peak intensity at 34° 2θ was observed, indicating the calcium sorption into the structure. The peak broadening reveals the change in lattice parameters of the input material due to intercalation of ions in the lattice of LDH, tension, and the occurrence of defects, as well as a decrease in the crystallite dimension.

Figures 14 and 15 present the SEM images and EDX analysis of LDH after 24 h of sorption with the S1 and RW solutions, respectively. The SEM images show that the morphology of the LDH is similar after sorption from both solutions. By comparing the initial LDH morphology (before sorption, as shown in Figures 6 and 7) with the structure after sorption (Figure 15), it is evident that the microstructure and chemical composition of the LDH have changed. The initial calcined LDH sample consists of aggregates sized around 100–300 nm, with distinct individual layers. In contrast, the LDH, after sorption, loses its organized aggregate structure and instead exhibits a swollen network of seemingly mismatched clusters, with a larger size ranging from 1 to 2 μm. These clusters also contain visible whisker-like particles. Hydrotalcites can absorb water into their interlayer spaces. Water molecules help expand the interlayer and participate in the ion-exchange process, as the LDH structure becomes more hydrated and stable. Initial ion absorption (including water) typically involves hydration followed by ion-exchange processes of cations (such as Ca^{2+}, Pb^{2+}, Zn^{2+}, and Cr^{3+}) facilitated by the Mg^{2+} and Al^{3+} ions in the hydrotalcite (HT) structure. Subsequently, anions (such as Cl^- and SO_4^{2-}) are bound through intercalation into the interlayer space of hydrotalcite [41,42]. The nanometric scale of the particles and LDH layers, the presence of OH^- groups, the high internal surface area due to the fluffy, multi-layered, and three-dimensional structure, and the combination of active, exchangeable cations (like Mg and Al) with their amphoteric character work synergistically during the sorption process. These characteristics enable the LDH material to achieve remarkable efficiency in the removal of both cations and anions from wastewater.

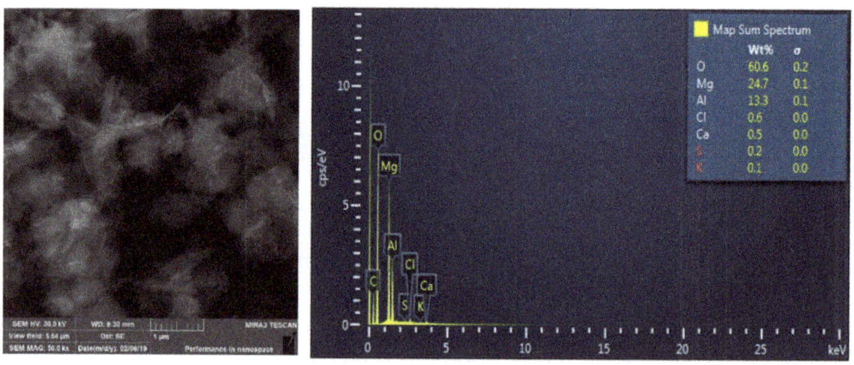

(a) (b)

Figure 14. (**a**) SEM image, (**b**) EDX analysis of LDH (100 g L^{-1}) after 24 h of sorption from S1.

EDX analysis revealed comparable compositions in the adsorbents from the S1 and RW solutions after the sorption process. The main LDH elements (Mg, Al and O, marked with white in Figure 14) were present in similar amounts, as were the levels of Ca, Cl and S. However, the percentage of Ca, Cl and S in the LDH from the RW solution is slightly lower. The presence of sulphur and potassium (marked with red in Figure 14) cannot be ruled out, but is difficult to confirm due to the low content and detection limits of the method used. Silicon was only detected in the LDH sample after sorption from the RW solution, which suggests the possible presence of silicon in the RW solution and the incorporation of silica (silicate) species into the LDH structure during sorption from the RW solution.

Conversely, the Si content is quite low (0.2%) (marked with red in Figure 15), which requires further investigation. As in Figure 6, carbon (C) was excluded from the EDX analysis in Figures 14 and 15 due to the improvement in the electrical conductivity of the sample.

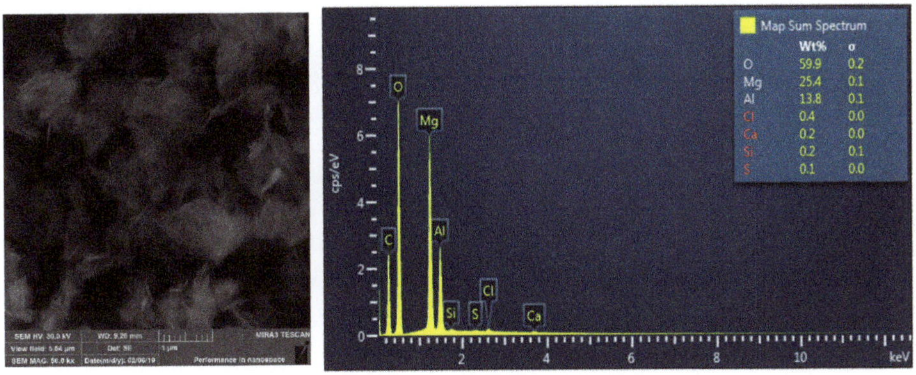

(a) (b)

Figure 15. (**a**) SEM image, (**b**) EDX analysis of LDH (100 g L^{-1}) after 24 h of sorption from RW.

The results obtained from sorption experiments using the S1 and RW solutions with an HT dosage of 100 g L^{-1} are summarized and compared in Table 2, which shows the total ion removal efficiency (%) after 24 h of sorption.

Table 2. Ion removal with hydrotalcite-2H (LDH) calcined at 500 °C. The summary of ionic species in S1 and RW solutions after sorption.

Ionic Species	Symbol	Ion Removal (%)	
		S1 Solution	RW Solution
Chloride	Cl^-	98 ± 0.9	100 ± 0.91
Sulphate	SO_4^{2-}	100 ± 0.7	100 ± 0.12
Sodium	Na^+	0 ± 9.7	0 ± 0.11
Potassium	K^+	0 ± 14.2	0 ± 15.1
Calcium	Ca^{2+}	100 ± 0.21	100 ± 2.1
Carbonates	CO_3^{2-}	-	0 ± 0.01
Zinc	Zn^{2+}	100 ± 0.09	95.6 ± 3.2
Lead	Pb^{2+}	100 ± 0.01	88.7 ± 11.8

The comparison of the results in Table 2 shows that sorption from the synthetic solution is 100% for most elements, except for chlorides. In contrast, the sorption from the RW source is also 100%, although certain cations, such as zinc and lead, are not fully sorbed. Additionally, sodium (Na) and potassium (K) are only sorbed in minimal amounts for both solutions, and hydrotalcite is generally not utilized for their removal.

3.6. Electrodialysis Tests

Ion removal from S1 and RW solutions was also evaluated by means of electrodialysis. Before the electrodialysis experiments, it is inherent to determine the limiting current density (i_{lim}) for both working solutions, which can be obtained using current–voltage curves. It can be seen in Figure 16 that the current–voltage curves showed a conventional shape, with three regions (except for the CEM, due to power supply limiting) and a limiting state [43]. The first region presents a quasi-ohmic behavior, with a linear dependence between the current density and membrane potential. The second region is associated with a limiting state, with a significant increase in the resistance of the membrane (an inclined "plateau") and where the i_{lim} is obtained. The i_{lim} value for AEM and CEM in contact with the S1 solution was 6.4 mA cm^{-1} and 8.6 mA cm^{-1}, respectively, while for the RW solution the i_{lim} values were 20.0 mA cm^{-1} and 30.0 mA cm^{-1}. The third region is linked to the overlimiting current density condition, which can be attributed to water splitting and electroconvection mechanisms [44].

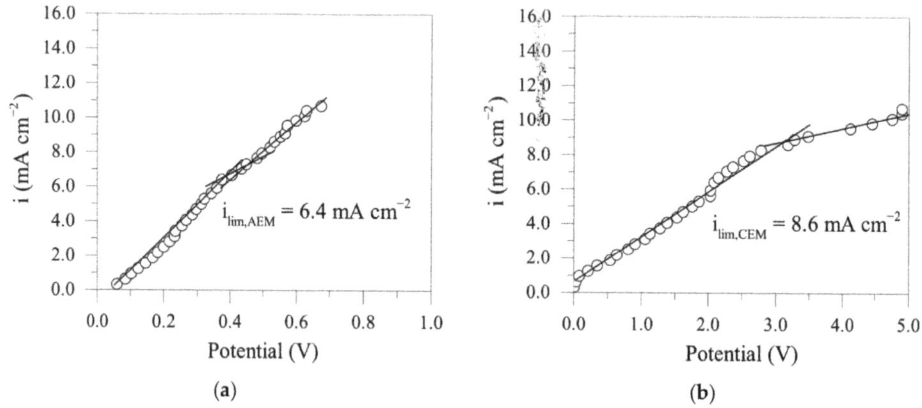

Figure 16. *Cont.*

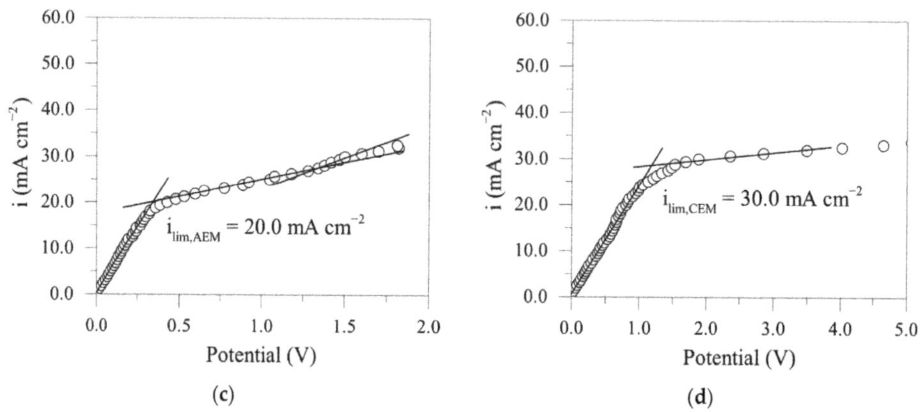

Figure 16. Current–voltage curves of (**a**) AEM and (**b**) CEM for S1 solution, as well as (**c**) AEM and (**d**) CEM for RW solution.

Based on the current–voltage curve results, the current density value to be applied in the electrodialysis experiments with both solutions was limited by the AEM. According to the literature [45], when operating under galvanostatic mode and underlimiting conditions, it is convenient to apply a current density value that corresponds to 70–80% of the limiting one, in order to mitigate the concentration polarization effects. In this sense, for the ED tests with the S1 solution, it was applied at 5.3 mA cm^{-2} for approximately 840 min, while for the RW solution, the experiments were carried out for 240 min with an imposed current density of 15 mA cm^{-2}. The electric conductivity (k) of the solution in the diluted and concentrated reservoirs was monitored each hour, and the values are presented in Figure 17. Comparing both ED setups for the S1 solution (Figure 17a) and for the RW solution (Figure 17b), it can be noted that this parameter presented a similar behavior, with k values decreasing in the diluted reservoir along the treatment time, while increasing in the concentrated reservoir. This is an expected performance in the electrodialysis process [46] since the ionic species are transferred from the diluted to the concentrated reservoir. Moreover, it indicates a successful demineralization of the working solution, corresponding to a reduction of approximately 98% of the electric conductivity for the S1 solution and 95% for the RW solution.

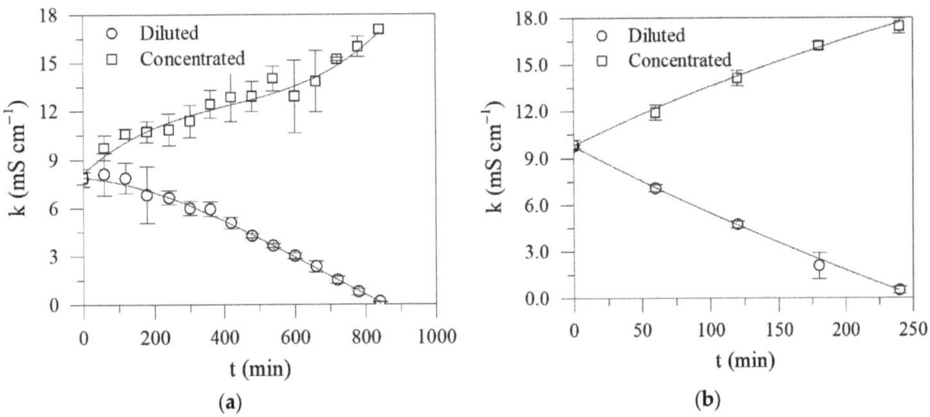

Figure 17. Values of electric conductivity for diluted and concentrated reservoir solution for ED experiments with (**a**) S1 solution and (**b**) RW solution.

The diluted and concentrated solution pH values for both ED setups were also recorded hourly. As can be observed, the pH of the concentrated reservoir solution remained nearly unchanged during the electrodialysis experiments at a value around 12 and 12.5 for the S1 (Figure 18a) and RW (Figure 18b) solutions, respectively. However, a different behavior was reported for the pH values of the diluted reservoir solution, which presented a decreasing tendency, especially after half of the treatment time. This behavior was also reported in [23] and may be associated with coupled effects of concentration polarization at the interface between the solution and the anion exchange membrane. Since ion concentration is one of the directly proportional parameters in the determination of the i_{lim} value [47], the demineralization of the diluted solution may lead to the system reaching or exceeding the i_{lim} value, and the generation of OH^- and H^+ due to water splitting in the membrane/solution region may appear. Thus, the OH^- may pass through the anion exchange membrane, while the H^+ ion may influence the pH value of the solution in the diluted reservoir [48].

The removal of chloride, sulphate, and calcium ions from the diluted solution is shown in Figure 19. With the S1 solution, it can be noted in Figure 19a that electrodialysis performed better than sorption in removing chlorides, reaching more than 99% extraction of these ions in 800 min, whereas at sorption this value was achieved after more than 1400 min with 150 g L^{-1} of LDH. However, this behavior is not observed when it comes to the removal of SO_4^{2-} and Ca^+ ions, since the removal rate of these ions was slower than sorption, despite reporting a percent extraction of 99.35% for sulphate and 98.91% for calcium ions. For the RW solution, none of the specified ions reached more than 90% of removal at the end of the ED tests, as can be seen in Figure 19b. The percentage removal of sulphate ions presented a singular behavior, where a considerable increase in its values can be seen after only 150 min. This can be explained by the competition of coexisting anions to pass through the AEM, where the transport of monovalent ions and those with a smaller ionic radius, such as chloride, is favored over divalent ions, such as sulphate [49]. Moreover, sulphate ions species had the lowest value of percentage removal, with less than 70%, compared to calcium and chlorides (approximately 73% and 83%, respectively).

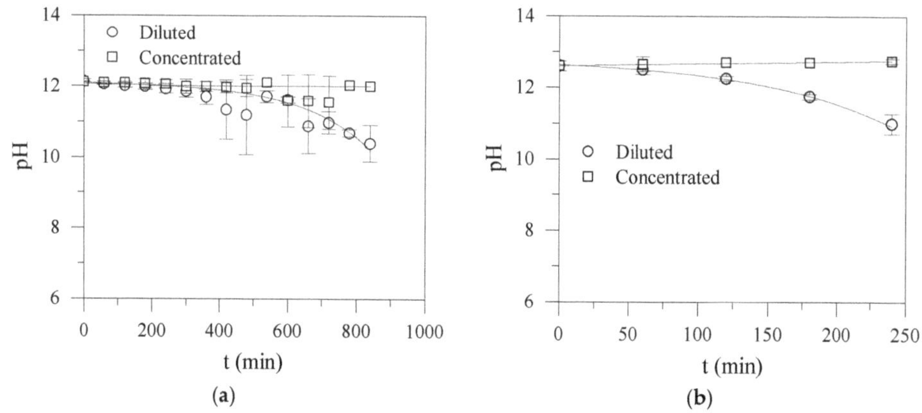

Figure 18. Values of pH for diluted and concentrated reservoir solution for ED experiments with (**a**) S1 solution and (**b**) RW solution.

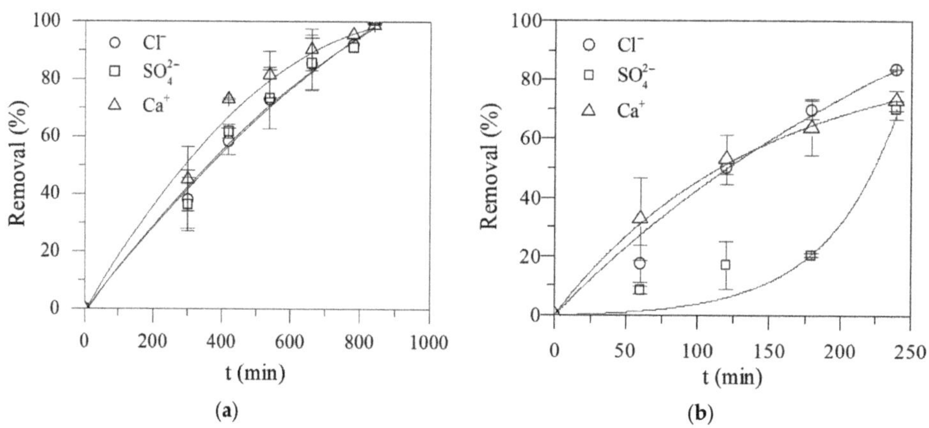

Figure 19. Values of chloride, sulphate, and calcium ion removal from diluted reservoir solution for ED experiments with (**a**) S1 solution and (**b**) RW solution.

From the comparison of the kinetic curves in Figure 19, it is evident that the sulphate removal rates differ significantly between S1 and RW, indicating that the mechanisms of sulphate removal are not the same. Although the initial concentrations of calcium, sulphate, and chloride were similar in S1 and RW (with RW having slightly higher conductivity), it can be concluded that sulphate removal from the RW solution is influenced by other ions and species present. These species may include carbonates, hydroxides, and heavy metal cations, which can significantly affect the rate of sulphate removal. This is also reflected in the higher removal efficiencies observed for Pb and Zn compared to other studied ions.

Interestingly, carbonates showed the lowest removal efficiency in the RW solution (56.46%). Conversely, the removal rates of Ca^{2+} and Cl^- within the 0–240 min range are higher in RW than in S1, despite the overall lower removal efficiency of RW at 240 min.

This behavior can be explained by the fact that the synthetic solution (S1) is more sensitive to polarization phenomena. In contrast, real water (RW), which contains a wider variety of ionic species, provides a greater number of charge carriers across the membranes. As a result, polarization is suppressed (or at least delayed), leading to improved process efficiency and a higher ion removal rate. It is therefore believed that the difference in transport rates arises from the ability of the RW system to operate at higher current densities, as illustrated by the polarization curve in Figure 16. Higher current density increases the amount of energy available to drive ion migration toward the concentrate compartment, which accounts for the enhanced performance observed in RW compared to S1. On the other hand, extending the duration of the experiment could further improve the overall ion removal efficiency in the RW system, allowing the process to take fuller advantage of its higher transport capacity.

The values of ion removal for all ionic species contained in S1 and RW solutions are summarized in Table 3. Overall, electrodialysis performed well in the desalination of both solutions, emphasizing the removal of sodium, potassium, and carbonate ions, which were not affected by the sorption process. However, it is important to highlight the occurrence of salt precipitation, especially in the concentrated reservoir, probably due to the high concentration of calcium and sulphate during the experiments. This phenomenon may cause severe damage to the integrity of the ion exchange membrane, reducing its mechanical properties and lifetime, as well as the efficiency of the electrodialysis process [50]. To overcome this situation, it is recommended to use a pre-treatment process that removes precipitable components.

Table 3. Ion removal summary of ionic species in S1 and RW solutions by electrodialysis.

Ionic Species	Symbol	Ion Removal (%)	
		S1 Solution	RW Solution
Chloride	Cl^-	99.60 ± 0.11	83.49 ± 0.20
Sulphate	SO_4^{2-}	99.36 ± 0.21	69.84 ± 3.12
Sodium	Na^+	98.28 ± 0.28	75.18 ± 12.50
Potassium	K^+	99.47 ± 0.08	82.77 ± 7.78
Calcium	Ca^{2+}	98.91 ± 0.04	72.83 ± 3.31
Carbonates	CO_3^{2-}	-	56.46 ± 5.35
Zinc	Zn^{2+}	-	83.10 ± 5.20
Lead	Pb^{2+}	90.32 ± 3.82	95.12 ± 1.10

3.7. Sorption and Electrodialysis as a Two-Stage Treatment

Considering the non-removal of monovalent ions like Na^+ and K^+ from sorption and the salt precipitation occurring in electrodialysis, the integration of both techniques in a two-stage treatment process with wastewater simulating the composition of RW, named S2, was studied. The solution S2 was prepared synthetically, reflecting the output of the sorption process with the RW solution. The composition of S2 is presented in Table 1. As a first step, it is suggested that LDH sorption be used since this technique presents a high capacity for the removal of calcium, chlorides, and sulphate ions, suppressing the formation of precipitates in the ED process. As a second step, the use of electrodialysis may reduce the salinity of the solution, removing monovalent ions, such as Na^+ and K^+, together with CO_3^{2-}.

The current–voltage curves are shown in Figure 20, where it can be seen that the AEM limited the current density value to be applied, with an i_{lim} = 29.1 mA cm^{-2}.

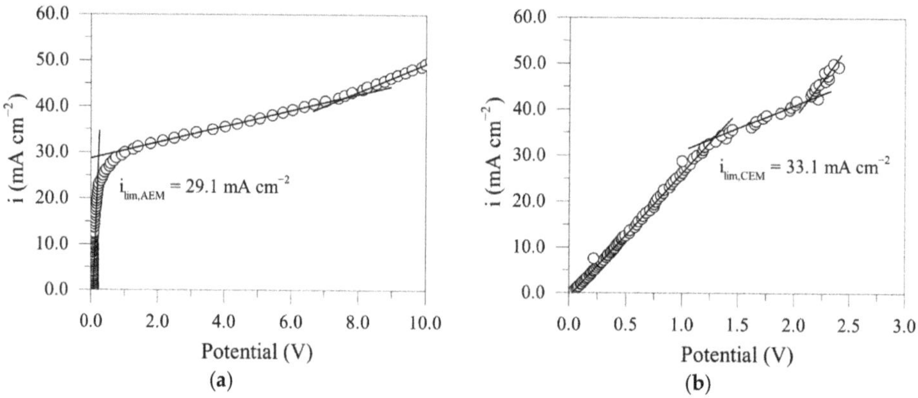

Figure 20. Current–voltage curves of (**a**) AEM and (**b**) CEM for S2 solution.

At this process stage, electrodialysis experiments were carried out with an applied current density of 21.8 mA cm^{-2} for 210 min. As in previous ED experiments, conductivity and pH values were monitored, and a similar behavior was reported in previous experiments. The conductivity value of the solution in the diluted reservoir decreased from (8.98 $\pm$ 0.42) mS cm^{-1} to (0.84 $\pm$ 0.06) mS cm^{-1}, representing a reduction higher than 90% at the end of the experiments. The pH value of the concentrated solution remained nearly unchanged (pH = 12.8) throughout the experimental time. At the same time, a sharp decrease was observed in the diluted reservoir solution between 100 and 150 min, from around 12.5 to 2.8, being kept virtually constant after this period. As mentioned before, this is probably due to the coupled effects of concentration polarization that occur according to the working solution's demineralization. Considering that the solution resulted in the ED diluted compartment having a low pH value, such produced water would be suitable for usage in the acid route in the hydrometallurgical process of Zn recovery. The percentage of ion removal from the S2 solution is shown in Table 4. Except for sulphate, which was added only to meet the concentrations of potassium and sodium ions, it can be seen that this step presented an ion removal higher than 90% for all ionic species in the working solution. The Na^+, K^+, and CO_3^{2-} ions, unable to be removed by sorption, reported a final concentration of (14.72 $\pm$ 2.21) mg L^{-1}, (12.10 $\pm$ 4.00) mg L^{-1}, and (5.36 $\pm$ 1.19) mg L^{-1} in the diluted reservoir. Additionally, after the ED experiments the concentration of Pb^+ and Zn^+ ions, partially removed by sorption, presented values lower than the limit detection of the method, that is, less than 0.02 mg L^{-1}; in this sense, sorption–electrodialysis as a two-stage treatment may be suitable to reduce the high salinity of the real EAFD wastewater. With this experimental setup, obtaining water with a low salinity of the final EAF wastewater solution was possible, and it may be reused in hydrometallurgical operations as process water. In the absence of calcium and chloride (after their removal), the results of electrodialysis (ED) showed significantly higher removal efficiencies for K^+, Na^+, and CO_3^{2-}, and also for sulphates. The calcium content appears to be one of the key parameters in the treatment of real EAFD wastewater, not only for ED but also for sorption using LDH.

Table 4. Ion removal efficiency in S2 solution.

Ionic Species	Symbol	Ion Removal (%) S2 Solution
Sulphate	SO_4^{2-}	76.98 ± 2.95
Sodium	Na^+	97.85 ± 0.21
Potassium	K^+	98.86 ± 0.27
Carbonates	CO_3^{2-}	98.31 ± 0.45
Zinc	Zn^+	>90%
Lead	Pb^{2+}	>93%

From the comparison of the calculated EC values for S1 (38.55 kWh·m^{-3}), RW (52.13 kWh·m^{-3}), and S2 (61.36 kWh·m^{-3}), it is evident that purifying the S1 solution requires less energy than treating the real RW solution. This difference is primarily attributed to the higher applied current density for RW and the higher impurity content in the RW solution. Integrating a sorption step before electrodialysis in the two-stage process slightly increases the EC, as observed for S2. On the other hand, the difference is insignificant, and the proposed two-stage process could be proven sustainable, in addition to preventing calcium and sulphate precipitation inside the electrodialysis cell.

Although the specific energy consumption values remain relatively high, several strategies can be adopted to mitigate this drawback. Operating under potentiostatic control constrains the applied potential, thereby limiting energy consumption even when treatment times are extended. In addition, the use of advanced low-resistivity membranes can significantly reduce ohmic losses, enhancing overall process efficiency and lowering the energy demand per cubic meter of the treated solution. Finally, electrodialysis and related electrochemical separation technologies are inherently compatible with decentralized renewable energy sources, such as solar and wind power. Their integration not only reduces dependence on conventional electricity grids but also enhances process sustainability and supports global decarbonization objectives [51].

4. Conclusions

This study demonstrated the successful application of a two-stage treatment system for electric arc furnace dust (EAFD) wastewater and synthetic solutions (S1, S2), combining sorption with Mg-Al layered double hydroxides (LDHs) and subsequent electrodialysis (ED). Sorption tests with synthetic solution (S1) showed high LDH efficiency in removing multivalent ions and heavy metals: Cr and Pb were removed almost immediately with >99.9% efficiency, while Ca^{2+} and SO_4^{2-} achieved >95% removal within 1–12 min at doses of 50–150 g L^{-1}. Cl^- required up to 24 h to reach equilibrium at 150 g L^{-1}. Monovalent ions (Na^+, K^+) were minimally removed, confirming the low LDH affinity. Similar trends were observed in real EAFD wastewater (RW), where Ca^{2+} and SO_4^{2-} were removed >99% within 5 min, Cl^- reached equilibrium after 240 min, and toxic metals were removed almost immediately. Electrodialysis achieved significant demineralization of the synthetic (S1) and real EAFD wastewater (RW) solutions. The electric conductivity of synthetic wastewater (S1) decreased by 98% after 840 min, while real wastewater (RW) decreased by 95% after 240 min. Ion removal efficiency in S1 reached >99% for Cl^-, 99.3% for SO_4^{2-}, and 98.9% for Ca^{2+}, whereas removal efficiency in RW was lower ($Cl^- \approx 83\%$, $Ca^{2+} \approx 73\%$, $SO_4^{2-} < 70\%$) due to competing ions. The worst removable ion in ED for RW was CO_3^{2-}; however, by implementing the sorption process before ED, the efficiency of carbonate removal increased substantially in the S2 solution from 56% to 98%. As a result of the combination of sorption–ED treatment from S2 (simulated wastewater after sorption from RW), >90% removal of Na^{+}, K^{+}, and CO_3^{2-} and reduced Pb and Zn below detection limits were observed. Despite challenges such as pH decrease in the dilute compartment and salt precipitation in the concentrate, the integrated approach was highly effective in producing low-salinity effluents suitable for reuse in hydrometallurgical processes. Overall, the combination of LDH sorption and electrodialysis is an effective strategy for removing heavy metals, Ca^{2+}, SO_4^{2-}, and, partially, Cl^{-} ions from EAFD wastewater. An optimized LDH dosage and contact time significantly affect treatment efficiency.

Future work should focus on the study of the role of specific species, especially Ca or carbonates and sulphate interactions, for such systems and the mechanism of ion removal in ED, optimizing LDH regeneration and reusability, long-term ED membrane performance using real wastewater and proving the proposed two-stage process from a long-term perspective, and strategies to mitigate scaling in concentrates. Extending the system to pilot or industrial scale will be essential for assessing energy efficiency, cost-effectiveness, and environmental impact. Additionally, exploring LDH modifications for improved affinity toward monovalent ions could enhance treatment performance and broaden applicability across diverse wastewater streams.

Author Contributions: Conceptualization, H.H., A.M. and A.M.B.; methodology, H.H., A.M., A.M.B., E.H.R. and T.B.; investigation, H.H., A.M.B., E.H.R. and T.B.; validation, H.H., A.M., Z.T. and A.M.B.; formal analysis, H.H., A.M. and A.M.B.; resources, H.H., A.M. and Z.T.; data curation, H.H., E.H.R. and A.M.; writing—original draft preparation, H.H., E.H.R., T.B. and A.M.B.; writing—review and editing, H.H., A.M., A.M.B., E.H.R., T.B. and Z.T.; visualization, H.H. and Z.T.; supervision, A.M. and A.M.B.; project administration, A.M. and A.M.B.; funding acquisition, A.M. and A.M.B. All authors have read and agreed to the published version of the manuscript.

Funding: This research was funded by the Slovak Research and Development Agency under the contracts No. APVV-23-0051 and No. APVV-14-0591, and Brazilian funding agencies CAPES, CNPq, FINEP, and FAPERGS.

Data Availability Statement: The original contributions presented in this study are included in the article. Further inquiries can be directed to the corresponding author.

Acknowledgments: The authors wish to thank the Slovak Research and Development Agency, and Brazilian funding agencies CAPES, CNPq, FINEP, and FAPERGS, for their support.

Conflicts of Interest: The authors declare no conflicts of interest.

References

1. Takacova, Z.; Piroskova, J.; Miskufova, A.; Vindt, T.; Hezelova, M.; Orac, D. Removal of Impurities from EAFD Ammonium Carbonate Leachate and Upgrading the Purity of Prepared ZnO. *Materials* **2023**, *16*, 5004. [CrossRef]
2. Lin, X.; Peng, Z.; Yan, J.; Li, Z.; Hwang, J.Y.; Zhang, Y.; Li, G.; Jiang, T. Pyrometallurgical Recycling of Electric Arc Furnace Dust. *J. Clean. Prod.* **2017**, *149*, 1079–1100. [CrossRef]
3. Steel Data Viewer-Worldsteel. Available online: https://worldsteel.org/data/steel-data-viewer/ (accessed on 11 June 2025).
4. Kukurugya, F.; Vindt, T.; Havlík, T. Behavior of Zinc, Iron and Calcium from Electric Arc Furnace (EAF) Dust in Hydrometallurgical Processing in Sulfuric Acid Solutions: Thermodynamic and Kinetic Aspects. *Hydrometallurgy* **2015**, *154*, 20–32. [CrossRef]
5. Kirkelund, G.M.; Magro, C.; Guedes, P.; Jensen, P.E.; Ribeiro, A.B.; Ottosen, L.M. Electrodialytic Removal of Heavy Metals and Chloride from Municipal Solid Waste Incineration Fly Ash and Air Pollution Control Residue in Suspension–Test of a New Two Compartment Experimental Cell. *Electrochim. Acta* **2015**, *181*, 73–81. [CrossRef]
6. Ezugbe, E.O.; Rathilal, S. Membrane Technologies in Wastewater Treatment: A Review. *Membranes* **2020**, *10*, 89. [CrossRef]
7. Shahid, M.K.; Mainali, B.; Rout, P.R.; Lim, J.W.; Aslam, M.; Al-Rawajfeh, A.E.; Choi, Y. A Review of Membrane-Based Desalination Systems Powered by Renewable Energy Sources. *Water* **2023**, *15*, 534. [CrossRef]
8. Hidalgo, A.M.; Charcosset, C. Classical and Recent Developments of Membrane Processes for Desalination and Natural Water Treatment. *Membranes* **2022**, *12*, 267. [CrossRef] [PubMed]

9. Aljumaily, M.M.; Alshami, A.W.; Ismael, B.H.; Hameed, M.M.; AlOmar, M.K.; Hussain, I.R.; Hameed, M.S.; Alsalhy, Q.F.; Alsaadi, M.A. A Review On Membrane Desalination Process in Water Treatment. *IOP Conf. Ser. Earth Environ. Sci.* **2022**, *1120*, 012035. [CrossRef]
10. Li, Y.; Yang, Z.; Yang, K.; Wei, J.; Li, Z.; Ma, C.; Yang, X.; Wang, T.; Zeng, G.; Yu, G.; et al. Removal of Chloride from Water and Wastewater: Removal Mechanisms and Recent Trends. *Sci. Total Environ.* **2022**, *821*, 153174. [CrossRef]
11. Dron, J.; Dodi, A. Comparison of Adsorption Equilibrium Models for the Study of CL^-, $NO3^-$ and $SO_4{}^{2-}$ Removal from Aqueous Solutions by an Anion Exchange Resin. *J. Hazard. Mater.* **2011**, *190*, 300–307. [CrossRef]
12. Hilal, N.; Kochkodan, V.; Al Abdulgader, H.; Mandale, S.; Al-Jlil, S.A. A Combined Ion Exchange–Nanofiltration Process for Water Desalination: I. Sulphate–Chloride Ion-Exchange in Saline Solutions. *Desalination* **2015**, *363*, 44–50. [CrossRef]
13. Darracq, G.; Baron, J.; Joyeux, M. Kinetic and Isotherm Studies on Perchlorate Sorption by Ion-Exchange Resins in Drinking Water Treatment. *J. Water Process Eng.* **2014**, *3*, 123–131. [CrossRef]
14. Dutta, R.; Ahmed, S.P.; Dolui, S.; Chandra Ray, B. Desalination of Oil Field Produced Water Using Ion Exchange System: As a Remediation Environmental Hazard. *Nat. Hazards Res.* **2024**, *4*, 596–603. [CrossRef]
15. Hashem, E.; El-Halim, A.; El-Gayar, D.A.; Farag, H.A. Treatment of Wastewater by Ion Exchange Resin Using a Pulsating Disc. *Desalination Water Treat.* **2020**, *193*, 133–141. [CrossRef]
16. Zhou, D.; Zhu, L.; Fu, Y.; Zhu, M.; Xue, L. Development of Lower Cost Seawater Desalination Processes Using Nanofiltration Technologies—A Review. *Desalination* **2015**, *376*, 109–116. [CrossRef]
17. Fu, C.; Li, F.; Li, H.; Yi, X.; Xie, B.; Fu, C.; Li, F.; Li, H.; Yi, X. Performance Study on Brackish Water Desalination Efficiency Based on a Novel Coupled Electrodialysis–Reverse Osmosis (EDRO) System. *Water* **2024**, *16*, 794. [CrossRef]
18. Zhao, X.; Zheng, Y.; Zhu, Q.; Cheng, M.; Zhang, Y.; Zhou, Z. Removal of Chloride Ions from Aqueous Phase as Layered Double Hydroxides via Uptake Strategies: A Critical Review. *Desalination* **2025**, *595*, 118317. [CrossRef]
19. Wang, Q.; Zhao, H.; Tian, Z.; Zhao, J.; Shen, X.; Lu, L. Variations in Chloride Ion Sorption within Layered Double Hydroxides Engineered with Different Cation Types. *Appl. Surf. Sci.* **2024**, *655*, 159610. [CrossRef]
20. Chen, Q.B.; Wang, J.; Liu, Y.; Zhao, J.; Li, P. Novel Energy-Efficient Electrodialysis System for Continuous Brackish Water Desalination: Innovative Stack Configurations and Optimal Inflow Modes. *Water Res.* **2020**, *179*, 115847. [CrossRef]
21. Alrashidi, A.; Aleisa, E.; Alshayji, K. Life Cycle Assessment of Hybrid Electrodialysis and Reverse Osmosis Seawater Desalination Systems. *Desalination* **2024**, *578*, 117448. [CrossRef]
22. Monat, L.; Liu, R.; Elimelech, M.; Nir, O. Integrating Divalent-Selective Electrodialysis in Brackish Water Desalination. *Environ. Sci. Technol. Lett.* **2024**, *11*, 172–178. [CrossRef]
23. Rotta, E.H.; Bitencourt, C.S.; Marder, L.; Bernardes, A.M. Phosphorus Recovery from Low Phosphate-Containing Solution by Electrodialysis. *J. Memb. Sci.* **2019**, *573*, 293–300. [CrossRef]
24. Benvenuti, T.; Krapf, R.S.; Rodrigues, M.A.S.; Bernardes, A.M.; Zoppas-Ferreira, J. Recovery of Nickel and Water from Nickel Electroplating Wastewater by Electrodialysis. *Sep. Purif. Technol.* **2014**, *129*, 106–112. [CrossRef]
25. Mahendra, C.; Satya Sai, P.M.; Anand Babu, C. Current–Voltage Characteristics in a Hybrid Electrodialysis–Ion Exchange System for the Recovery of Cesium Ions from Ammonium Molybdophosphate-Polyacrylonitrile. *Desalination* **2014**, *353*, 8–14. [CrossRef]
26. Mitko, K.; Rosiński, A.R.; Turek, M. Energy Consumption in Membrane Capacitive Deionization and Electrodialysis of Low Salinity Water. *Desalination Water Treat.* **2021**, *214*, 294–301. [CrossRef]
27. Mokhtar, M.; Inayat, A.; Ofili, J.; Schwieger, W. Thermal Decomposition, Gas Phase Hydration and Liquid Phase Reconstruction in the System Mg/Al Hydrotalcite/Mixed Oxide: A Comparative Study. *Appl. Clay Sci.* **2010**, *50*, 176–181. [CrossRef]
28. Santos, P.S.; Santos, H.S.; Toledo, S.P. Standard Transition Aluminas. Electron Microscopy Studies. *Mater. Res.* **2000**, *3*, 104–114. [CrossRef]
29. Mališová, M.; Horňáček, M.; Mikulec, J.; Hudec, P.; Jorík, V. FTIR Study of Hydrotalcite. *Acta Chim. Slovaca* **2018**, *11*, 147–166. [CrossRef]
30. Zhang, X.; Huestis, P.L.; Pearce, C.I.; Hu, J.Z.; Page, K.; Anovitz, L.M.; Aleksandrov, A.B.; Prange, M.P.; Kerisit, S.; Bowden, M.E.; et al. Boehmite and Gibbsite Nanoplates for the Synthesis of Advanced Alumina Products. *ACS Appl. Nano Mater.* **2018**, *1*, 7115–7128. [CrossRef]
31. Inside Front Cover (Ed Board). *Spectrochim. Acta A Mol. Biomol. Spectrosc.* **2014**, *123*, IFC. [CrossRef]
32. Theiss, F.; López, A.; Frost, R.L.; Scholz, R. Spectroscopic Characterisation of the LDH Mineral Quintinite Mg4Al2(OH)12CO3·3H2O. *Spectrochim. Acta A Mol. Biomol. Spectrosc.* **2015**, *150*, 758–764. [CrossRef]
33. Busca, G. The Surface of Transitional Aluminas: A Critical Review. *Catal. Today* **2014**, *226*, 2–13. [CrossRef]
34. Frost, R.L.; Scholz, R.; López, A.; Theiss, F.L. Vibrational Spectroscopic Study of the Natural Layered Double Hydroxide Manasseite Now Defined as Hydrotalcite-2H–$Mg_6A_{12}(OH)_{16}[CO_3]\cdot 4H_2O$. *Spectrochim. Acta A Mol. Biomol. Spectrosc.* **2014**, *118*, 187–191. [CrossRef] [PubMed]

35. Jauffret, G.; Morrison, J.; Glasser, F.P. On the Thermal Decomposition of Nesquehonite. *J. Therm. Anal. Calorim.* **2015**, *122*, 601–609. [CrossRef]
36. Van, K.V.; Habashi, F. Identification and Thermal Stability of Copper(I) Sulfate. *Can. J. Chem.* **2011**, *50*, 3872–3875. [CrossRef]
37. Babilas, D.; Dydo, P. Zinc Salt Recovery from Electroplating Industry Wastes by Electrodialysis Enhanced with Complex Formation. *Sep. Sci. Technol.* **2020**, *55*, 2250–2258. [CrossRef]
38. Kameda, T.; Yoshioka, T.; Hoshi, T.; Uchida, M.; Okuwaki, A. Treatment of Hydrochloric Acid with Magnesium–Aluminum Oxide at Ambient Temperatures. *Sep. Purif. Technol.* **2006**, *51*, 272–276. [CrossRef]
39. Horváthová, H.; Miškufová, A.; Takáčová, Z.; Bernardes, A.M.; Bureš, R.; Fáberová, M.; Oráč, D. Simultaneous Removal of Chlorides and Calcium from EAF Dust Wastewater. *Minerals* **2025**, *15*, 239. [CrossRef]
40. Xavier, L.D.; Yokoyama, L.; de Oliveira, V.R.; Ribeiro, G.T.; Araújo, O. The Role of Coagulation-Flocculation in the Pretreatment of Reverse Osmosis in Power Plant. *J. Sustain. Dev. Energy Water Environ. Syst.* **2020**, *8*, 118–131. [CrossRef]
41. Rojas, R. Layered double hydroxides applications as sorbents for environmental remediation. In *Hydroxides: Synthesis, Types and Applications*; Nova Science Publishers, Inc.: Hauppauge, NY, USA, 2012; Chapter 2; pp. 39–71. ISBN 978-1-62081-021-7. Available online: https://www.researchgate.net/publication/230634843_Layered_double_hydroxides_applications_as_sorbents_for_environmental_remediation#fullTextFileContent/ (accessed on 10 May 2025).
42. Vaccari, A. Layered Double Hydroxides: Present and Future: V. Rives (Ed.), Nova Science Publishers, Inc., New York, 2001, IX+439 Pp., ISBN 1-59033-060-9. *Appl. Clay Sci.* **2002**, *22*, 75–76. [CrossRef]
43. Káňavová, N.; Machuča, L.; Tvrzník, D. Determination of Limiting Current Density for Different Electrodialysis Modules. *Chem. Pap.* **2014**, *68*, 324–329. [CrossRef]
44. Nikonenko, V.V.; Kovalenko, A.V.; Urtenov, M.K.; Pismenskaya, N.D.; Han, J.; Sistat, P.; Pourcelly, G. Desalination at Overlimiting Currents: State-of-the-Art and Perspectives. *Desalination* **2014**, *342*, 85–106. [CrossRef]
45. Chandramowleeswaran, M.; Palanivelu, K. Treatability Studies on Textile Effluent for Total Dissolved Solids Reduction Using Electrodialysis. *Desalination* **2006**, *201*, 164–174. [CrossRef]
46. Moura Bernardes, A.; Zoppas Ferreira, J.; Siqueira Rodrigues, M.A. *Electrodialysis and Water Reuse: Novel Approaches*; Springer: Berlin/Heidelberg, Germany, 2014; pp. 1–144. [CrossRef]
47. Lee, H.J.; Strathmann, H.; Moon, S.H. Determination of the Limiting Current Density in Electrodialysis Desalination as an Empirical Function of Linear Velocity. *Desalination* **2006**, *190*, 43–50. [CrossRef]
48. Marder, L.; Sulzbach, G.O.; Bernardes, A.M.; Zoppas Ferreira, J. Removal of Cadmium and Cyanide from Aqueous Solutions through Electrodialysis. *J. Braz. Chem. Soc.* **2003**, *14*, 610–615. [CrossRef]
49. Kabay, N.; Arar, Ö.; Samatya, S.; Yüksel, Ü.; Yüksel, M. Separation of Fluoride from Aqueous Solution by Electrodialysis: Effect of Process Parameters and Other Ionic Species. *J. Hazard. Mater.* **2008**, *153*, 107–113. [CrossRef] [PubMed]
50. Andreeva, M.A.; Gil, V.V.; Pismenskaya, N.D.; Dammak, L.; Kononenko, N.A.; Larchet, C.; Grande, D.; Nikonenko, V.V. Mitigation of Membrane Scaling in Electrodialysis by Electroconvection Enhancement, PH Adjustment and Pulsed Electric Field Application. *J. Memb. Sci.* **2018**, *549*, 129–140. [CrossRef]
51. Cournoyer, A.; Bazinet, L. Electrodialysis Processes an Answer to Industrial Sustainability: Toward the Concept of Eco-Circular Economy?—A Review. *Membranes* **2023**, *13*, 205. [CrossRef]

water

MDPI

Article

Optimization of *Auxenochlorella pyrenoidosa* Photobioreactor Parameters for Low Carbon-to-Nitrogen Ratio Wastewater Treatment: Effects of Inoculum Density, Aeration, Light Intensity, and Photoperiod

Lin Zhao [1,*], Yuwei Xu [1], Tian Tian [1], Yifan Zhang [1], Guanqin Huang [2] and Jun Tang [1]

[1] Anhui Province Key Laboratory of Pollution Damage and Biological Control for Huaihe River Basin, College of Biology and Food Engineering, Fuyang Normal University, Fuyang 236037, China

[2] Shenzhen Collaborative Innovation Public Service Platform for Marine Algae Industry, College of Life Sciences and Oceanography, Shenzhen University, Shenzhen 518060, China

* Correspondence: linzhao@fynu.edu.cn

Abstract

Treating wastewater with a low carbon-to-nitrogen (C/N) ratio remains a major challenge for conventional biological processes because insufficient organic carbon limits heterotrophic denitrification. To address this issue, microalgae-based photobioreactors offer a sustainable alternative that couples nutrient removal with biomass valorization. This study systematically evaluated the effects of four key operational parameters—initial inoculum density, aeration rate, light intensity, and photoperiod—on nutrient removal, biomass productivity, and metabolite accumulation of *Auxenochlorella pyrenoidosa* (*A. pyrenoidosa*) treating synthetic low C/N wastewater. Optimal operating conditions were identified as an initial OD_{680} of 0.1, aeration rate of 2 L air min^{-1}, light intensity of 112 μmol m^{-2} s^{-1}, and a 16L:8D photoperiod. Under these conditions, the photobioreactor achieved 86.35% total nitrogen and 98.43% total phosphorus removal within 11 days while producing biomass rich in proteins, polysaccharides, and lipids. Metagenomic analysis revealed a metabolic transition from denitrification-driven pathways during early operation to assimilation-dominated nitrogen metabolism under optimized conditions, emphasizing the synergistic interactions within algal–bacterial consortia. These findings demonstrate that optimized *A. pyrenoidosa*-based photobioreactors can effectively recover nutrients and produce valuable biomass, offering a viable and sustainable solution for the treatment of low C/N wastewater.

Keywords: wastewater treatment; algal-bacterial consortium; nutrient assimilation; metagenome; functional genes

1. Introduction

The rapid expansion of global population and industrialization has placed mounting pressure on water resources and wastewater management systems. A particular challenge lies in the treatment of effluents with low carbon-to-nitrogen (C/N) ratios, commonly originating from advanced biological nitrogen removal processes in municipal wastewater treatment plants [1]. These effluents contain insufficient organic carbon to sustain conventional heterotrophic denitrification [2], necessitating alternative treatment strategies that are both environmentally sustainable and economically viable. Recent studies have highlighted that low-C/N wastewater treatment requires integrated biological processes capable of coupling nutrient removal with biomass recovery. For instance, Satiro et al. [3] demonstrated in a pilot-scale high-rate pond system that inoculation of a microalgae–bacteria consortium significantly improved nitrogen and phosphorus removal, providing practical evidence of the role of initial inoculum and community composition in driving system performance. This underscores the importance of operational and biological factors in shaping overall treatment efficiency.

Among the proposed solutions, microalgae-based photobioreactors have attracted significant attention for their dual role in nutrient removal and biomass valorization. Microalgae such as *Chlorella* spp. can directly assimilate inorganic nitrogen and phosphorus, eliminating the need for an external organic carbon [4]. Additionally, photosynthetic CO_2 fixation during algal growth offers potential for carbon sequestration, enhancing the sustainability of algal biotechnology [5]. In particular, *Chlorella* species exhibit robust adaptability, high nutrient uptake efficiency, and tolerance to variable wastewater matrices [4,6,7]. Prior pilot-scale work has verified the feasibility of such systems for integrated nutrient removal and biomass production [8]. The industrial scalability and real-world relevance of algal-based treatment technologies have also been demonstrated in hybrid systems. Recent reported that integrating photobioreactors with constructed wetlands effectively treated complex paper-pulp industry effluents, achieving simultaneous nutrient recovery and effluent polishing [9]. These advances reinforce the potential of photobioreactors as core units in circular, nature-based wastewater management frameworks.

However, under low C/N conditions, limited organic carbon severely constrains cellular carbon flux and alters the metabolic balance between autotrophic, heterotrophic, and mixotrophic pathways. Insufficient carbon skeletons restrict the tricarboxylic acid cycle and amino acid biosynthesis, leading to reduced protein formation and a compensatory shift toward energy storage compounds such as lipids or carbohydrates. This imbalance disrupts the coordination between photosynthetic carbon fixation and nitrogen assimilation, ultimately diminishing growth and nutrient removal efficiency. Such conditions also intensify metabolic competition between algal and bacterial partners in mixed consortia, as both rely on overlapping carbon sources for growth and redox balance.

The interplay of key physical parameters—including aeration rate, light intensity, photoperiod, and inoculum density—plays a decisive role in mitigating these carbon-limiting effects. Aeration influences oxygen transfer, CO_2 dissolution, and microbial respiration; light regulates energy supply and photosynthetic activity; while inoculum density governs self-shading, nutrient competition, and community stability. Prior studies have examined these parameters independently, yet few have comprehensively investigated their coupled effects under carbon-deficient conditions. Gao et al. [10] demonstrated that *Chlorella* growth and lipid accumulation vary with the total organic carbon to total nitrogen (TN) ratio, reinforcing the direct link between carbon availability and metabolic partitioning. Likewise, Diaz-Hernandez et al. [11] used data-driven modeling to predict optimal photobioreactor performance, underscoring the need for holistic parameter optimization. Furthermore, Chu et al. [12] recently showed that regulating operational symmetry between microalgae and bacteria in a membrane photobioreactor markedly enhanced nitrogen and phosphorus removal in low-C/N domestic wastewater while reducing antibiotic resistance gene propagation. Their findings emphasize that optimized physical parameters (light intensity, aeration, and photoperiod) can control microbial interactions and maintain consortium stability—principles directly relevant to the current study's optimization strategy.

From a metabolic standpoint, low C/N stress amplifies competition for carbon skeletons between energy storage (lipid synthesis) and biomass growth (protein formation). This trade-off dictates whether assimilated nutrients are directed toward structural cell growth or stored macromolecules. Fine-tuning external conditions can steer this metabolic balance toward efficient nutrient assimilation and balanced biomass composition rather than excessive lipid accumulation. Understanding this metabolic plasticity provides the theoretical foundation for the multi-parameter optimization strategy employed herein. Light, in particular, is a dominant regulatory factor in algal–bacterial systems. Wang et al. [13] demonstrated that light intensity directly shapes community structure and metabolic pathways in algal–bacterial biofilm reactors, influencing both nitrogen conversion and photosynthetic capacity. Furthermore, recent proteomic and transcriptomic studies have revealed the molecular mechanisms underlying these physiological shifts. For instance, *Chlorella vulgaris* under nitrogen starvation exhibits upregulation of lipid biosynthesis genes while downregulating photosystem II and ribosomal protein pathways, reflecting reallocation of carbon and energy toward lipid accumulation [14, 15]. Similarly, proteomic analyses of *Auxenochlorella protothecoides* (*A. protothecoides*) and *Chromochloris zofingiensis* (*C. zofingiensis*) have revealed the upregulation of triacylglycerol (TAG)–synthesizing enzymes and carbon fixation–related proteins under nutrient-limited conditions. In *A. protothecoides*, nitrogen starvation significantly upregulated diacylglycerol acyltransferase genes (ApDGAT1 and ApDGAT2b), which was closely correlated with enhanced TAG accumulation [16]. Likewise, in *C. zofingiensis*, nitrogen deprivation stimulated extensive TAG synthesis, accompanied by the enrichment of lipid metabolism–associated proteins within lipid droplets, including major lipid droplet proteins, caleosins, and lipases that facilitate TAG assembly [17]. These molecular insights provide a robust physiological basis for understanding carbon–nitrogen metabolic interactions in microalgae, reinforcing their suitability as model organisms for low C/N wastewater treatment.

Within algal–bacterial consortia, operational conditions further influence these molecular processes. Aeration and mixing affect oxygen dynamics and CO_2 mass transfer, modulating bacterial nitrification–denitrification and algal photosynthesis coupling. Light intensity and photoperiod govern energy supply and circadian regulation, while inoculum density shapes the autotrophic–heterotrophic balance. Together, these parameters determine the degree of metabolic synergy within the consortium, ultimately affecting nutrient removal efficiency and biomass productivity under carbon limitation. From an application perspective, integration of optimized photobioreactors into wastewater treatment facilities offers a sustainable path toward nutrient recovery and resource reuse. *Auxenochlorella pyrenoidosa* (*A. pyrenoidosa*) stands out due to its high nutrient assimilation efficiency, stress tolerance, and capacity for simultaneous nutrient removal and value-added biomass generation [18,19]. This aligns with the recent shift toward nature-based engineered systems for industrial and municipal effluents [9], highlighting the relevance of optimizing operational parameters for scalability and environmental resilience.

This study addresses this knowledge gap by testing the hypothesis that under low C/N stress, microalgal cells experience intensified competition for carbon skeletons between energy storage (lipid synthesis) and biomass growth (protein formation). We further hypothesize that optimizing key operational parameters—including light intensity, aeration rate, inoculum density, and photoperiod—can modulate this metabolic balance, steering nutrient assimilation pathways and determining biomass composition and productivity. To validate this hypothesis, we systematically evaluate the interactive effects of these parameters on nutrient removal, biomass yield, and metabolite accumulation of *A. pyrenoidosa* treating synthetic low C/N wastewater. The results provide mechanistic insights into how environmental regulation can alleviate carbon limitation, enhance nutrient recovery, and promote sustainable wastewater treatment.

2. Materials and Methods

2.1. Microalgae Culture and Photobioreactors Construction

In this study, *Auxenochlorella pyrenoidosa* (*A. pyrenoidosa*, formerly *Chlorella pyrenoidosa* [20]) was obtained from the Algal Germplasm Bank at Shenzhen University. Seed cultures were maintained in Blue-Green 11 (BG-11) medium (250 mL Erlenmeyer flasks with 100 mL working volume) at 25 ± 1 °C under a 12L:12D photoperiod and 83 μmol m^{-2} s^{-1} illumination from daylight fluorescent lamps. Scaled-up cultivation was performed in 1 L sterile glass bottles containing 800 mL BG-11, continuously aerated with filtered air through electromagnetic pumps. Log-phase cultures (The optical density measured at a wavelength of 680 nm (OD_{680}) ≈ 1.5) were harvested by centrifugation (5000 rpm, 5 min, 4 °C), washed twice with ultrapure water, and resuspended in synthetic wastewater to the desired OD_{680}. To ensure realistic microbial interactions, all operations were intentionally conducted under non-sterile conditions, following the rationale of Satiro et al. [3] and Chu et al. [12], who emphasized that open or semi-open algal–bacterial systems better replicate environmental consortia and enable natural co-development of heterotrophic bacteria.

The photobioreactor consisted of transparent acrylic cylinders (20 cm diameter, 30 cm height), equipped with three adjustable 15 W LED tubes. Light intensity and photoperiod were monitored using an MQ-500 light meter (Apogee Instruments, Logan, UT, USA) and controlled by a programmable timer. Aeration was provided from the bottom via diffusers connected to glass rotameters and electromagnetic compressors (ACO-318, HAILEA, Raoping, China), ensuring homogeneous mixing and CO_2 supply.

2.2. Operating Conditions

Photobioreactors were operated indoors under controlled temperature (25 ± 1 °C) to evaluate the effects of initial inoculum density, aeration rate, light intensity, and photoperiod on system performance. Synthetic wastewater simulating low C/N effluent was prepared by dissolving analytical-grade NH_4Cl, $NaNO_3$, KH_2PO_4, K_2HPO_4, and glucose in tap water, yielding ≈ 5 mg L^{-1} NH_4^+–N, 10 mg L^{-1} NO_3^-–N, 1 mg L^{-1} TP, and 30 mg L^{-1} COD. Although typical secondary effluents usually contain low ammonia and relatively high nitrate concentrations, the inclusion of a measurable NH_4^+–N fraction (5 mg/L) in this study was intentional. This adjustment aimed to simulate post-denitrification or partial nitrification effluents, where residual ammonium frequently coexists with nitrate, providing a more representative nitrogen spectrum for assessing algal–bacterial nitrogen transformation dynamics. Similar synthetic wastewater compositions have been employed in previous laboratory-scale studies investigating nitrogen metabolism in photobioreactor and algal–bacterial systems [12,21]. Each reactor contained 5.0 L working volume at start-up. The algal suspension was evenly distributed into triplicate reactors at target OD_{680} values (0.1, 0.2, 0.4) according to experimental design. Aeration rates (1, 2, 4 L min^{-1}) were set using calibrated rotameters (LZB-3WB, Xinghua Xiangjin Flow Instrument Co., Ltd., Taizhou, Jiangsu, China). Light intensities (28 ± 1, 68 ± 4, 112 ± 4 μmol m^{-2} s^{-1}) and photoperiods (8L:16D, 12L:12D, 16L:8D) were controlled by LED timers. The parameter effects were examined using a one-factor-at-a-time approach to isolate independent impacts under constant conditions. The chosen illumination range and photoperiod were supported by Wang et al. [13], who demonstrated that light intensity between 30 and 120 μmol m^{-2} s^{-1} optimizes photosynthetic activity and nitrogen removal in algal–bacterial reactors. Similarly, Satiro et al. [9] validated controlled photoperiod operation (12L:12D and 16L:8D) in industrial wastewater photobioreactors, providing an engineering precedent. The aeration rates of 1, 2, and 4 L min^{-1} were chosen based on their influence on CO_2 availability and mass transfer efficiency. Previous studies have highlighted that moderate aeration rates (1–4 L min^{-1}) optimize algal growth and nutrient removal, avoiding excessive CO_2 stripping while maintaining sufficient oxygen supply for photosynthesis [22–24].

Although the wastewater medium was prepared using tap water and analytical-grade reagents, no sterilization was performed prior to inoculation. Continuous aeration through 0.22 μm filters minimized particle entry while allowing native microorganisms to colonize the system. Similar approaches have been reported in previous studies investigating algal–bacterial photobioreactors operated under open or semi-open conditions [21,25]. This design choice, mirroring Chu et al. [12], facilitates the spontaneous formation of algal–bacterial consortia typical of real treatment plants and supports dynamic co-metabolic processes such as nitrification and denitrification. The detailed construction and operation of the photobioreactor are described in our previously published study [26]. However, unlike that study, which focused on the microalga *Tetradesmus obliquus*, the present work investigates *Auxenochlorella pyrenoidosa* as the functional strain. The two species differ markedly in cellular morphology, metabolic regulation, and nitrogen assimilation capacity. Moreover, all data presented in this study are newly obtained, and the work specifically highlights the optimization of operational parameters and the metabolic transitions associated with algal–bacterial interactions under low C/N conditions, thereby providing new mechanistic insights distinct from our previous publication.

2.3. *Sampling and Analytical Methods*

2.3.1. Physicochemical Analysis

Daily, 100 mL of mixed liquor was collected from each reactor. Fifty milliliters were filtered (0.45 μm) for analysis of NH_4^+–N, NO_2^-–N, NO_3^-–N, TN, TP, and COD using Chinese Standard Methods [27]; the remaining portion was used to determine dry weight and chlorophyll *a* content (via freeze–thaw acetone extraction). OD_{680} was measured daily with a UV-5100 spectrophotometer (Metash, Shanghai, China). All tests were performed in triplicate. Dissolved organic carbon (DOC) was quantified by high-temperature catalytic oxidation (680 °C) using a Shimadzu TOC-LCPH/CPN analyzer (Kyoto, Japan). pH, DO, and conductivity were monitored in situ (Orion Star A329, Thermo Fisher, Waltham, MA, USA).

2.3.2. Biomass and Biochemical Composition

A of algal suspension were filtered (0.45 μm membrane). The filter was folded and stored at −20 °C in darkness. Chlorophyll *a* was extracted in 90% acetone and quantified spectrophotometrically [28]. Dry weight was measured gravimetrically after drying filters at 65 °C to constant weight [29]. At the end of each run, biomass was centrifuged (4000 rpm, 5 min), oven-dried (60 °C), and ground. Protein content was determined via BCA assay (TransGen Biotech, Beijing, China) [30]; carbohydrates via phenol–sulfuric acid method [31]; lipids by chloroform–methanol extraction [32]. Specific growth rate (μ, d^{-1}) of *T. obliquus* was calculated according to Equation (1):

$$\mu = (\ln N_2 - \ln N_1)/(T_2 - T_1) \quad (1)$$

where N_1 and N_2 represent the biomass (dry weight) o at times T_1 and T_2 (days), respectively.

2.4. Metagenomic Analyses

Biomass samples collected on days 1, 5, and 11 were centrifuged, flash-frozen in liquid nitrogen (60 s), and stored at −80 °C. DNA was extracted and sequenced (Majorbio Bio-pharm Technology Co., Ltd., Shanghai, China). Sequencing integrity was checked via 1% agarose gel electrophoresis, and data were analyzed on the Majorbio Cloud Platform (https://cloud.majorbio.com/page/tools/, accessed on 2 August 2025).

2.5. Data Processing

Differences in nutrient removal, biomass production, and community composition across operational phases were analyzed by one-way ANOVA (SPSS v26.0, IBM, Armonk, NY, USA). Microbial taxonomic variance was evaluated using R (v3.3.1). Sequence clustering (CD-HIT v4.6.1) and KEGG annotation (DIAMOND v0.8.35) were used to calculate relative enzyme and KO abundances for N and P metabolism. Gene abundance was expressed as RPKM. Mantel tests (Spearman method) quantified correlations between pathway gene sets and seven environmental variables (pH, NH_4^+–N, NO_3^-–N, NO_2^-–N, TN, TP, DOC). Differential expression analyses were performed using the stats package in R (v3.3.1).

3. Results and Discussion

3.1. Effects of Different Environmental Factors on the Algal Growth and Nutrient Removals

3.1.1. Initial Inoculation Concentration and Aeration Rates

In this study, the effects of inoculum density and aeration rate on the growth of *A. pyrenoidosa* were evaluated. Parameters such as dry weight, specific growth rate, chlorophyll content, and pH were used to assess the microalgae's response to these conditions (Figure 1). As shown in Figure 1a, both initial inoculum density and aeration rate significantly affected biomass accumulation. The initial dry weights were 0.018 g/L (OD_{680} = 0.1), 0.028 g/L (OD_{680} = 0.2), and 0.057 g/L (OD_{680} = 0.4). During cultivation, total biomass increased in all groups, with the highest final dry weight observed at OD_{680} = 0.4 (0.204 g L^{-1} on day 11), followed by OD_{680} = 0.2 (0.168 g L^{-1}) and OD_{680} = 0.1 (0.153 g L^{-1}). However, this higher final biomass mainly reflects the larger initial cell load rather than superior growth performance, as supported by the specific growth rate analysis described below. Aeration also influenced biomass productivity: cultures aerated at 2 and 4 L min^{-1} reached comparable dry weights (~0.168 g L^{-1}), whereas 1 L min^{-1} yielded lower biomass (0.159 g L^{-1}). This suggests that moderate aeration (2 L min^{-1}) ensured adequate CO_2 transfer and mixing without imposing shear stress or unnecessary energy use, indicating an optimal aeration threshold. The specific growth rate data (Figure 1b) revealed distinct physiological responses among inoculum groups. At day 1, the OD_{680} = 0.1 and 0.2 cultures exhibited rapid growth rate of 0.79 d^{-1} and 0.69 d^{-1}, respectively, whereas the OD_{680} = 0.4 culture showed a negative growth rate (−0.13 d^{-1}). The initial decline at high density indicates a lag phase caused by self-shading, oxygen supersaturation, and nutrient competition, which temporarily suppressed photosynthetic activity. After day 4, all groups approached similar apparent rates due to the onset of stationary conditions; however, the OD_{680} = 0.1 culture maintained a higher steady-state growth rate (0.20 d^{-1} on day 11) and more consistent nutrient removal efficiency, indicating sustained metabolic activity compared to high-density cultures. These results collectively indicate that while higher inoculum densities may accelerate early biomass accumulation through a larger initial population, they eventually experience growth inhibition due to restricted light penetration, local nutrient depletion, and oxygen oversaturation [33]. Therefore, the convergence in growth rates after day 4 likely represents a balance between nutrient limitation and self-shading effects across treatments, rather than a uniform physiological response. Moreover, the apparent increase in total biomass at OD_{680} = 0.4 could also partly arise from the proliferation of heterotrophic bacteria supported by organic exudates from stressed algal cells, rather than active algal growth per se. In contrast, lower inoculum densities facilitate uniform light distribution and sustained carbon and nitrogen uptake, thereby prolonging the exponential phase and enhancing long-term system stability.

Chlorophyll *a* content (Figure 1c) followed a similar pattern. The OD_{680} = 0.4 group showed the highest absolute chlorophyll *a* concentration (2.31 mg L^{-1} on day 11), yet its normalized chlorophyll *a* per unit biomass declined in later stages, reflecting a shift from active photosynthesis to maintenance metabolism under nutrient depletion. This supports the interpretation that elevated chlorophyll levels at high inoculum density result from cell concentration rather than higher photosynthetic efficiency. For aeration, higher rates generally enhanced chlorophyll *a* accumulation owing to improved CO_2 availability and gas–liquid mass transfer. At 4 L min^{-1}, chlorophyll *a* reached 2.13 mg L^{-1}, compared with 1.58 mg L^{-1} at 2 L min^{-1} and 2.09 mg L^{-1} at 1 L min^{-1}. Beyond 2 L min^{-1}, further aeration offered negligible benefits, indicating diminishing returns once CO_2 saturation was achieved. This suggests that improved CO_2 availability and enhanced mass transfer under suitable aeration rates contribute to more efficient photosynthesis [34]. The pH data (Figure 1d) increased gradually across all treatments, consistent with CO_2 uptake and bicarbonate formation during photosynthesis. The rise was more pronounced in reactors with higher inoculum densities, indicating higher CO_2 fixation rates early in cultivation. Conversely, cultures with lower aeration exhibited larger pH shifts, likely due to limited CO_2 buffering. These findings align with prior reports that CO_2 concentration modulates key enzymes such as carbonic anhydrase and Rubisco, thereby influencing photosynthetic efficiency [35,36]. As cultivation progressed, the pH values of all experimental groups gradually converged during the later phase. This convergence can be attributed to the stabilization of algal metabolic activity and the dynamic equilibrium established between CO_2 fixation and release as the system approached steady-state conditions. In the early and middle stages, differences in inoculum density and aeration rate caused distinct CO_2 dynamics and pH fluctuations; however, once nutrient availability declined and algal growth slowed, respiration and photosynthetic CO_2 uptake became balanced, leading to similar final pH levels across treatments. This convergence therefore reflects a transition from active exponential growth to a quasi-stationary metabolic equilibrium rather than identical physiological behavior under all conditions.

Taken together, these observations demonstrate that an initial inoculum density of OD_{680} = 0.1, combined with an aeration rate of 2 L min^{-1}, provides the most balanced condition—ensuring sufficient light and nutrient availability while minimizing mass-transfer limitations. This combination supports steady biomass growth, efficient nutrient removal, and prolonged photobioreactor stability under low C/N conditions.

At OD_{680} = 0.2, NH_4^+–N was rapidly reduced to 0.02 mg L^{-1} by day 3, accompanied by a marked decrease in NO_3^-–N from an initial 10.17 mg L^{-1} to 1.02 mg L^{-1} on day 11, resulting in the highest total nitrogen (TN) removal efficiency of 87.54%. In contrast, the OD_{680} = 0.1 group showed delayed nitrification, with NO_2^-–N accumulation reaching 1.39 mg/L on day 11 (Figure 2b). TP removal was fastest at an inoculum density of OD_{680} = 0.2, with a residual concentration of 0.12 mg L^{-1} on day 2, although all groups ultimately achieved removal efficiencies greater than 93%. COD removal was greatest at OD_{680} = 0.1, suggesting that higher inoculum densities may have caused intracellular release. With respect to aeration, higher aeration (4 L min^{-1}) accelerated NH_4^+–N removal, reducing its concentration to 0.48 mg/L by day 2, whereas 1 and 2 L min^{-1} still retained 1.42 and 1.30 mg/L, respectively. NO_2^-–N remained low (0–0.15 mg/L) under low (1 L min^{-1}) and high (4 L min^{-1}) aeration, but in the 2 L min^{-1} group it gradually increased to 0.86 mg/L by day 11. NO_3^-–N consumption was slower at low aeration, while medium and high aeration (2 and 4 L min^{-1}) enhanced nitrogen conversion, resulting in the highest TN removal efficiencies (88.02% and 89.03%). Phosphorus removal was similar across treatments (~0.05 mg/L), whereas COD removal was less effective at 1 and 4 L min^{-1}, with residuals up to 12.33–14.50 mg/L and rebound likely linked to cell lysis or extracellular release [37].

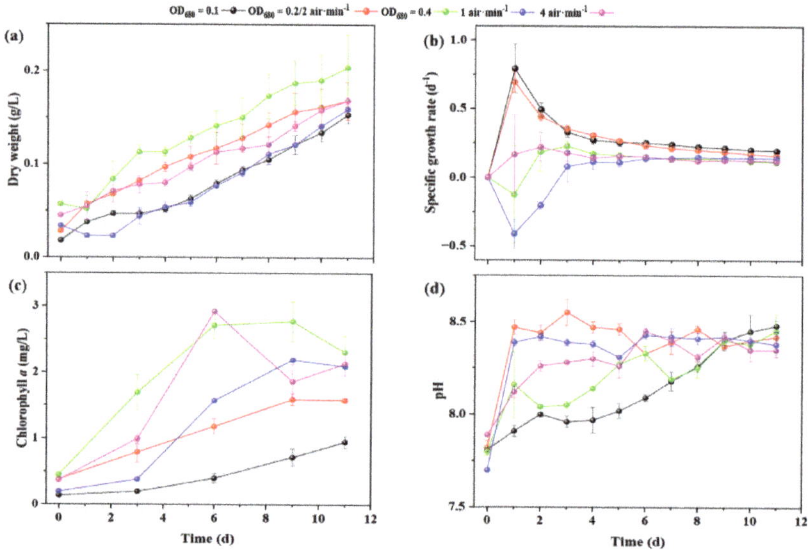

Figure 1. Temporal dynamics of biomass, specific growth rate, chlorophyll *a*, and pH in the algal–bacterial symbiotic system operated under varying initial inoculum densities and aeration rates. (**a**) Variations in dry weight; (**b**) variations in specific growth rate; (**c**) variations in chlorophyll *a* content; (**d**) variations in pH. "Time (d)" denotes the duration of reactor operation in days. The red line corresponds to the treatment with an initial inoculum density of OD_{680} = 0.2 and an aeration rate of 2 L min^{-1}.

It should be noted that at higher aeration rates and under alkaline conditions (pH > 8), a fraction of total ammonia nitrogen may exist as free ammonia (NH_3) gas, which can be stripped from the culture medium due to increased air flow and turbulence [38,39]. Although this physicochemical mechanism was not explicitly considered in the present analysis, it may have contributed partially to the observed reduction in NH_4^+–N, particularly in the 4 L min^{-1} group. Accordingly, future studies should incorporate quantitative evaluation of ammonia volatilization under varying pH and aeration conditions to better distinguish biological assimilation from physicochemical stripping effects.

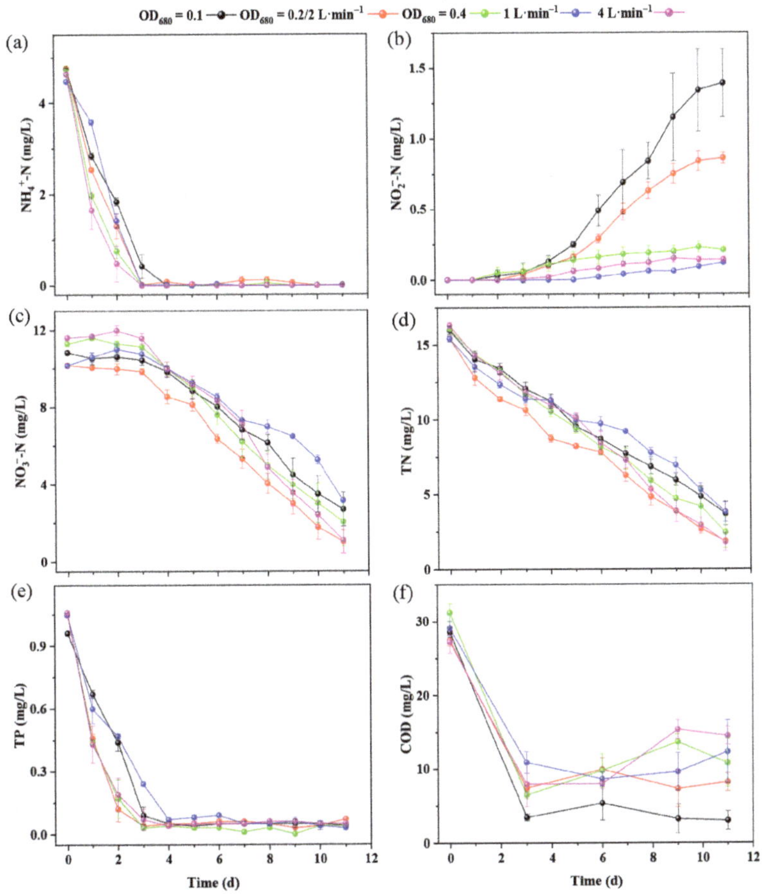

Figure 2. Variations in inorganic nitrogen, TN, TP and COD concentrations in the algae–bacteria symbiotic photobioreactor system under different initial inoculation densities and aeration rates. (**a**) Variations in $NH_4{}^+$–N concentration; (**b**) variations in $NO_2{}^-$–N concentration; (**c**) variations in $NO_3{}^-$–N concentration; (**d**) variations in TN concentration; (**e**) variations in TP concentration; (**f**) variations in COD concentration. "Time (d)" denotes the duration of reactor operation in days. The red line corresponds to the treatment with an initial inoculum density of OD_{680} = 0.2 and an aeration rate of 2 L min^{-1}.

3.1.2. Light Intensity and Photoperiods

In the light intensity experiment, 112 µmol m^{-2} s^{-1} favored biomass accumulation, reaching a dry weight of 0.19 g/L on day 11, approximately 90% and 12% higher than the 28 µmol m^{-2} s^{-1} (0.10 g/L) and 68 µmol m^{-2} s^{-1} (0.17 g/L) treatments, respectively (Figure 3a). The specific growth rate peaked on day 3 in the high light intensity group (0.36 d^{-1}) (Figure 3b), indicating a pronounced stimulatory effect of elevated light intensity during the mid-growth phase. In the photoperiod experiment, the longest light duration (16L:8D) yielded the highest biomass (0.20 g/L on day 11), 2.02- and 1.18-fold higher than the 8L:16D and 12L:12D treatments, respectively. Notably, the 12L:12D group exhibited the highest early-stage specific growth rate (0.69 d^{-1} on day 1), suggesting that a moderate light–dark cycle promotes growth initiation. Chlorophyll *a* content peaked under the 12L:12D photoperiod, reaching 1.58 mg/L on day 11 (Figure 3c), indicating that moderate light intensity combined with a balanced light–dark cycle favors pigment synthesis. pH increased steadily across all treatments, stabilizing between 8.40 and 8.50 (Figure 3d). Overall, high light intensity (112 μmol m^{-2} s^{-1}) and extended illumination (16L:8D) promoted biomass accumulation, whereas moderate light (68 μmol m^{-2} s^{-1}) with a regular light–dark cycle (12L:12D) enhanced pigment content and stabilized growth rates. These trends align with the typical microalgal response of boosting photosynthetic capacity under strong light while avoiding photoinhibition from continuous illumination [40,41].

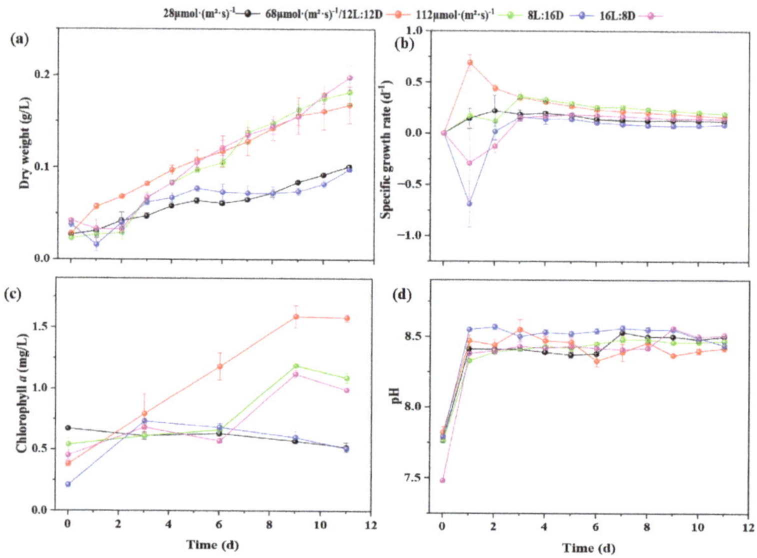

Figure 3. Temporal dynamics of biomass, specific growth rate, chlorophyll *a*, and pH in the algal–bacterial symbiotic system operated under varying light intensities and photoperiods. (**a**) Variations in dry weight; (**b**) variations in specific growth rate; (**c**) variations in chlorophyll *a* content; (**d**) variations in pH. "Time (d)" denotes the reactor operation period in days. The red line corresponds to the treatment condition with a light intensity of 68 μmol m^{-2} s^{-1} and a photoperiod of 12L:12D.

This study systematically examined the combined effects of light intensity and photoperiod on nutrient removal in an *A. pyrenoidosa* photobioreactor treating low C/N wastewater (Figure 4). Under 112 μmol m^{-2} s^{-1}, TN removal was highest, reaching 0.90 mg/L on day 11, with NO_3^-–N fully consumed (0.63 mg/L); however, significant NO_2^-–N accumulation occurred, peaking at 2.53 mg/L on day 8, indicating that high light intensity may exacerbate nitrification imbalance. In contrast, 28 and 68 μmol m^{-2} s^{-1} treatments achieved rapid NH_4^+–N removal without notable NO_2^-–N accumulation, but final TN remained at 9.96 and 1.85 mg/L, respectively, suggesting that low to moderate light may limit coordinated nitrogen metabolism. In the photoperiod experiment, the 12L:12D treatment exhibited the most balanced denitrification, with lower NO_2^-–N accumulation (0.86 mg/L on day 11) and near-complete NH_4^+–N removal by day 3, highlighting that a moderate light–dark cycle stabilizes nitrification–denitrification. Although the 16L:8D group exhibited higher NO_2^-–N accumulation (2.79 mg L^{-1} on day 11), the TN and NO_3^-–N removal rates between the 16L:8D and 12L:12D groups showed no significant differences ($p > 0.05$). TP was effectively removed in all treatments, with concentrations below 0.07 mg/L by day 11, indicating that light conditions had minimal impact on phosphorus removal. However, during the initial phase (0–4 days), TP reduction was slower under the lower light intensity (28 μmol m^{-2} s^{-1}) and milder operating conditions, which can be attributed to lower biomass accumulation and limited microbial activity at the early stage. Under these conditions, both algal and bacterial populations had not yet reached stable growth, leading to reduced phosphorus uptake through assimilation. As the cultures developed and photosynthetic activity increased, the algal–bacterial consortium became metabolically balanced, resulting in the accelerated TP removal observed in the later stage of operation. In contrast, COD removal varied more noticeably among treatments. All groups showed some COD rebound in the later stage; this effect was especially pronounced under 112 μmol m^{-2} s^{-1} and 16L:8D conditions, likely due to cell lysis and the release of dissolved organic matter under high light intensity or prolonged illumination.

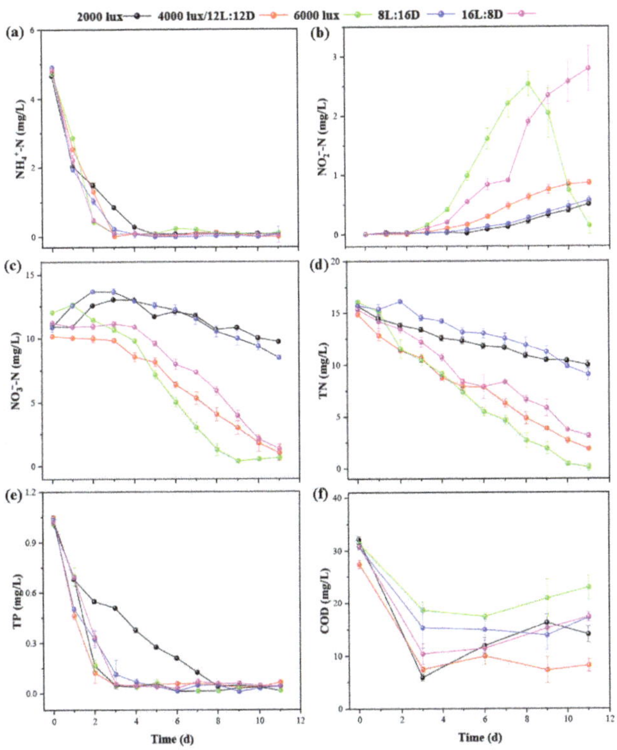

Figure 4. Variations in inorganic nitrogen, TN, TP and COD concentrations in the algae–bacteria symbiotic photobioreactor system under different light intensities and photoperiods. (**a**) Variations in NH_4^+–N concentration; (**b**) variations in NO_2^-–N concentration; (**c**) variations in NO_3^-–N concentration; (**d**) variations in TN concentration; (**e**) variations in TP concentration; (**f**) variations in COD concentration. "Time (d)" denotes the reactor operation period in days. The red line corresponds to the treatment condition with a light intensity of 68 μmol m^{-2} s^{-1} and a photoperiod of 12L:12D.

3.2. Chemical Compositions of Algal Biomass

This study systematically evaluated the effects of operational parameters on protein, polysaccharide, and total lipid production in an *A. pyrenoidosa* photobioreactor treating low C/N wastewater (Figure 5a). A higher initial inoculum (OD_{680} = 0.4) promoted total lipid accumulation, reaching 44.03 mg/L, while protein and polysaccharide yields were also highest at 22.23 mg/L and 17.24 mg/L, respectively, indicating that a higher biomass starting point favors energy-rich compound synthesis. Even at OD_{680} = 0.1, substantial biomass accumulation occurred, with protein and polysaccharide reaching 19.51 mg/L and 14.42 mg/L, slightly lower than the high-inoculum group. Aeration effects were more complex: total lipid production peaked at 4 L min^{-1} (37.98 mg/L), whereas protein and polysaccharide yields were higher at 2 L min^{-1} (17.77 mg/L and 13.83 mg/L), suggesting that moderate oxygen availability favors lipid synthesis, while protein and polysaccharide accumulation depends on a balanced metabolic state provided by intermediate aeration.

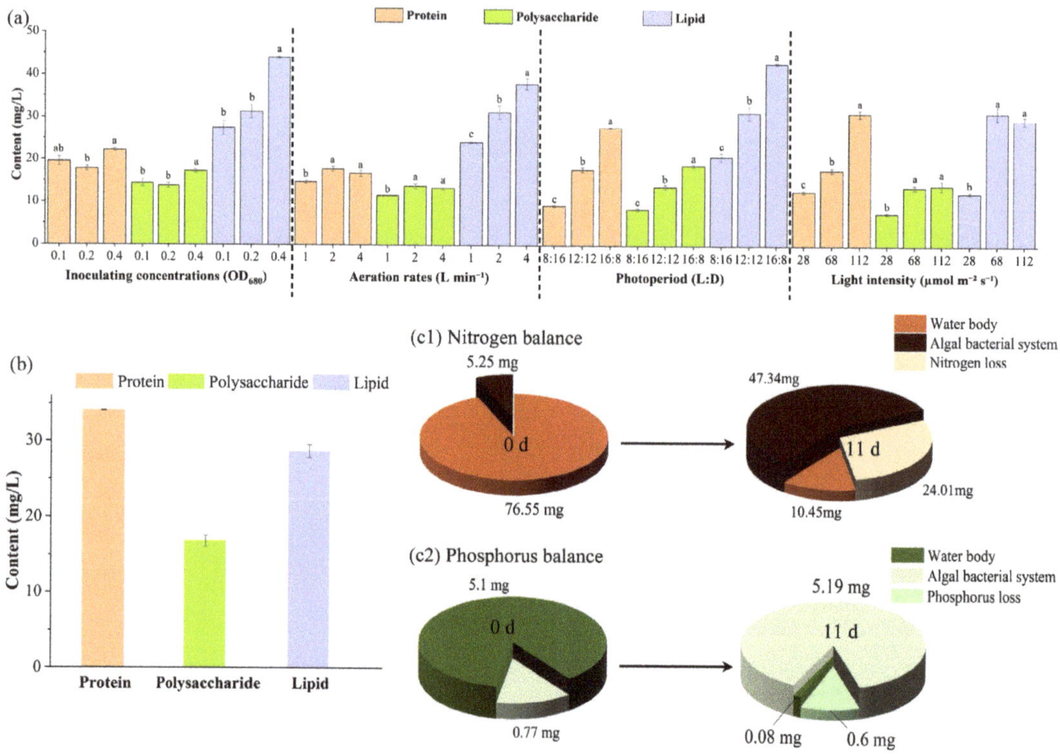

Figure 5. Variations in protein, polysaccharide, and lipid contents, and nitrogen–phosphorus balance analysis in the algal–bacterial symbiotic system under single-factor and optimal operating conditions. (**a**) Changes in protein, polysaccharide, and lipid contents under different initial inoculum densities, aeration rates, and light conditions. (**b**) Temporal variations in active biomass content under optimal operating conditions. (**c1**) Nitrogenmass balance analysis under optimal operating conditions. (**c2**) phosphorus mass balance analysis under optimal operating conditions. Different letters indicate statistically significant differences ($p < 0.05$).

Regarding light conditions, 112 μmol m^{-2} s^{-1} significantly enhanced protein synthesis, reaching 31.18 mg/L ($p < 0.05$), while total lipid and polysaccharide yields were also higher than those under moderate or low light, indicating that high light intensity promotes nitrogenous and energy-storage compound production. Extended illumination (16L:8D) will also significantly increased all target products, with protein, polysaccharide, and total lipid reaching 27.64 mg/L, 18.76 mg/L, and 42.79 mg/L, respectively, suggesting that longer light periods favor overall photosynthetic carbon fixation and energy metabolism. In contrast, low light (28 μmol m^{-2} s^{-1}) and short photoperiods (8L:16D) markedly reduced the yields of all three compounds. These results indicate that higher light intensity and longer illumination enhance the accumulation of bioactive compounds in *A. pyrenoidosa*, likely by supplying sufficient energy to drive carbon flux and biosynthetic pathways [13], whereas inoculum density and aeration modulate product synthesis indirectly by affecting biomass accumulation and oxygen transfer efficiency, consistent with the metabolic characteristics of photoautotrophic microorganisms [22,37].

3.3. Performance of the A. pyrenoidosa Photobioreactor Under Optimal Operating Parameters

Considering both nitrogen and phosphorus removal as well as biomass production, the optimal operating conditions were identified as an initial inoculum of OD_{680} = 0.1, aeration rate of 2 L min^{-1}, light intensity of 112 μmol m^{-2} s^{-1}, and a photoperiod of 16L:8D. Under these conditions, the *A. pyrenoidosa* photobioreactor demonstrated outstanding nitrogen and phosphorus removal performance alongside efficient biomass production during low C/N wastewater treatment (Figure 5(c1,c2)). After 11 days of operation, TN and TP in the water phase decreased sharply, with TN dropping from 76.55 mg to 10.45 mg (86.35% removal) and TP from 5.10 mg to 0.08 mg (98.43% removal), demonstrating the system's excellent nitrogen and phosphorus removal performance. At the same time, a clear mass balance was established between the aqueous and solid phases. Specifically, TN and TP accumulated within the algal–bacterial biomass increased from 5.25 mg to 47.34 mg (an accumulation of 42.09 mg) and from 0.77 mg to 5.19 mg (an accumulation of 4.42 mg), respectively. These increases in biomass-associated nitrogen and phosphorus closely matched the reductions observed in the aqueous phase, confirming that most of the removed nutrients were assimilated into the algal–bacterial biomass rather than lost through volatilization or precipitation. This consistency supports the overall process performance and validates the system's nutrient removal efficiency through mass balance closure. Moreover, the strong correlation between nutrient removal and biomass accumulation highlights the superior inorganic nutrient fixation capacity of *A. pyrenoidosa* under optimized operating conditions. It should be noted that the optimized conditions for low C/N wastewater treatment provide a reliable and efficient strategy for resource recovery. Therefore, the simultaneous removal and conversion of nitrogen and phosphorus into algal–bacterial biomass demonstrate the system's potential as a sustainable and integrated approach for the treatment and resource recovery of low C/N ratio wastewater.

Additionally, high light intensity (112 μmol m^{-2} s^{-1}) and extended illumination (16L:8D) markedly enhanced photosynthetic efficiency and carbon–nitrogen flux, providing sufficient energy and carbon skeletons for protein, polysaccharide, and lipid synthesis. Moderate aeration (2 L min^{-1}) maintained dissolved oxygen levels while avoiding excessive stripping of soluble nitrogen sources. A lower inoculum density (OD_{680} = 0.1) helped prolong the exponential growth phase, improving the conversion efficiency of nutrients into biomass. These findings are consistent with previous reports that microalgae enhance nutrient assimilation and biomass accumulation under high light intensity and extended illumination [13,42,43]. It should be noted that although the system's performance under high C/N wastewater or other wastewater conditions requires further investigation, the results of the present study—particularly those obtained for low C/N wastewater—demonstrate the system's strong potential for resource recovery and nutrient removal. These findings provide a solid foundation for future research to further evaluate and optimize the system's applicability under diverse wastewater characteristics. They confirm that parameter optimization enables the simultaneous treatment of wastewater and production of high-value biomass, providing a reliable strategy for resource recovery from low C/N wastewater.

3.4. Metagenomic Analysis

3.4.1. Microbial Community Structure Changes

Under the optimized operating conditions, the microbial community in the *A. pyrenoidosa* photobioreactor exhibited pronounced dynamics at the phylum level (Figure 6a). At the start of operation (C1), Chlorophyta overwhelmingly dominated (83.7%), indicating rapid niche occupation by the inoculated algae and active photosynthetic carbon fixation. By day 5 (C5), its relative abundance sharply decreased to 6.97%, while Pseudomonadota (46.5%) and Actinomycetota (38.6%) became predominant, suggesting a strong bacterial proliferation phase driven by the increasing availability of algal-released organic exudates. This transient bacterial dominance is a typical feature of early-stage algal–bacterial symbiosis, in which heterotrophic bacteria utilize photosynthate as carbon substrates while regenerating inorganic nitrogen and phosphorus for algal reuse [44]. The subsequent recovery of *Chlorella* by day 11 (C11)—with Chlorophyta abundance increasing to 46.7%—was closely linked to the optimized light and aeration conditions. The moderate light intensity (112 μmol m^{-2} s^{-1}) and 16L:8D photoperiod ensured adequate photosynthetically active radiation without photoinhibition, maintaining effective carbon fixation and oxygen evolution. Simultaneously, the aeration rate of 2 L min^{-1} supplied sufficient CO_2 and prevented oxygen oversaturation, thereby stabilizing pH and creating a favorable redox environment for both algae and aerobic bacteria (*Pseudomonas*, *Sphingopyxis*). This balanced gas exchange supported bacterial mineralization of organic compounds during the mid-phase, which in turn replenished dissolved inorganic nitrogen and phosphorus, facilitating *Chlorella*'s resurgence and biomass accumulation in the late phase. Notably, the moderate increase in Bacillota (4.2%) and the persistence of Uroviricota (0.19%) indicate the activation of spore-forming and viral populations, respectively, which are known to regulate microbial turnover and horizontal gene transfer in photobioreactor systems [45–48]. This observation aligns with recent findings that phage-driven lytic cycles and auxiliary metabolic gene transfer significantly enhance microbial adaptability in nutrient-limited environments [49,50].

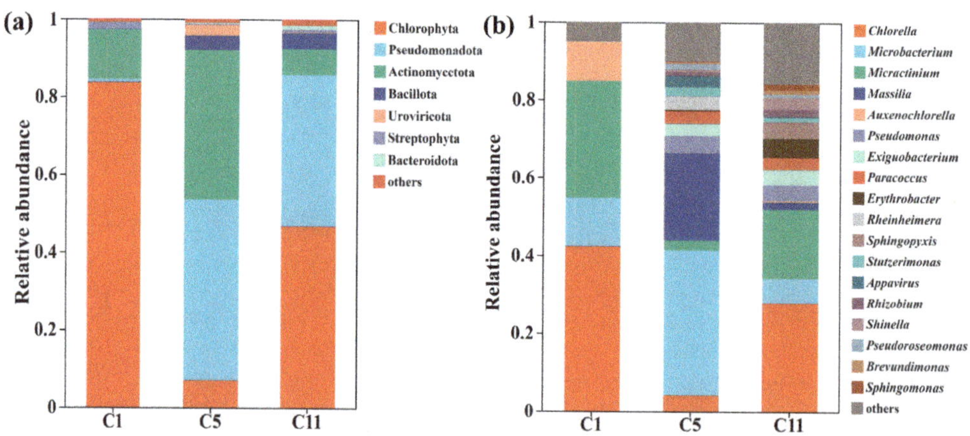

Figure 6. Microbial community composition at the (**a**) phylum level and (**b**) genus level across different operational stages.

At the genus level, community succession revealed distinct ecological shifts (Figure 6b). Initially, *Chlorella* (42.2%) and *Micractinium* (29.9%) dominated, consistent with the algal inoculum composition and early autotrophic growth. By day 5, green algal abundance dropped markedly (*Chlorella* to 4.1%), whereas bacterial genera such as *Microbacterium* (37.2%), *Massilia* (22.5%), and *Pseudomonas* (4.3%) proliferated, reflecting a transition toward bacterial dominance. These genera are known participants in organic matter degradation, nitrification, and heterotrophic denitrification, facilitating nitrogen turnover and maintaining nutrient availability under low C/N conditions [51–53]. By day 11, *Chlorella* recovered to 27.9% and *Micractinium* to 17.6%, accompanied by the persistence of denitrification-associated genera (*Erythrobacter, Sphingopyxis, Paracoccus*) at moderate abundance. These coexisting taxa indicate a stabilized metabolic partnership wherein oxygenic photosynthesis by algae sustained aerobic nitrogen oxidation, while bacterial mineralization and denitrification regenerated nutrients for continued algal assimilation. The observed rebound of *Chlorella* thus reflects the establishment of a cooperative steady state driven by light–aeration–nutrient coupling.

The microbial succession therefore represents a transition from an algae-dominated startup phase to a bacterial proliferation phase and finally to a balanced, mutually reinforcing consortium. The optimized physical parameters fostered a finely tuned exchange of oxygen, dissolved organic carbon, and nitrogen species between algae and bacteria, enabling sustained nutrient removal and biomass production. Such parameter-driven ecological stabilization underscores the feasibility of steering microbial community structure and function through controlled environmental conditions, thereby achieving efficient nutrient recovery and enhanced photobioreactor performance under low C/N constraints.

3.4.2. Correlation Analysis of Environmental Factors and Nitrogen–Phosphorus Metabolism

(1) Correlation between microbe communities and environmental factors

Under optimal operating conditions, microbial community structure showed significant correlations with water-quality parameters (Figure 7), reflecting coordinated metabolic interactions in the algae–bacteria system treating low C/N wastewater. *Chlorella* and *Micractinium* were significantly negatively correlated with pH ($r = -0.83$, $p < 0.01$; $r = -0.88$, $p < 0.01$) and positively correlated with dissolved organic carbon (DOC) ($r = 0.88$, $p < 0.01$; $r = 0.93$, $p < 0.001$), indicates active photosynthetic CO_2 uptake and the concurrent release of organic exudates, which elevate pH and DOC, consistent with algae-driven autotrophic metabolism. Both genera were also correlated positively with NH_4^+–N ($r = 0.68$, $p < 0.05$; $r = 0.73$, $p < 0.05$), supporting their strong NH_4^+–N assimilation capacity that contributes directly to protein synthesis and nitrogen removal.

Several bacterial genera exhibited linkages to nitrogen transformations. *Pseudomonas, Rhizobium, Sphingomonas, Brevundimonas*, and *Stutzerimonas* were negatively correlated with NO_3^-–N, suggesting active denitrification. Notably, the relative abundance of *Sphingopyxis* increased 3.7-fold from phase C1 to C11, coinciding with a 78% decline in NO_3^-–N and 87% reduction in TN, demonstrating its quantitative contribution to nitrate and total-nitrogen removal. Similarly, *Shinella* abundance rose in parallel with TP removal (97.6%), consistent with its reported polyphosphate-accumulating function. Furthermore, pH correlated positively ($r > 0.67$, $p < 0.05$) with denitrification-related genera such as *Pseudomonas* and *Paracoccus*, indicating that slightly alkaline conditions favored bacterial nitrogen metabolism. Algae fix inorganic carbon and assimilate NH_4^+–N through photosynthesis to promote biomass accumulation, while bacteria remove nitrogen and phosphorus to improve water quality, together achieving both wastewater treatment and resource recovery [21,25]. These relationships substantiate a cooperative division of labor: photosynthetic algae fix inorganic carbon and assimilate NH_4^+–N, while heterotrophic bacteria mediate complementary nitrate reduction and phosphorus uptake, jointly enhancing effluent quality.

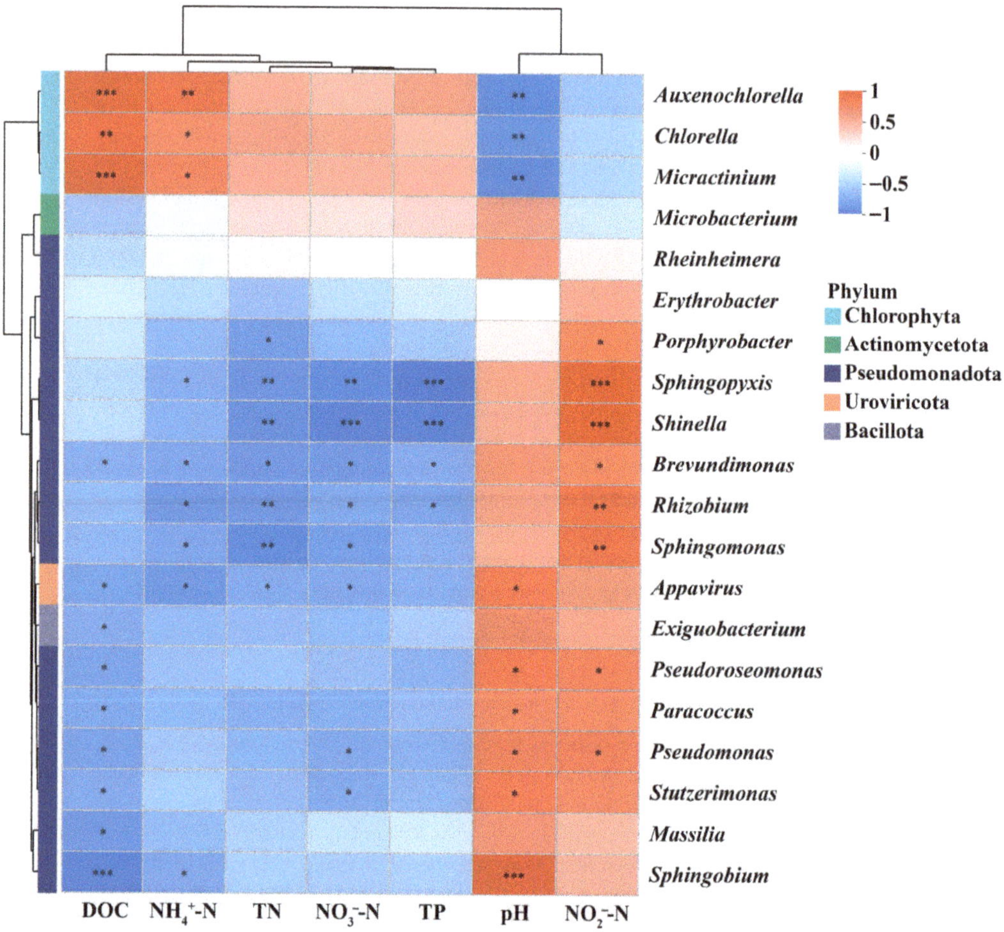

Figure 7. Heatmap of correlations between environmental factors and microbial community. Red boxes represent the positive correlations, while blue boxes represent the negative correlations. Significance codes are represented by asterisks: '***' for $p < 0.001$, '**' for $p < 0.01$, '*' for $p < 0.05$.

(2) Mechanistic analysis of nitrogen and phosphorus metabolism

Mantel tests confirmed significant correlations between functional pathways and environmental variables (Figure 8a). Nitrogen metabolism correlated positively with pH ($r = 0.527$, $p = 0.004$), NH_4^+–N ($r = 0.493$, $p = 0.024$), NO_2^-–N ($r = 0.573$, $p = 0.002$), TP ($r = 0.456$, $p = 0.026$), and DOC ($r = 0.651$, $p = 0.011$), highlighting the importance of carbon availability and alkalinity in promoting nitrification–assimilation coupling. Phosphorus and carbon metabolism were also significantly correlated with pH ($r = 0.537$, $p = 0.015$; $r = 0.541$, $p = 0.008$) and DOC ($r = 0.504$, $p = 0.018$; $r = 0.466$, $p = 0.024$), suggesting that pH and organic carbon availability in the photobioreactor can enhance microbial P and C metabolism by regulating microbial activity. These results highlight the critical role of pH and nutrient availability (NH_4^+–N, DOC) in shaping microbial C–N–P metabolic processes.

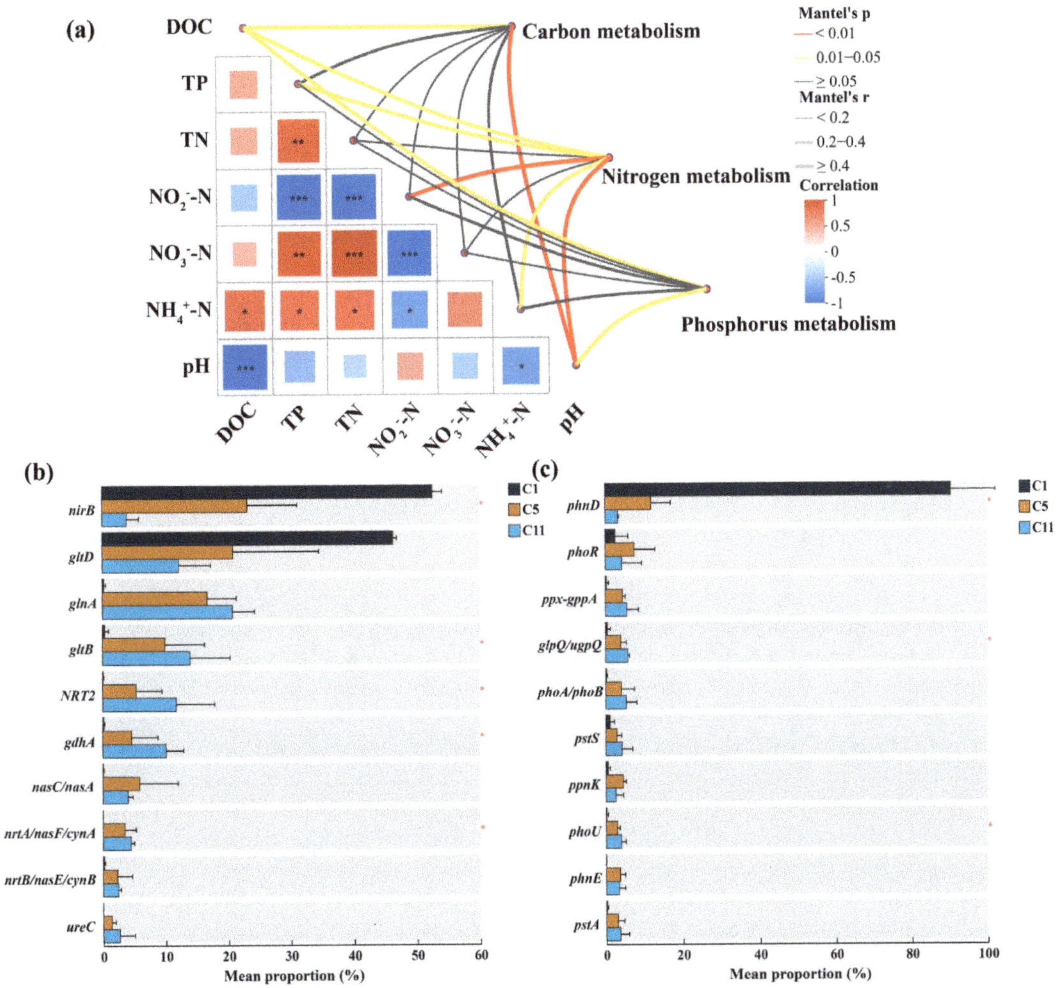

Figure 8. Metagenomic analysis of algae-bacteria symbiotic system for treatment of low C/N ratio wastewater. (**a**) Correlation analysis of environmental factors and carbon-nitrogen-phosphorus metabolism and relative abundances of the top 10 key functional genes involved in (**b**) nitrogen and (**c**) phosphorus transformation based on metagenomic analysis in algae-bacteria symbiotic system. '***' for $p < 0.001$, '**' for $p < 0.01$, '*' for $p < 0.05$.

Metagenomic data revealed a temporal shift in nitrogen-transformation dominance (Figure 8b). During the initial phase (C1), denitrification-related *nirB* accounted for 52.61% of nitrogen genes, whereas by phase C11 its abundance declined to 3.87%. In contrast, assimilatory genes glnA and gltB increased from 0.24% → 20.75% and 0.35% → 13.99%, respectively. The elevated expression of glnA (glutamine synthetase) and gltB (glutamate synthase) directly aligns with the observed 97.60% NH_4^+–N and 87.0% TN removal, confirming that algal assimilation became the dominant nitrogen-removal mechanism. Meanwhile, denitrification genes (*narG*, *nirK*, *norB*) remained below 2.50% abundance in C11, indicating minor bacterial denitrification. The co-occurrence of *Pseudomonas* and *Sphingopyxis* with low-level denitrification genes suggests their auxiliary role in reducing residual nitrate and maintaining redox balance, complementing algal nitrogen uptake.

This metabolic shift underscores that under optimized light and aeration conditions, algal assimilation became the dominant nitrogen removal pathway [54], closely linked to the biomass synthesis requirements of photosynthetic microorganisms such as *Chlorella* and *Micractinium*.

Phosphorus-metabolism genes displayed a parallel functional transition (Figure 8c). In the initial phase, the key organic phosphorus mineralization gene *phnD* was highly dominant (90.35%), indicating that organic phosphorus decomposition was the primary mechanism of phosphorus mobilization. As operation progressed, its abundance decreased sharply (to 3.34% in C11), while several genes related to inorganic phosphorus uptake and transport were significantly upregulated. These included the high-affinity phosphate transport system genes *pstS* (increasing from 1.13% to 4.25%), *pstA* (0.29% to 3.71%), and *pstC* (0.33% to 1.92%), as well as the alkaline phosphatase gene *phoA* (0.25% to 5.40%) and the polyphosphate kinase gene *ppk1* (0.26% to 2.80%). These shifts suggest a transition in metabolic emphasis from organic phosphorus degradation to active inorganic phosphorus uptake and intracellular storage [5], correlating with the consistently high removal efficiency of TP (97.6%) observed in water quality data. The enrichment of Shinella and Sphingopyxis—known polyphosphate-accumulating bacteria—provides a direct microbial explanation for these macroscopic results. Furthermore, increased abundances of *glpQ* (0.48% to 5.83%) and *ppx* (0.31% to 5.63%) indicated enhanced recycling capacity of organic phosphorus substrates under phosphorus-limited conditions. This metabolic strategy shift—from decomposition to absorption, and from inorganic transport to intracellular storage—reflects the efficient phosphorus utilization by the algal-bacterial system in a low-carbon-nitrogen-ratio wastewater environment and provides a molecular basis for phosphorus accumulation in microalgal biomass [55].

Together, these quantitative linkages between microbial composition, functional-gene abundance, and nutrient-removal data substantiate the synergistic role of algal–bacterial consortia: algae supplied organic carbon and oxygen, sustaining heterotrophic denitrifiers and phosphorus-accumulating bacteria, while bacterial mineralization regenerated inorganic nutrients for algal assimilation. This mutualistic exchange maintained high NH_4^+–N, TN, and TP removal and stabilized reactor performance under low-C/N conditions. The coordinated succession of nitrogen- and phosphorus-metabolism genes thus provides a molecular basis for the observed improvements in water quality and biomass productivity.

4. Conclusions

This study elucidated how operational parameter optimization governs the metabolic balance and functional dynamics of an *Auxenochlorella pyrenoidosa*-based photobioreactor treating low C/N wastewater. Inoculum density, aeration rate, light intensity, and photoperiod each exerted distinct but interdependent effects on algal growth, nutrient removal, and system stability. Among these, the combination of an initial OD_{680} of 0.1, aeration rate of 2 L min^{-1}, light intensity of 112 μmol m^{-2} s^{-1}, and a 16L:8D photoperiod achieved the best balance between efficient nutrient removal and sustainable biomass production. Metagenomic and water-quality analyses together revealed that optimized physical parameters did not merely improve reactor performance but actively redirected nitrogen metabolism—from bacterial denitrification toward algal nitrogen assimilation—by reshaping the algal–bacterial community and metabolic gene expression (notably *glnA*, *gltB*). This metabolic transition underpins the system's high nitrogen removal efficiency and carbon utilization under limited organic-C conditions. These findings provide a broader conceptual insight: controlling operational parameters can strategically modulate microbial community function, allowing the co-optimization of wastewater treatment and biomass valorization in a single process. This demonstrates the feasibility of regulating microbial metabolism through engineering design, enabling carbon-efficient, resource-recovering wastewater treatment systems. Future research should integrate metabolomic and kinetic modeling approaches to further quantify carbon–nitrogen fluxes and evaluate the downstream valorization potential of the produced biomass in biorefinery applications.

Author Contributions: Methodology, Data analysis and visualization, Writing—original draft, Writing—review & editing, Funding acquisition, L.Z.; Investigation, Data analysis and visualization, Y.X.; Investigation, T.T.; Investigation, Y.Z.; Investigation, G.H.; Supervision, Writing—review & editing, J.T. All authors have read and agreed to the published version of the manuscript.

Funding: This work was supported by the National Natural Science Foundation of China (32401412), Opening Foundation of Shenzhen Collaborative Innovation Public Service Platform for Marine Algae Industry (SQ202403), Key Project of Natural Science Research for Higher Education Institutions of Anhui Province (2024AH051458), Fuyang Normal University 2023 Doctoral Research Launch Fund (2023KYQD0025) to Lin Zhao.

Data Availability Statement: The original contributions presented in this study are included in the article. Further inquiries can be directed to the corresponding author.

Conflicts of Interest: The authors declare no conflicts of interest.

References

1. Zhao, L.; Zhang, Y.; Tang, J. Application of multi-electron donor autotrophic denitrification (MEDAD) substrate in constructed wetlands for treating low carbon-to-nitrogen ratio wastewater: A feasible framework. *Ecol. Eng.* **2025**, *217*, 107648. [CrossRef]
2. Guo, H.; Zhai, X.; Hu, M.; Chang, J.S.; Lee, D.J. Atypical removals of nitrogen and phosphorus with biochar-pyrite vertical flow constructed wetlands treating wastewater at low C/N ratio. *Bioresour. Technol.* **2025**, *422*, 132219. [CrossRef]
3. Sátiro, J.; dos Santos Neto, A.; Tavares, J.; Marinho, I.; Magnus, B.; Kato, M.; Albuquerque, A.; Florencio, L. Impact of inoculum on domestic wastewater treatment in high-rate ponds in pilot-scale: Assessment of organic matter and nutrients removal, biomass growth, and content. *Algal Res.* **2025**, *86*, 103923. [CrossRef]
4. Zhu, S.; Qin, L.; Feng, P.; Shang, C.; Wang, Z.; Yuan, Z. Treatment of low C/N ratio wastewater and biomass production using co-culture of *Chlorella vulgaris* and activated sludge in a batch photobioreactor. *Bioresour. Technol.* **2019**, *274*, 313–320. [CrossRef]
5. Wang, S.; Zhang, C.; Zhang, K.; Zhang, L.; Bi, R.; Zhang, Y.; Hu, Z. One-step bioremediation of hypersaline and nutrient-rich food industry process water with a domestic microbial community containing diatom *Halamphora coffeaeformis*. *Water Res.* **2024**, *254*, 121430. [CrossRef] [PubMed]
6. Kusuma, H.S.; Illiyanasafa, N.; Jaya, D.E.C.; Darmokoesoemo, H.; Putra, N.R. Utilization of the microalga *Chlorella vulgaris* for mercury bioremediation from wastewater and biomass production. *Sustain. Chem. Pharm.* **2024**, *37*, 101346. [CrossRef]
7. Lugo, L.A.; Thorarinsdottir, R.I.; Bjornsson, S.; Palsson, O.P.; Skulason, H.; Johannsson, S.; Brynjolfsson, S. Remediation of Aquaculture Wastewater Using the Microalga *Chlorella sorokiniana*. *Water* **2020**, *12*, 3144. [CrossRef]
8. Ayatollahi, S.Z.; Esmaeilzadeh, F.; Mowla, D. Integrated CO_2 capture, nutrients removal and biodiesel production using *Chlorella vulgaris*. *J. Environ. Chem. Eng.* **2021**, *9*, 104763. [CrossRef]
9. Sátiro, J.; Marchiori, L.; Morais, M.V.; Marinho, T.; Florencio, L.; Gomes, A.; Muñoz, R.; Albuquerque, A.; Simões, R. Integrating photobioreactors and constructed wetlands for paper pulp industry wastewater treatment: A nature-based system approach. *J. Water Process Eng.* **2025**, *71*, 107237. [CrossRef]
10. Gao, F.; Yang, H.L.; Li, C.; Peng, Y.Y.; Lu, M.M.; Jin, W.H.; Bao, J.J.; Guo, Y.M. Effect of organic carbon to nitrogen ratio in wastewater on growth, nutrient uptake and lipid accumulation of a mixotrophic microalgae *Chlorella* sp. *Bioresour. Technol.* **2019**, *282*, 118–124. [CrossRef]
11. Diaz-Hernandez, M.M.; Krebss-Kleingezinds, E.; Parra-Saldivar, R.; Alfaro-Ponce, M.; Chairez, I.; Melchor Martinez, E.M.; Kjellberg, K.; Soheil Mansouri, S. Bioprocessing sidestream valorization as culture stream for *Chlorella vulgaris* biomass accumulation: Neural data-driven design of experiments. *Biofuels* **2025**, *16*, 579–591. [CrossRef]
12. Chu, W.; Li, X.; Li, P.; Li, J.; Wang, Z.; Zhou, H.; Yang, X.; Chen, S.; Zhou, M.; Wang, S.; et al. Enhanced treatment of low C/N domestic wastewater in a membrane photobioreactor: Operational control of microalgal-bacterial symbiosis for synergistic pollutant and antibiotic resistance genes removal. *J. Environ. Manag.* **2025**, *394*, 127398. [CrossRef]

13. Wang, Q.; Chu, G.; Gao, C.; Tian, T.; Zhang, W.; Chen, W.; Gao, M. Effect of light intensity on performance, microbial community and metabolic pathway of algal-bacterial symbiosis in sequencing batch biofilm reactor treating mariculture wastewater. *Bioresour. Technol.* **2025**, *433*, 132726. [CrossRef]
14. Ikaran, Z.; Suárez-Alvarez, S.; Urreta, I.; Castañón, S. The effect of nitrogen limitation on the physiology and metabolism of *Chlorella vulgaris* var L3. *Algal Res.* **2015**, *10*, 134–144. [CrossRef]
15. Shen, X.F.; Chu, F.F.; Lam, P.K.; Zeng, R.J. Biosynthesis of high yield fatty acids from *Chlorella vulgaris* NIES-227 under nitrogen starvation stress during heterotrophic cultivation. *Water Res.* **2015**, *81*, 294–300. [CrossRef] [PubMed]
16. Liu, K.; Li, J.; Xing, C.; Yuan, H.; Yang, J. Characterization of *Auxenochlorella protothecoides* acyltransferases and potential of their protein interactions to promote the enrichment of oleic acid. *Biotechnol Biofuels Bioprod* **2023**, *16*, 69. [CrossRef] [PubMed]
17. Wang, X.; Wei, H.; Mao, X.; Liu, J. Proteomics Analysis of Lipid Droplets from the Oleaginous Alga *Chromochloris zofingiensis* Reveals Novel Proteins for Lipid Metabolism. *Genom. Proteom. Bioinform.* **2019**, *17*, 260–272. [CrossRef] [PubMed]
18. Li, J.; Wang, Y.; Fang, Y.; Lyu, X.; Zhu, Z.; Wu, C.; Xu, Z.; Li, W.; Liu, N.; Du, C.; et al. *Phycospheric Bacteria* Alleviate the Stress of Erythromycin on *Auxenochlorella pyrenoidosa* by Regulating Nitrogen Metabolism. *Plants* **2025**, *14*, 121. [CrossRef]
19. Liu, Q.; Jia, J.; Hu, H.; Li, X.; Zhao, Y.; Wu, C. Nitrogen and phosphorus limitations promoted bacterial nitrate metabolism and propagation of antibiotic resistome in the phycosphere of *Auxenochlorella pyrenoidosa*. *J. Hazard. Mater.* **2024**, *468*, 133786. [CrossRef]
20. Lou, M.; Zhang, Y.; Zhang, M.; Zhou, H.; Zhang, Y.; Sheng, Q.; Zhao, J.; Xu, Q.; Zhang, R. Effects of Intracellular Polysaccharides and Proteins of *Auxenochlorella pyrenoidosa* on Water Quality, Floc Formation, and Microbial Composition in a Biofloc System. *Microorganisms* **2025**, *13*, 1704. [CrossRef]
21. Luo, L.; Xiaojuan, H.; Qin, Y.; Liu, Y.; Wu, Y.; Chen, M.; Liao, Y.; Zhang, L.; Li, X. Optimizing nutrient removal of algal-bacterial symbiosis system for treating low C/N farmland drainage. *Agric. Water Manag.* **2024**, *306*, 109149. [CrossRef]
22. Xue, C.; Gao, K.; Qian, P.; Dong, J.; Gao, Z.; Liu, Q.; Chen, B.; Deng, X. Cultivation of *Chlorella sorokiniana* in a bubble-column bioreactor coupled with cooking cocoon wastewater treatment: Effects of initial cell density and aeration rate. *Water Sci. Technol.* **2021**, *83*, 2615–2628. [CrossRef]
23. Huang, J.; Cheng, S.; Zhang, Y.; Teng, J.; Zhang, M.; Lin, H. Optimizing aeration intensity to enhance self-flocculation in algal-bacterial symbiosis systems. *Chemosphere* **2023**, *341*, 140064. [CrossRef] [PubMed]
24. Tang, C.C.; Zuo, W.; Tian, Y.; Sun, N.; Wang, Z.W.; Zhang, J. Effect of aeration rate on performance and stability of algal-bacterial symbiosis system to treat domestic wastewater in sequencing batch reactors. *Bioresour. Technol.* **2016**, *222*, 156–164. [CrossRef] [PubMed]
25. Tang, C.C.; Tian, Y.; Liang, H.; Zuo, W.; Wang, Z.W.; Zhang, J.; He, Z.W. Enhanced nitrogen and phosphorus removal from domestic wastewater via algae-assisted sequencing batch biofilm reactor. *Bioresour. Technol.* **2018**, *250*, 185–190. [CrossRef]
26. Zhao, L.; Xu, Y.; Zhang, Y.; Huang, Y.; Huang, G.; Hu, Z.; Wu, C.; Tang, J. Performance and mechanisms of an algae-bacteria symbiotic system for treating low carbon-to-nitrogen ratio wastewater. *Biochem. Eng. J.* **2026**, *226*, 109995. [CrossRef]
27. State Environmental Protection Administration of China. *The Monitoring and Analysis Methods of Water and Wastewater (4th Version)*; China Environmental Science Press: Beijing, China, 2002; pp. 210–213.
28. Jeffrey, S.W.; Humphrey, G.F. New spectrophotometric equations for determining chlorophylls a, b, c1 and c2 in higher plants, algae and natural phytoplankton. *Biochem. Physiol. Pflanz.* **1975**, *167*, 191–194. [CrossRef]
29. Mousavi, S.; Najafpour, G.D.; Mohammadi, M.; Seifi, M.H. Cultivation of newly isolated microalgae *Coelastrum* sp. in wastewater for simultaneous CO_2 fixation, lipid production and wastewater treatment. *Bioprocess Biosyst. Eng.* **2018**, *41*, 519–530. [CrossRef]
30. Wang, Q.; Zhuang, J.; Ni, S.; Luo, H.; Zheng, K.; Li, X.; Lan, C.; Zhao, D.; Bai, Y.; Jia, B.; et al. Overexpressing CrePAPS Polyadenylate Activity Enhances Protein Translation and Accumulation in *Chlamydomonas reinhardtii*. *Mar. Drugs* **2022**, *20*, 276. [CrossRef]
31. Ben--Amotz, A.; Tornabene, T.G.; Thomas, W.H. Chemical profile of selected species of microalgae with emphasis on lipids. *J. Phycol.* **1985**, *21*, 72–81. [CrossRef]
32. Deng, X.-Y.; Gao, K.; Zhang, R.-C.; Addy, M.; Lu, Q.; Ren, H.-Y.; Chen, P.; Liu, Y.-H.; Ruan, R. Growing *Chlorella vulgaris* on thermophilic anaerobic digestion swine manure for nutrient removal and biomass production. *Bioresour. Technol.* **2017**, *243*, 417–425. [CrossRef] [PubMed]
33. Bohutskyi, P.; Kligerman, D.; Byers, N.; Nasr, L.; Cua, C.; Chow, S.; Su, C.; Tang, Y.; Betenbaugh, M.; Bouwer, E. Effects of inoculum size, light intensity, and dose of anaerobic digestion centrate on growth and productivity of Chlorella and Scenedesmus microalgae and their poly-culture in primary and secondary wastewater. *Algal Res.* **2016**, *19*, 278–290. [CrossRef]
34. Kryvenda, A.; Tischner, R.; Steudel, B.; Griehl, C.; Armon, R.; Friedl, T. Testing for terrestrial and freshwater microalgae productivity under elevated CO_2 conditions and nutrient limitation. *BMC Plant Biol.* **2023**, *23*, 27. [CrossRef]

35. Mohsenpour, S.F.; Willoughby, N. Effect of CO_2 aeration on cultivation of microalgae in luminescent photobioreactors. *Biomass Bioenergy* **2016**, *85*, 168–177. [CrossRef]
36. Yang, Y.; Gao, K. Effects of CO_2 concentrations on the freshwater microalgae, *Chlamydomonas reinhardtii*, *Chlorella pyrenoidosa* and Scenedesmus obliquus (Chlorophyta). *J. Appl. Phycol.* **2003**, *15*, 379–389. [CrossRef]
37. Zhao, L.; Tang, J.; Xu, Y.; Zhang, Y.; Song, Z.; Fu, G.; Hu, Z. A vertical-flow constructed wetland–microalgal membrane photobioreactor integrated system for treating high-pollution-load marine aquaculture wastewater: A lab-scale study. *Sci. Total Environ.* **2024**, *919*, 170465. [CrossRef] [PubMed]
38. Zhao, Q.-B.; Ma, J.; Zeb, I.; Yu, L.; Chen, S.; Zheng, Y.-M.; Frear, C. Ammonia recovery from anaerobic digester effluent through direct aeration. *Chem. Eng. J.* **2015**, *279*, 31–37. [CrossRef]
39. Guštin, S.; Marinšek-Logar, R. Effect of pH, temperature and air flow rate on the continuous ammonia stripping of the anaerobic digestion effluent. *Process Saf. Environ. Prot.* **2011**, *89*, 61–66. [CrossRef]
40. Gao, P.; Guo, L.; Gao, M.; Zhao, Y.; Jin, C.; She, Z. Regulation of carbon source metabolism in mixotrophic microalgae cultivation in response to light intensity variation. *J. Environ. Manag.* **2022**, *302 Pt B*, 114095. [CrossRef]
41. Ardo, F.M.; Khoo, K.S.; Ahmad Sobri, M.Z.; Suparmaniam, U.; Ethiraj, B.; Anwar, A.F.; Lam, S.M.; Sin, J.C.; Shahid, M.K.; Ansar, S.; et al. Modelling photoperiod in enhancing hydrogen production from *Chlorella vulgaris* sp. while bioremediating ammonium and organic pollutants in municipal wastewater. *Environ. Pollut.* **2024**, *346*, 123648. [CrossRef]
42. Iasimone, F.; Panico, A.; De Felice, V.; Fantasma, F.; Iorizzi, M.; Pirozzi, F. Effect of light intensity and nutrients supply on microalgae cultivated in urban wastewater: Biomass production, lipids accumulation and settleability characteristics. *J. Environ. Manag.* **2018**, *223*, 1078–1085. [CrossRef]
43. Parveen, A.; Bhatnagar, P.; Gautam, P.; Bisht, B.; Nanda, M.; Kumar, S.; Vlaskin, M.S.; Kumar, V. Enhancing the bio-prospective of microalgae by different light systems and photoperiods. *Photochem. Photobiol. Sci.* **2023**, *22*, 2687–2698. [CrossRef]
44. Zhang, X.; Ji, B.; Tian, J.; Liu, Y. Development, performance and microbial community analysis of a continuous-flow microalgal-bacterial biofilm photoreactor for municipal wastewater treatment. *J. Environ. Manag.* **2023**, *338*, 117770. [CrossRef]
45. Li, Y.; Yan, X.; Qin, T.; Gan, Y.; Li, N.; Zheng, C. Applicability of blue algae as an activator for microbial enhanced coal bed methane technologies. *J. Environ. Manag.* **2024**, *370*, 123063. [CrossRef]
46. Liao, H.; Li, H.; Duan, C.S.; Zhou, X.Y.; An, X.L.; Zhu, Y.G.; Su, J.Q. Metagenomic and viromic analysis reveal the anthropogenic impacts on the plasmid and phage borne transferable resistome in soil. *Environ. Int.* **2022**, *170*, 107595. [CrossRef]
47. Wu, R.; Cao, Z.; Jiang, Y.; Chen, W.; Sun, Y.; Li, Q.; Mi, J.; Deng, L.; Liao, X.; Feng, Y.; et al. Early life dynamics of ARG and MGE associated with intestinal virome in neonatal piglets. *Vet. Microbiol.* **2022**, *274*, 109575. [CrossRef] [PubMed]
48. Zhang, J.; Tang, A.; Jin, T.; Sun, D.; Guo, F.; Lei, H.; Lin, L.; Shu, W.; Yu, P.; Li, X.; et al. A panoramic view of the virosphere in three wastewater treatment plants by integrating viral-like particle-concentrated and traditional non-concentrated metagenomic approaches. *Imeta* **2024**, *3*, e188. [CrossRef] [PubMed]
49. Liang, X.; Yang, S.; Radosevich, M.; Wang, Y.; Duan, N.; Jia, Y. Bacteriophage-driven microbial phenotypic heterogeneity: Ecological and biogeochemical importance. *NPJ Biofilms Microbiomes* **2025**, *11*, 82. [CrossRef] [PubMed]
50. Hassen, B.; Hammami, S. Environmental phages: Ecosystem dynamics, biotechnological applications and their limits, and future directions. *J. Appl. Microbiol.* **2025**, *136*, lxaf136. [CrossRef]
51. Xie, F.; Thiri, M.; Wang, H. Simultaneous heterotrophic nitrification and aerobic denitrification by a novel isolated *Pseudomonas mendocina* X49. *Bioresour. Technol.* **2021**, *319*, 124198. [CrossRef]
52. Gong, X.; Peng, Q.; Jiang, R.; Yang, N.; Xing, C.; Wang, R. Mn-oxidizing microalgae and woodchip-denitrifying bioreactor system for recovering manganese and removing nitrogen from electrolytic manganese metal industrial tailwater. *J. Hazard. Mater.* **2025**, *488*, 137383. [CrossRef]
53. Zhao, Y.; Wang, Q.; Yang, Z.; Jia, X.; Cabrera, J.; Ji, M. Bio-capture of Cr(VI) in a denitrification system: Electron competition, long-term performance, and microbial community evolution. *J. Hazard. Mater.* **2022**, *432*, 128697. [CrossRef] [PubMed]
54. Li, Q.; Xu, Y.; Liang, C.; Peng, L.; Zhou, Y. Nitrogen removal by algal-bacterial consortium during mainstream wastewater treatment: Transformation mechanisms and potential N_2O mitigation. *Water Res.* **2023**, *235*, 119890. [CrossRef] [PubMed]
55. Xiong, W.; Wang, S.; Jin, Y.; Wu, Z.; Liu, D.; Su, H. Insights into nitrogen and phosphorus metabolic mechanisms of algal-bacterial aerobic granular sludge via metagenomics: Performance, microbial community and functional genes. *Bioresour. Technol.* **2023**, *369*, 128442. [CrossRef] [PubMed]

water

MDPI

Article

Utilization of Coal-Based Activated Carbon (JA) for the Adsorption of Methyl Orange Azo Dye in Wastewater

Lihua Dong *, Qianwen Xu, Yang Yao, Yingjie Han and Guanghui Zhang *

School of Environmental Science and Engineering, Tianjin University, Tianjin 300072, China; 18503235688@163.com (Q.X.); yy123@tju.edu.cn (Y.Y.); jackhan_141@tju.edu.cn (Y.H.)
* Correspondence: donglihua12@tsinghua.org.cn (L.D.); zgh@tju.edu.cn (G.Z.); Tel.: +86-228-740-2072 (G.Z.)

Abstract

Currently, the mainstream methods for dye removal internationally include advanced oxidation, catalytic degradation, and adsorption. Catalytic and oxidation methods are costly and unsuitable for large-scale application. While adsorption is straightforward, selecting and modifying raw materials poses significant challenges. Therefore, identifying readily available and inexpensive adsorbents is crucial for dye removal. This study utilized Type A coal as raw material to prepare a series of specialized activated carbon (JA) for adsorbing methyl orange from wastewater, followed by optimization. The optimized screening results indicated that JA-12 exhibited the highest methyl orange removal rate (90.54%). This performance is attributed to its larger micropore structure and increased pore volume. Further analysis revealed that the adsorption process follows pseudo-second-order kinetics and the Langmuir adsorption isotherm model ($R^2 \approx 0.999$). Compared to the theoretical adsorption capacity calculated based on specific surface area, the adsorption capacity calculated based on pore volume (270.66 mg/g) was closer to the actual adsorption capacity, indicating that the pore structure of JA-12 plays a dominant role in the adsorption process. Combined with the Langmuir adsorption model, it can be inferred that dye molecules in solution adsorb onto the inner surface of JA-12 in a monolayer form. Surface functional group analysis revealed that protonation enhances JA-12's adsorption capacity for the azo dye methyl orange. Collectively, our findings elucidate the removal mechanism of methyl orange using readily available coal as raw material to prepare low-cost specialty activated carbon, providing a framework for cost-effective, large-scale dye removal.

Keywords: coal-based activated carbon; physical activation; azo dye methyl orange; activated carbon characterization

1. Introduction

The textile printing and dyeing industry has long been a pillar of national economies worldwide, serving diverse sectors such as printing, textiles, food processing, and leather manufacturing [1]. Dyes also play an indispensable role in daily life, enriching our world with vibrant colors. However, this industry consumes vast amounts of water and energy, a problem that is becoming increasingly prominent.

Over 100,000 types of dyes are currently known, with annual dye production exceeding 700,000 tons [2]. According to statistics, the world's only textile industry consumes more than 10,000 tons of dyes each year, and about 100 tons of dyes are discharged into the water body each year [3]. The discharge of textile industrial wastewater in China is about 20×10^8 t/a, and the industrial wastewater produced in the printing and dyeing process can reach 3~4 million tons per day. Research indicates that over 12% of synthetic dyes are lost annually during their preparation and application, while 20% remain in water [4]. The discharge of large volumes of dye-containing wastewater into water bodies will cause severe pollution to the surrounding aquatic environment [5,6]. Azo dyes is the collective term for all molecules containing azo groups. They are the most commonly used synthetic dyes, accounting for over 70% of all synthetic dyes [7]. Azo dyes have high chemical inertness [8,9], the resulting wastewater composition is complex, and it is more difficult to decolorize than natural dyes. Its biological toxicity is greater, and a large number of azo dyes enter the water due to its complex structure. It cannot degrade naturally within a short period of time. Therefore, the treatment of wastewater contaminated with azo dyes has become an urgent problem to be solved. Some studies have employed chemical methods to treat wastewater containing azo dyes, but chemical treatment is costly, inefficient, and produces highly toxic intermediate products, making it difficult to apply widely [10].

Currently, there are various methods for treating dyes. (See Figure 1 for the methods and their respective advantages and disadvantages.) The most commonly used methods are membrane separation, physical adsorption [11], chemical oxidation [12], and microbial degradation [13]. Although dye treatment methods are diverse, each possesses specific applicability and limitations, with different approaches offering distinct advantages and disadvantages. For chemical treatment processes, while the oxidation process is straightforward to implement, the aqueous reagents require activation treatment (and the activation steps and methods are cumbersome). When using hydrogen peroxide and ferrous salts to remove dyes, the Fenton reagent exhibits the highest efficiency, but it also generates sludge during the dye removal process [13]. Ozonation does not increase the volume of wastewater and sludge [14], but the disadvantage is that the half-life is too short, usually only 20 min [15]. Photochemical reactions are also commonly used as a method for dye removal. This approach similarly does not produce sludge, but byproducts are still generated. The sodium hypochlorite treatment method [16] can quickly break the azo bond, but aromatic amines will be released during the removal process. The electrochemical treatment method [17] has no drug consumption or sludge deposition, but the increase in flow rate will lead to a decrease in dye removal rate. The biological treatment method [13] is also the mainstream treatment method at present, which is mainly divided into decolorization of white rot fungi, mixed culture of other microorganisms, and adsorption of live/dead microbial biomass. However, each of these three methods has its own advantages and disadvantages. White rot fungi can break down dyes using enzymes, but enzyme production is not entirely reliable [18]. Other microbial mixed cultures can rapidly decolorize under aerobic conditions, but the drawback lies in the fact that azo dyes themselves are extremely difficult to decolorize. Live/dead microbial biomass adsorption can specifically bind dyes, demonstrating good treatment efficacy; however, not all dyes possess specific binding sites. Regarding physical treatment, activated carbon adsorption is currently the most mainstream, simple, and efficient method, with its strong adsorption properties yielding favorable treatment results for various dyes. The drawback lies in the high cost of activated carbon [19]. Membrane filtration can remove all types of dyes [20], but it also produces concentrated sludge. Ion exchange [21] can avoid the loss of the adsorbent and can be regenerated, but this method is also not effective for all dyes. Electrocoagulation technology is less expensive and more economical, but it produces a larger volume of sludge.

From the above analysis, it can be seen that among various physical, chemical, and biological treatment methods, activated carbon adsorption in physical treatment not only produces no byproducts or sludge but also delivers stable adsorption performance with excellent removal efficiency for various dyes. Its sole drawback lies in its relatively high

https://doi.org/10.3390/w18020289

cost [22]. Therefore, it is of great practical significance to develop a low-cost and high-efficiency activated carbon.

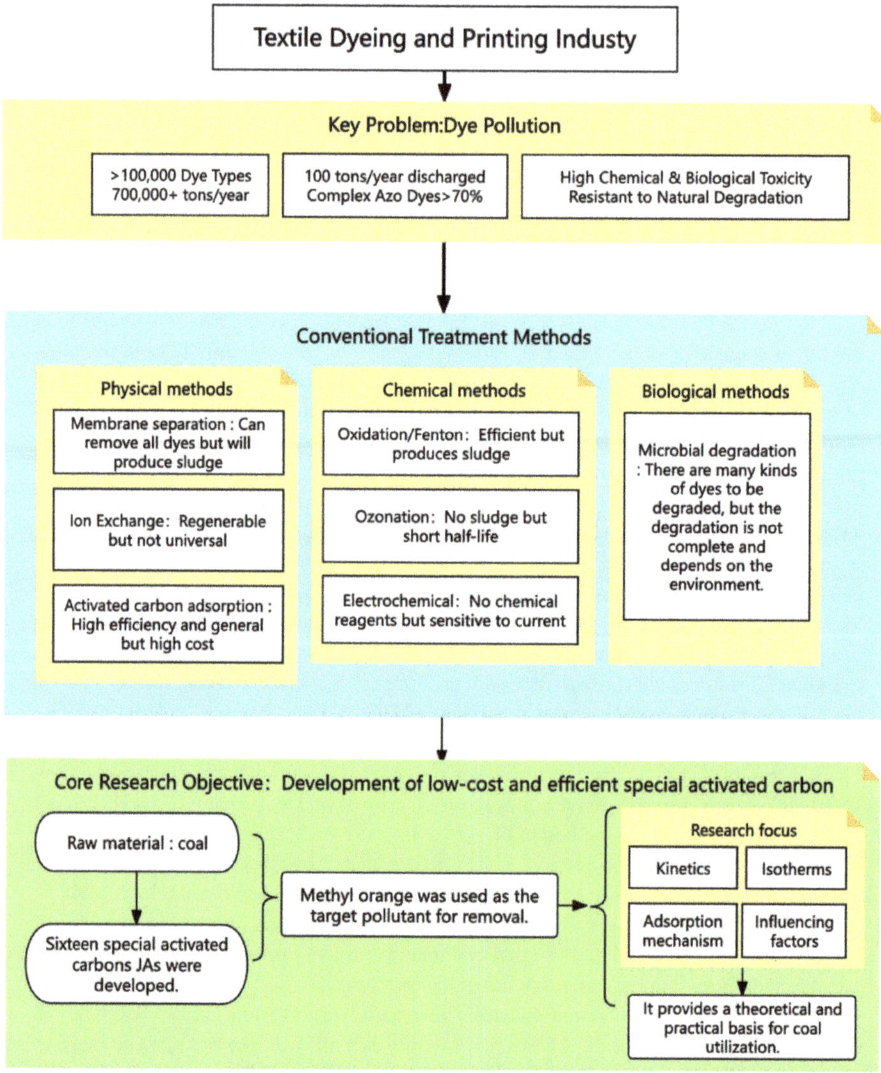

Figure 1. Current dye removal technologies and their challenges.

This study utilized Type A coal as raw material to prepare specialized activated carbon for adsorbing azo dyes in wastewater. Sixteen JA-type activated carbons were synthesized via physical activation, with comprehensive investigations conducted on their adsorption kinetics, adsorption isotherms, influencing factors, and mechanisms for adsorbing methyl orange. The findings provide theoretical and practical foundations for coal utilization while demonstrating the potential of coal-based activated carbon for treating dye wastewater. This approach offers a readily available and cost-effective solution for large-scale purification of dye wastewater, particularly in the textile industry.

https://doi.org/10.3390/w18020289

2. Material and Methods

2.1. Materials

A-type coal belongs to the discarded Uygur Autonomous Region and is supplied by the National Energy Group (Urumqi, Xinjiang, China) (moisture (%): 4.83, ash (%): 1.76, volatiles (%): 39.15).

The drugs and reagents used in the experiment are as follows: coal (discarded Uygur Autonomous Region, China), methyl orange ($C_{14}H_{14}N_3NaO_3S$) (*AR*, purity of 99%, Tianjin Damao Chemical Reagent Factory, Tianjin, China), disodium hydrogen phosphate dodecahydrate ($Na_2HPO_4 \bullet 12H_2O$) (AR, purity of 99%, Tianjin No.1 Chemical Reagent Factory, Tianjin, China), potassium dihydrogen phosphate (KH_2PO_4) (AR, purity of 99%, Tianjin Institute of Photorefining Fine Chemicals, Tianjin, China), and methylene blue ($C_{16}H_{18}N_3ClS \bullet 3H_2O$) (AR, purity of 99%, Beijing Chemical Plant, Beijing, China).

2.2. Preparation of JA

JA was prepared by water vapor activation [23]. Firstly, the raw coal was crushed, and then the JA was prepared by carbonization and an activation process in a tube furnace (BTF-1200C, Anhui Bayik Equipment Technology Co., Ltd., Hefei, Anhui, China), with the carbonization process involving heating under isolated air and the activation process involving activating the carbonized material with water vapor [24]. During the preparation of JA, different temperatures and times of carbonization and activation were set. After carbonization, the carbonized material was continuously heated to the activation temperature and then JA was obtained. Finally, the JA was reduced in particle size by ball mill so that 95% passes through a 325-mesh sieve and dried 3 h at 150 °C for use.

2.3. Characterization of JA

2.3.1. Determination of Water Capacity

The water capacity of coal-based activated carbon was determined according to the test method of coal-based granular activated carbon [25].

The activated carbon sample of (25 ± 1) mL was weighed and placed in a beaker with a volume of 250 mL. The sample was recorded as m_2. A total of 100 mL of water was added to the beaker, heated to 100 degrees, and kept boiling for 15 min. During this period, water was injected into the beaker to keep the liquid level basically unchanged. After heating, it was cooled to room temperature. Filter paper was placed in the suction funnel and moistened with water, the suction filter was used to vacuum, and the filter paper was placed close to the suction funnel. The vacuum pump was opened, and the vacuum degree of the vacuum pump was adjusted to (8.0 ± 0.6) kPa. After 5 min of filtration, the vacuum degree was maintained and the sample was moved to the balance and weighed immediately. The interval was not more than 3 min, and the mass of the sample was recorded as m_1.

The mass fraction of water capacity was calculated according to the following equation:

$$V_1(\%) = \frac{m_1 - m_2}{m_2} \times 100 \quad (1)$$

where V_1 (%) is the mass fraction of water capacity, m_1 (g) is the quality of the sample after boiling with water suction filtration, and m_2 (g) is the mass of the sample.

2.3.2. Methylene Blue Adsorption Value Test

The methylene blue adsorption value (MB) of coal-based activated carbon was determined according to the "determination of methylene blue adsorption value of wood-based activated carbon test method".

https://doi.org/10.3390/w18020289

MB is defined as the mg of methylene blue adsorbed by one gram of AC in equilibrium with a solution of methylene blue with a concentration of 1.5 mg/L, which indicates the level of development of sub-micropores in AC and is an important indicator of AC porosity characteristics in water treatment [26].

2.3.3. N_2 Adsorption and Desorption, FT-IR Characterization, and Isoelectric Point Analysis

N_2 adsorption and desorption of activated carbon [27]: Using an automatic specific surface area porosity analyzer (ASAP2460, Shanghai, China), the total pore volume of activated carbon was calculated according to the single-point method. The micropore volume and micropore surface area of activated carbon were calculated by the t-method. The specific surface area and average pore size of activated carbon were calculated. The surface functional groups of the samples were characterized by Fourier transform infrared spectroscopy (FT-IR) using an infrared spectrometer (Nicolet iS 10,Thermo Fisher Scientific, Waltham, MA, USA). The pH value of the solid surface potential in the aqueous solution is zero, which is called the isoelectric point. The pH value of the carbon suspension and the ZETA point of the carbon particles were determined. The ZETA potential value was plotted against pH, and the pH value when the ZETA potential was zero was obtained by the interpolation method, that is, the isoelectric point was obtained.

2.4. *Experimental and Analytical Methods*

2.4.1. Adsorption Test

The adsorption experiment was conducted according to Ref. [28].

According to the experimental results, the initial concentration of the methyl orange solution was approximately 150 mg/L, the JA dosage was set to 0.5 mg/L, and the adsorption of methyl orange was conducted at 25 °C for 120 min. The adsorption capacity of JA for methyl orange was calculated according to Equation (2).

$$q_e = \frac{(C_0V - CV)}{M} \tag{2}$$

where C_0 (mg/L) is the initial concentration of the adsorbate solution, C (mg/L) is the equilibrium concentration of the adsorbate solution, M (g) is the weight of carbon, q_e (mg/g) is the volume of the adsorbate solution, and V (L) is the volume of the adsorbate solution.

2.4.2. Adsorption Dynamics

The quasi-first-order kinetic adsorption equation, quasi-second-order kinetic adsorption equation [29], and intraparticle diffusion model [30] were used to fit the experimental data, and their expressions are shown in Equations (1)–(3), respectively [31].

$$\frac{dq_t}{dt} = k_1(q_e - q_t) \tag{3}$$

$$\frac{t}{qt} = \frac{1}{k_2q_e^2} + \frac{1}{q_e}t \tag{4}$$

$$q_t = k_p t_p^{1/2} \tag{5}$$

where q_e (mg g^{-1}) is the equilibrium concentration of the solid phase, q_t (mg g^{-1}) is the average solid concentration at time t, k_1 (min^{-1}) is the quasi-first-order rate constant, k_2 ($g(mg\ min)^{-1}$) is the quasi-second-order rate constant, and k_p (mg g^{-1} $min^{-0.5}$) is the intraparticle diffusion rate constant.

https://doi.org/10.3390/w18020289

If the adsorption process conforms to the quasi-first-order rate equation, then the relationship between $\ln(q_e - q_t)$ and time is a straight line.

If the adsorption conforms to the pseudo-second-order rate equation, then t/q_t is linear with time t.

2.4.3. Adsorption Isotherm

(1) Langmuir adsorption isotherm

The Langmuir equation [32,33] is expressed by the following relation:

$$q_e = \frac{q_m K_L C_e}{1 + K_L C_e} \quad (6)$$

where q_e (mg/g) is the equilibrium adsorption capacity, C_e (mg/L) is the equilibrium concentration of the adsorbate, q_m (mg/g) is the saturated capacity, and K_L (L·mg^{-1}) is the isotherm constant of Langmuir.

For the convenience of calculation, the above equation is often rewritten as the reciprocal Equation (7):

$$\frac{1}{q_e} = \frac{1}{K_L q_m}\frac{1}{C_e} + \frac{1}{q_m} \quad (7)$$

After rewriting, the values of q_m and K_L can be obtained by the above equation.

(2) Fruendlich adsorption isotherm [34,35]

The temperature is constant, the volume of the solution is controlled, the initial concentration of the adsorbate is constant, and the amount of carbon added is changed. The residual solute equilibrium concentration Ce and the equilibrium adsorption capacity q_e also change. The change law is a curve, which is expressed as an exponential function equation:

$$q_e = K_f C_e^{\frac{1}{n}} \quad (8)$$

The linear form of Freundlich can be expressed as:

$$\ln q_e = \ln K_f + \frac{1}{n}\ln C_e \quad (9)$$

where q_e (mg/g) is the equilibrium adsorption capacity, C_e (mg/L) is the equilibrium adsorption concentration, K_f is the constant of the adsorbent, and $1/n$ is the adsorption constants.

The adsorption constants K_f and $1/n$ can be obtained by the slope and intercept of the straight line.

3. Results and Discussion

3.1. Preparation of Discarded A-Type Coal-Based Activated Carbon (JA)

3.1.1. Orthogonal Experimental Design for the Preparation of JA

To quickly explore the best experimental conditions, based on the temperatures and times of carbonization and activation, an orthogonal experimental table was designed [36]. Four groups of carbonization temperature gradients (450 °C, 500 °C, 550 °C, 600 °C), three groups of carbonization time gradient (60 min, 90 min, 120 min), four groups of activation temperature gradient (850 °C, 900 °C, 950 °C, 1000 °C), and three groups of activation time gradient (90 min, 120 min, 150 min) were set [37]. During the experiments, excessive water vapor was introduced to fully react with the carbonized material, and the heating rate was less than 10 °C/min. As shown in Table 1, a total of 16 types of activated carbon samples needed to be prepared.

https://doi.org/10.3390/w18020289

Table 1. Orthogonal experimental design of JA preparation.

No.	Carbonization Temperature (°C)	Activation Temperature (°C)	Carbonization Time (min)	Activation Time (min)
JA-1	450	850	60	90
JA-2	450	900	90	120
JA-3	450	950	120	150
JA-4	450	1000	120	150
JA-5	500	850	120	150
JA-6	500	900	60	150
JA-7	500	950	60	120
JA-8	500	1000	90	90
JA-9	550	850	60	120
JA-10	550	900	120	90
JA-11	550	950	90	150
JA-12	550	1000	60	150
JA-13	600	850	90	150
JA-14	600	900	60	150
JA-15	600	950	60	90
JA-16	600	1000	120	120

3.1.2. The Yield of Different Processes

According to the method described in Section 2.1 and the conditions in Table 1, 16 types of JA were prepared, and the carbonization yield and activation yield of each group are respectively shown in Figure 2b,c.

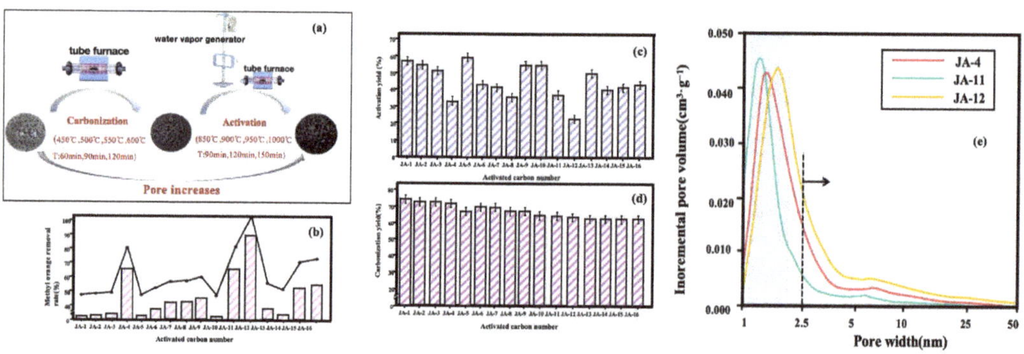

Figure 2. The carbonization rate and activation rate of JA, (**a**) Preparation of activated carbon, (**b**) JA removal rate, (**c**) JA activation yield, (**d**) JA carbonization yield, (**e**) Pore distribution of JA-4, JA-11, and JA-12.

As shown in Figure 2c,d, both temperature and time can affect the yield of carbonization and activation processes, especially the activation process. During the carbonization process, temperature is the main factor affecting the carbonization yield and is negatively correlated with the carbonization yield. Although the carbonization time will affect the carbonization yield and shows a weak negative correlation, the degree of influence is relatively small. When the temperature is 600 °C and the time is 120 min (JA-16), the carbonization yield is the lowest, which is 66.01%. When the temperature is 450 °C and the time is 60 min (JA-1), the carbonization yield is the highest, at 77.93%. This is because carbonization is a heating process under isolated air with the purpose of removing impurities such as volatiles in coal. Therefore, as the temperature and time increase, impurities such as volatiles in coal will be fully reacted, resulting in greater coal weight loss [38].

 https://doi.org/10.3390/w18020289

The activation process yielded similar conclusions to the carbonization process. Activation time and activation temperature exhibited a negative correlation with activation yield. At an activation temperature of 1000 °C and an activation time of 150 min, the activation rate was the lowest, reaching only 22.67%. The activation process primarily involves the reaction between water vapor and carbon, leading to the formation of a pore structure. Higher temperatures and longer reaction times promote the forward progression of the reaction, accelerating its rate. The more complete the reaction, the more developed the pore structure of the activated carbon becomes, resulting in better adsorption performance. However, this also leads to a decrease in the activation yield [39].

Compared to carbonization and activation conditions, longer carbonization times and higher carbonization temperatures do not result in lower activation yields. When the activation temperature is sufficiently high and the activation time is sufficiently long, the influence of carbonization conditions on activation yield is minimal, which should be attributed to further carbonization occurring during the activation heating process. Specific carbonization and activation conditions need to be comprehensively determined based on factors such as their performance in removing target pollutants and yield.

3.2. Determination of Optimal JA for the Adsorption of Azo Dye

3.2.1. Screening Based on the MB and Water Capacity of JA

In order to preliminarily determine the pore structure characteristics of JA, MB [40] and water capacity value were measured, with MB representing the amount of sub-micropores in AC and the water capacity reflecting the total pore volume of activated carbon to a certain extent. Figure 3 shows the relationship between MB and the water capacity of the 16 JAs and their active water.

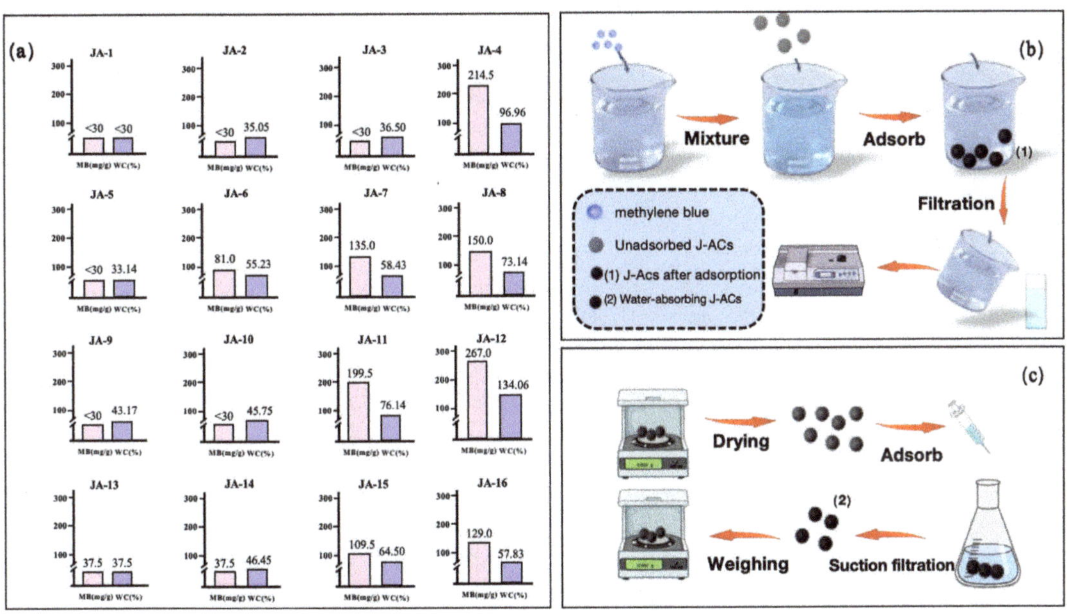

Figure 3. MB and water capacity of 16 types of JA. (**a**) JA's methylene blue value and water capacity value, (**b**) Methylene blue value test, (**c**) Water capacity value test.

As shown in Figure 3, the MB and water capacity of JA-12 exhibited a negative correlation with its activation yield, indicating that greater weight loss occurs with more developed pore structure. This also corroborates the earlier inferences regarding the carbonization

https://doi.org/10.3390/w18020289

and activation mechanisms. Among the 16 activated carbons, JA-12 demonstrated the lowest activation yield, followed by JA-4 and JA-11. Conversely, JA-12 exhibited the highest MB adsorption capacity and water-holding capacity, at 267.0 mg/g and 134.06%, respectively. JA-4 and JA-11 exhibited activation yields approximately 10–14% higher than JA-12, but their MB adsorption capacity and water-holding capacity were about 50–66 mg/g and 37–58% lower than JA-12, respectively. Overall, JA-12, JA-4, and JA-11 all exhibited well-developed pore structures.

3.2.2. Screening Based on the Adsorption Capacity for Methyl Orange

In order to further determine the optimal JA, methyl orange adsorption experiments were conducted on 16 types of JA, and the results are shown in Figure 3.

As shown in Figure 3, when the initial methyl orange concentration was 146.02 mg/L and the JA dosage was 0.5 g/L, JA-12 achieved the highest removal rate of 90.54%, followed by JA-4 and JA-11 with removal rates of 68.11% and 67.55%, respectively. The removal rates of other JA types were all below 60%. Although JA-12 exhibited the best adsorption performance, its yield was the lowest. In contrast, JA-4 and JA-11 maintained methyl orange adsorption capacities of 60–70% while ensuring yields exceeding 30%. Therefore, JA-4 and JA-11 also possess practical value.

3.2.3. Pore Structure of JA-4, JA-11, and JA-12

In order to further understand the role of different pore structures in the adsorption of methyl orange, nitrogen adsorption and desorption experiments were conducted on JA-4, JA-11, and JA-12 using an automatic specific surface area porosity analyzer [41] (ASAP2460, McMurray-Tec (Shanghai) Instrument Co., Ltd., Shanghai, China), and the results are shown in Table 2.

Table 2. Pore structure parameters of JA-4, JA-11, and JA-12.

No.	Specific Surface Area (m^2/g)	Average Pore Diameter (nm)	Average Pore Volume (cm^3/g)	Micropore Volume (cm^3/g)
JA-4	1050.86	2.20	0.555	0.246
JA-11	964.05	2.02	0.444	0.294
JA-12	1141.80	2.46	0.669	0.221

As mentioned earlier, JA-12 demonstrated the highest removal efficiency for methyl orange, followed by JA-4 and JA-11. This is evident from Table 2 and Figure 2e indicates that the specific surface area, pore size, and average pore volume of the three JA types all showed a positive correlation with removal efficiency, while micropore volume exhibited a negative correlation. Therefore, it can be inferred that activated carbon with a larger specific surface area and greater average pore volume demonstrates stronger removal capacity for methyl orange.

Based on the pore distribution of JA-4, JA-11, and JA-12, JA-12, JA-4, and JA-11 formed peaks in the 2.20–2.5 nm range, shifting successively to the right. The larger the peak, the stronger the adsorption capacity. When pore sizes exceeded 2.5 nm, the variation in pore volume also followed the pattern JA-12 > JA-4 > JA-11. This indicates that larger micropores are more conducive to JA adsorption, and the adsorption capacity increases with greater incremental pore volume. This finding aligns with the previous MB and water capacity characterization results, validating the reliability of these two metrics.

Based on the above analysis, JA-12 was identified as the optimal product for adsorbing methyl orange. Therefore, JA-12 will serve as the representative sample for subsequent mechanism studies.

https://doi.org/10.3390/w18020289

3.3. *Adsorption Performance of JA-12 for Methyl Orange*

3.3.1. Adsorption Kinetics Analysis

According to the method described in Section 2.4.1, it can be seen in Figure 4 that JA-12 was used for adsorption kinetics experiments. By fitting the experimental results with quasi-first-order, quasi-second-order, and intra-ion diffusion models, it is shown that the adsorption of methyl orange by the JA-12 is more in line with the second-order kinetic model, with a correlation coefficient of 0.9998 and an equilibrium adsorption capacity of 277.78 mg/g ($q(e_2)$) ($q(e_2)$) is the equilibrium adsorption capacity of the quasi-second-order model). This shows that the adsorption of methyl orange by JA-12 belongs to chemical adsorption, and the adsorption rate has a linear relationship with the dosage of JA-12 and the concentration of methyl orange.

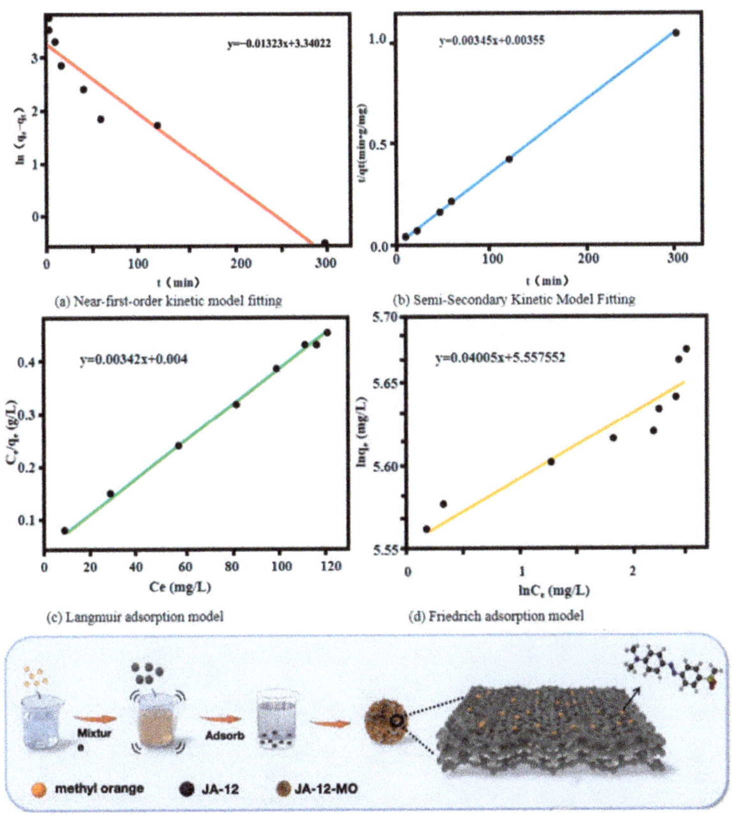

Figure 4. JA-12 adsorption kinetics and adsorption isotherm fitting curves. (**a**) Near-first-order kinetic model fitting, (**b**) Semi-Secondary Kinetic Model Fitting, (**c**) Langmuir adsorption model, (**d**) Friedrich adsorption model.

3.3.2. Adsorption Isotherm Analysis

The adsorption isotherm fitting results for methyl orange adsorption onto JA-12 indicate that the Langmuir model better fits the adsorption process compared to the Freundlich model (R^2 = 0.99). Based on the assumptions of the Langmuir model, it can be concluded that the adsorption of methyl orange onto JA-12 follows a monolayer adsorption mechanism [42].

https://doi.org/10.3390/w18020289

3.3.3. Influencing Factor on the Adsorption of Methyl Orange by JA-12

Effects of Temperature

According to the methods in Section 2.4.1, five portions of 50 mL methyl orange solution with an initial concentration of 150 mg/L were placed in a centrifuge tube, and JA-12 was added to each portion. Five gradients of 20 °C, 25 °C, 30 °C, 35 °C, and 40 °C were set.

It can be seen from Figure 5a when the temperature increased, the removal rates of JA-12 for methyl orange decreased slightly. The best effect was achieved at 25 °C with a removal rate of 90.89%. Compared with other coal-based activated carbons [43], JA-12 performed better at room temperature. At temperatures above 35 degrees Celsius, the removal rate decreased, and the reason for this may be that for JA-12 at pH < 6.65 (isoelectric point), surface hydroxyl and carboxyl groups underwent protonation, forming positively charged sites. The electrostatic attraction between these positive sites and SO_3^- released substantial Coulombic energy, which far exceeded the energy required for both disassembling the methyl orange solvation layer (hydrogen bonding between water molecules and SO_3^-), and diffusion of methyl orange molecules into activated carbon pores. The net energy change manifested as an "energy release," indicating that the adsorption process was exothermic ($\Delta H < 0$). For an exothermic reaction, an increase in temperature promotes the reverse direction of the reaction, which may be the reason for the decrease in adsorption rate.

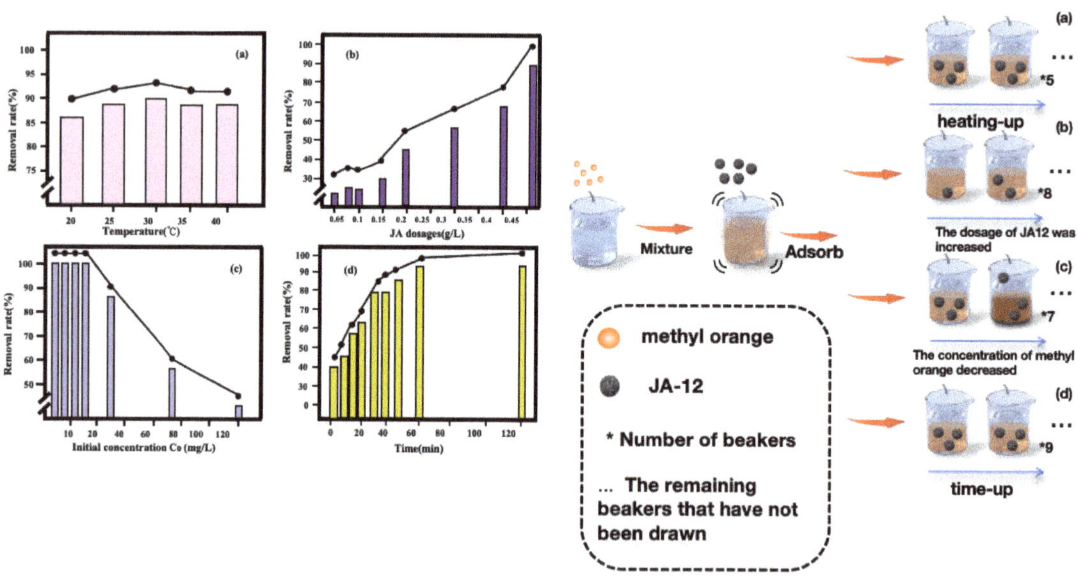

Figure 5. Influencing factor on the adsorption of methyl orange. (**a**) Effect of temperature on adsorption, (**b**) Effect of JA dosage on adsorption, (**c**) Effect of initial concentration Co on adsorption, (**d**) Effect on time on adsorption.

Effect of JA Dosages

The effect of JA dosage on the adsorption of methyl orange was studied using the method described in Section 2.4.1, with eight different dosages. The experimental results are shown in Figure 5b.

Eight samples of 50 mL methyl orange solution were taken at a constant concentration (146.43 mg/L) as experimental samples. Activated carbon samples were added with

https://doi.org/10.3390/w18020289

concentrations of 0.5 g/L, 0.4 g/L, 0.3 g/L, 0.2 g/L, 0.15 g/L, 0.1 g/L, 0.09 g/L, and 0.07 g/L to the eight samples.

From the data results in the Figure 5b, it can be seen that the removal rate of methyl orange was obviously improved when the amount of carbon was increased. When the amount of carbon added increased, the activated carbon had more active binding sites to bind to the methyl orange.

Effect of Initial Concentration

A total of 150 mg/L of methyl orange mother liquor was diluted step by step to obtain 120 mg/L, 75 mg/L, 30 mg/L, 18 mg/L, 12 mg/L, 6 mg/L, and 3 mg/L methyl orange solutions. The experiment results are shown in Figure 5c.

It can be seen from Figure 5c that the removal rate of methyl orange by JA-12 was negatively correlated with the initial concentration of methyl orange, but with the increase in initial concentration, the adsorption capacity of JA-12 for methyl orange increased. As the initial concentration increased, the mass transfer resistance between the solution and JA-12 also decreased accordingly, which increased the equilibrium solid-phase concentration. Therefore, JA-12 had a greater adsorption capacity at high concentrations. When the adsorption capacity of JA-12 reached saturation, it was necessary to increase the amount of JA-12 or regenerate it to ensure the effluent quality.

The Effect of Adsorption Time

Following the experimental method described in Section 2.4.1, nine different adsorption times were set (1 min, 3 min, 5 min, 10 min, 20 min, 45 min, 60 min, 120 min, 300 min). The experimental results are shown in Figure 5d.

It can be seen from Figure 5d that the removal rate of methyl orange gradually increased with the increase in adsorption time, but the growth rate of the removal rate gradually decreased and finally tended to reach equilibrium after 60 min, indicating that the adsorption process mainly occurred in the first 60 min.

3.4. *Adsorption Mechanism Analysis of Methyl Orange by JA-12*

At the initial concentration (C_0 = 146.02 mg/L, T = 25 °C, m/v = 0.5 g/L), the adsorption test was carried out. After adsorption, the equilibrium concentration of methyl orange was 14.67 mg/L, and the adsorption capacity of JA-12 for methyl orange was 270.66 mg/g. Characterization was performed on JA-12 before and after the adsorption of methyl orange, and they were labeled as JA-12 and JA-12-MO, respectively.

3.4.1. Zeta Potential Analysis

The Zeta potential measurement results of JA-12 are shown in Figure 6.

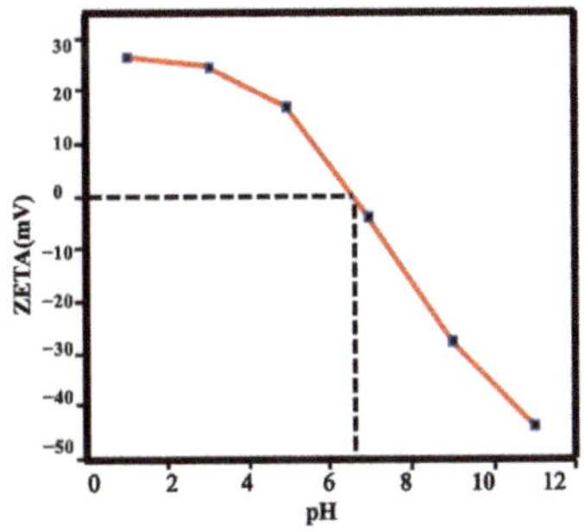

Figure 6. ZETA potential diagram for JA-12.

https://doi.org/10.3390/w18020289

As shown in Figure 6, the isoelectric point [44] of JA-12 was at pH = 6.65. When the pH of the solution was greater than 6.65, the surface of JA-12 was negatively charged. When the pH of the solution was less than 6.65, the surface of JA-12 was positively charged, which promoted the adsorption of methyl orange with the negatively charged sulfonic acid group [45].

3.4.2. The Changes in Pore Structures of JA-12 Before and After Adsorption of Methyl Orange

The Specific Surface Area and Pore Structure Characteristics

Specific surface area and pore structure tests were conducted on JA-12 and JA-12-MO to investigate the changes in the pore structure of JA-12 before and after methyl orange adsorption. Nitrogen adsorption–desorption isotherm tests were also performed. The results are shown in Table 3.

Table 3. The changes in specific surface area and pore structure of JA-12 before and after adsorption of methyl orange.

Activated Carbon Number	Specific Surface Area (m^2/g)	Total Pore Volume (cm^3/g)	Average Pore Size (nm)
JA-12	1141.80	0.669	2.46
JA-12-MO	614.92	0.387	2.48

It can be seen from Table 3 and Figure 7 that the specific surface area and total pore volume of JA-12-MO were significantly lower than those of JA-12. This can be attributed to the fact that methyl orange molecules entered the pores of JA-12 and occupied the internal adsorption sites [46]. In order to further explore the change in pore distribution when JA-12 adsorbs methyl orange, JA-12 was taken as a control, and the pore structure distribution of JA-12-MO is given in Figure 7c.

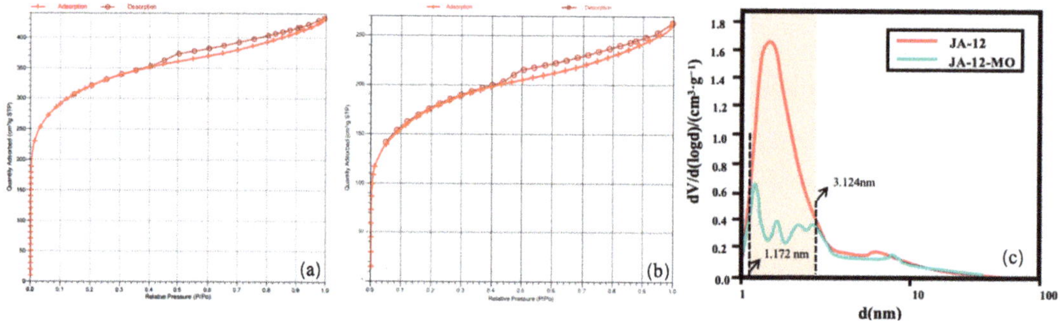

Figure 7. N_2 adsorption images and pore size distribution images before and after adsorption. (**a**) JA-12 Nitrogen adsorption image prior to adsorption, (**b**) JA-12 Nitrogen adsorption image after adsorption, (**c**) JA-12 pore size changes before and after adsorption.

As shown in Figure 7c, when d < 1.172 nm and d > 3.124 nm, the pore size distribution curves of JA-12-MO and JA-12 were nearly identical, indicating that pores within this range had no effect on adsorption. However, within the pore size range of 1.172 nm $\leq$ d $\leq$ 3.124 nm, the pore volume of JA-12-MO was significantly smaller than that of JA-12. This indicates that methyl orange molecules entering the pores were adsorbed by JA-12, meaning methyl orange molecules can access the sub-micropores (1.2 nm to 3.2 nm) of JA-12 [47]. However, the decrease in pore volume varied across different pore sizes, with the most pronounced reduction occurring in the 1.412 nm < d < 1.871 nm range.

https://doi.org/10.3390/w18020289

This indicates that JA-12 pores within this size range exhibited the highest adsorption efficiency for methyl orange. In summary, pore size directly influences methyl orange capture. JA-12's well-developed sub-micropore structure makes it an excellent adsorbent for methyl orange. Therefore, controlling the preparation conditions of activated carbon is essential to ensure sufficient sub-micropore volume.

To further investigate the contributions of pore volume and specific surface area to the adsorption of methyl orange by JA-12, the theoretical adsorption capacity of methyl orange on JA-12 was determined from the perspectives of changes in pore volume and specific surface area. These values were compared with the experimentally measured adsorption capacity to elucidate the adsorption mechanism.

Theoretical Adsorption Capacity Based on Pore Volume Change

It is assumed that when the adsorption equilibrium is reached, compared with JA-12, the reduced pore volume of JA-12-MO is filled with methyl orange molecules. Therefore, the theoretical adsorption capacity can be derived based on the reduction in pore volume according to Equation (10) [48].

$$q_{V-MO}q_{V-MO} = (V_0 - V) \times 1000\rho \quad (10)$$

where q_{V-MO} (mg/g) is the theoretical adsorption capacity based on pore volume change, V_0 (cm^3/g) is the initial pore volume of JA-12, V (cm^3/g) is the pore volume of JA-12-*MO*, and ρ (g/cm^3) is the density of methyl orange.

The density of methyl orange is 0.98 g/cm^3, and the pore volume of JA-12 before and after adsorption of methyl orange is shown in Table 3. Based on Equation (10), the theoretical calculated q_{V-MO} is 276.36 mg/g, which is consistent with the actual measured adsorption capacity of JA-12 for methyl orange (270.66 mg/g). It can be concluded that the pore structure of JA-12 plays a major role in the adsorption of methyl orange.

According to Section 3.2.2 above, the adsorption of methyl orange onto JA-12 conforms to the Langmuir monolayer model. To further investigate the distribution of methyl orange molecules within the JA-12 pores, the relationship between the change in specific surface area before and after adsorption onto JA-12 and the adsorption amount was analyzed.

Theoretical Adsorption Capacity Based on the Change in Specific Surface Area

Assuming that upon reaching adsorption equilibrium, methyl orange molecules densely cover the JA-12 surface in a monolayer configuration, the decrease in specific surface area of JA-12-MO relative to JA-12 is entirely attributable to the attachment of methyl orange molecules to the surface. According to the molecular sphere model, the occupied area of a single methyl orange molecule can be calculated based on its maximum cross-sectional circle. The theoretical adsorption capacity of JA-12 for methyl orange is then calculated using Equation (11).

$$q_A = \frac{4000\,(S_0 - S)\cdot\mu}{\pi d^2 N_A} q \quad (11)$$

where q_A (mg/g) is the theoretical adsorption capacity based on specific surface area change, S_0 (m^2/g) is the initial specific surface area of JA-12, S (m^2/g) is the specific surface area of JA-12-MO, μ (g/mol) is the molar mass of the adsorbate, d (m) is the molecular diameter of the adsorbate, and N_A (mol^{-1}) is the Avogadro constant.

According to Equation (11), Table 3 indicates that the theoretical adsorption capacity based on specific surface area variation is 431.050 mg/g, significantly higher than the actual adsorption capacity of 270.66 mg/g. This indicates that although the Langmuir isotherm model suggests methyl orange adsorbs as a monolayer on the JA-12 surface, the surface is

https://doi.org/10.3390/w18020289

not completely covered by a dense layer of methyl orange molecules. This conclusion can be inferred based on the Langmuir isotherm model.

3.4.3. Changes in Surface Functional Groups Before and After Adsorption of Methyl Orange

From Figures 8 and 9, it can be concluded that compared to JA-12, there was no significant deviation for the main characteristic peaks in JA-12-MO. This indicates that the adsorption process of methyl orange does not destroy the original functional group structure on the surface of JA-12. At the same time, the intensity of the three main characteristic peaks was enhanced to a certain extent, which was due to the formation of hydrogen bonds between methyl orange molecules and JA-12 surface functional groups during the adsorption. This is because hydrogen bonds were formed between the methyl orange molecules and the surface functional groups of JA-12, exhibiting weak absorption peaks at wave numbers of 1620 cm^{-1} and 842 cm^{-1}, respectively. The perks should be attributed to the stretching vibration peak of the carbon atom of the benzene ring in methyl orange and the hydrocarbon stretching on the benzene ring [49,50]. As described in Section 3.3.1 above, when the pH of the solution was less than 6.65, as shown in Figure 8a, the hydroxyl and carboxyl functional groups on the surface of JA-12 were protonated, resulting in a positive charge on its surface. While methyl orange is an anionic azo dye, its sulfonic acid group SO_3^- is negatively charged in aqueous solution [51,52]; therefore, under the action of electrostatic force, the methyl orange molecule is combined with the surface functional group of JA-12. The above characterization results further indicate that methyl orange molecules are adsorbed on the surface of JA-12. Namely, when JA-12 adsorbs methyl orange molecules, the pore structure of JA-12 plays a major role in adsorption, and the functional groups on its surface also promote the adsorption process [53].

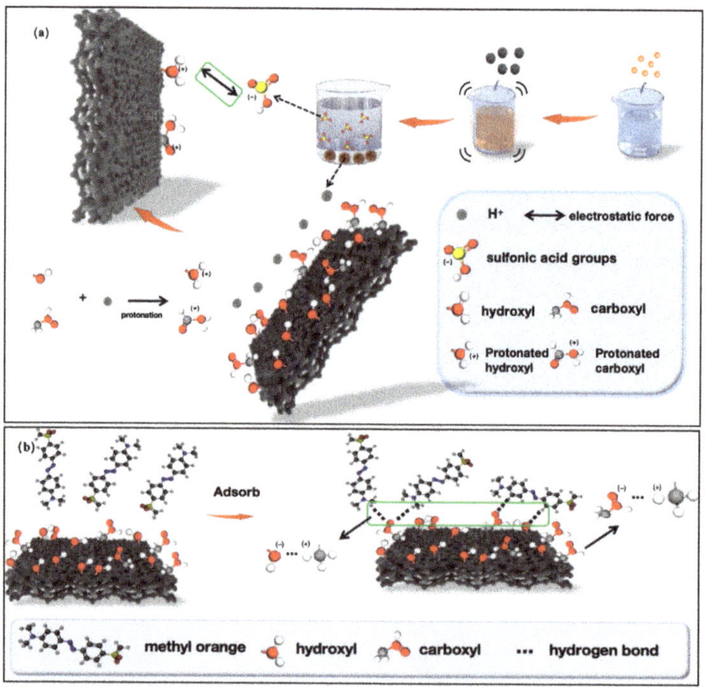

Figure 8. Adsorption mechanism analysis of JA-12. (**a**) Rotonation of functional groups, (**b**) Methyl orange molecules form hydrogen bonds with the surface of JA-12.

https://doi.org/10.3390/w18020289

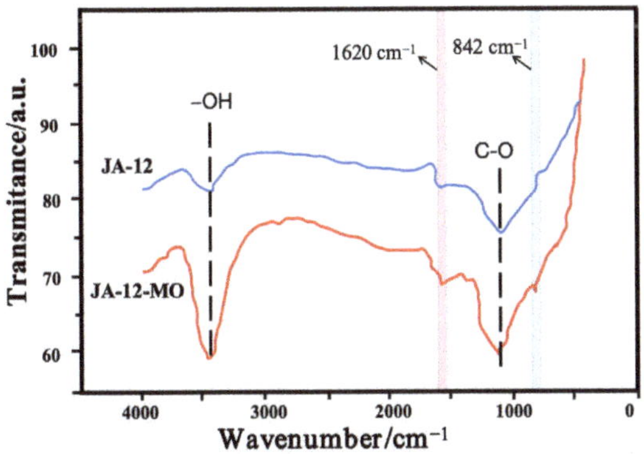

Figure 9. FT-IR characterization before and after JA-12 adsorption.

4. Conclusions

This study utilized A-type coal waste as raw material to prepare 16 types of activated carbon (JA) via physical activation. Based on preliminary optimization of the carbon's pore structure and methyl orange adsorption performance, JA-12, with a methylene blue value of 267 mg/g and a moisture capacity of 134.06%, was selected for subsequent research. Subsequently, systematic investigations were conducted on the adsorption kinetics, adsorption isotherms, influencing factors, and adsorption mechanism of methyl orange by JA-12. The main conclusions are as follows.

(1) At an addition rate of 0.5 g/L and an initial methyl orange concentration of 150 mg/L, JA-12 demonstrated excellent adsorption capacity, achieving a maximum removal rate of 90.54%. The adsorption process better conformed to the Langmuir isotherm model and pseudo-second-order kinetic model, indicating that JA-12's adsorption of methyl orange was a monolayer chemisorption.

(2) FT-IR test results indicate that after methyl orange adsorption, absorption peaks corresponding to the stretching vibration of benzene ring carbon atoms and the alkyl chain extension on the benzene ring appeared on the JA-12 surface. When the pH of the methyl orange solution was below the isoelectric point of JA-12 (pH = 6.65), electrostatic interactions between the hydroxyl and carboxyl functional groups on the JA-12 surface and the sulfonate group SO_3^- of methyl orange promoted the adsorption process.

(3) After adsorbing methyl orange, the specific surface area and total pore volume of JA-12 were significantly decreased. The pore volume within the pore size range of 1.412 nm < d < 1.871 nm showed the highest adsorption efficiency for methyl orange.

(4) The theoretical adsorption capacity calculated based on pore volume changes was closer to the actual adsorption capacity than the theoretical value based on specific surface area, indicating that JA-12's adsorption of methyl orange primarily occurs within the internal pore space. Therefore, conditions must be controlled during preparation to ensure the activated carbon possesses sufficient sub-micropore volume. In summary, the results of this study will provide a useful and simple idea for the development of discarded coal-based low-cost activated carbon, wastewater decolorization, and environmental protection applications.

Author Contributions: L.D.: conceptualization, methodology; Q.X.: writing—original draft, visualization; Y.Y.: data curation, investigation; Y.H.: validation, funding acquisition, supervision; G.Z.: writing—review and editing, funding acquisition. All authors have read and agreed to the published version of the manuscript.

https://doi.org/10.3390/w18020289

Funding: This research was funded by the National Energy Group Xinjiang Energy Co., Ltd., grant number "2020GKF-0924" and The APC was funded by "National Energy Group Xinjiang Energy Co., Ltd.".

Data Availability Statement: The original contributions presented in this study are included in the article. Further inquiries can be directed to the corresponding authors.

Acknowledgments: First and foremost, I extend my heartfelt gratitude to everyone who has contributed to this paper. My sincere thanks go first to my advisor, Zhang Guanghui. His guidance and encouragement in translation studies have been profoundly inspiring. I feel deeply honored and delighted to have completed my studies under his mentorship and care. Moreover, his personal integrity and diligent work ethic have provided me with lifelong lessons for which I am eternally grateful. I also extend my heartfelt gratitude to all friends and classmates who offered support and companionship throughout the writing process. Furthermore, I am profoundly grateful to my family for their unwavering love and support. Finally, sincere thanks go to all faculty members and colleagues who took the time to review this thesis and provide valuable feedback. Their insights will greatly enrich my future studies.

Conflicts of Interest: The authors declare that this study received funding from Xinjiang Energy Co., Ltd. The funder was not involved in the study design; collection, analysis, or interpretation of data; the writing of this article; or the decision to submit it for publication.

References

1. Lalnunhlimi, S.; Krishnaswamy, V. Decolorization of azo dyes (Direct Blue 151 and Direct Red 31) by moderately alkaliphilic bacterial consortium. *Braz. J. Microbiol.* **2016**, *47*, 39–46. [CrossRef] [PubMed]
2. Sen, T.K.; Afroze, S.; Ang, H.M. Equilibrium, Kinetics and Mechanism of Removal of Methylene Blue from Aqueous Solution by Adsorption onto Pine Cone Biomass of *Pinus radiata*. *Water Air Soil Pollut.* **2011**, *218*, 499–515. [CrossRef]
3. Yagub, M.T.; Sen, T.K.; Ang, H.M. Equilibrium, Kinetics, and Thermodynamics of Methylene Blue Adsorption by Pine Tree Leaves. *Water Air Soil Pollut.* **2012**, *223*, 5267–5282. [CrossRef]
4. Yu, Z.H.; Qi, T.; Qu, J.K.; Wang, L.N.; Chu, J.L. Removal of Ca(II) and Mg(II) from potassium chromate solution on Amberlite IRC 748 synthetic resin by ion exchange. *J. Hazard. Mater.* **2009**, *167*, 406–412. [CrossRef]
5. Shao, D.D.; Zhou, W.W.; Bouma, T.J.; Asaeda, T.; Wang, Z.B.; Liu, X.L.; Sun, T.; Cui, B.S. Physiological and biochemical responses of the salt-marsh plant *Spartina alterniflora* to long-term wave exposure. *Ann. Bot.* **2020**, *125*, 291–299. [CrossRef]
6. Fan, H.Y.; Xia, Y.Z.; Sun, C.Z.; Liu, R.P.; Chen, F.Y.; Li, M.; Zhu, W.C.; Yang, X.P.; Zhang, Z. Degradation of Anthraquinone dye wastewater by sodium percarbonate with CoO heterogeneous activation. *J. Chem. Technol. Biotechnol.* **2024**, *99*, 1821–1836. [CrossRef]
7. Tan, L.; He, M.Y.; Song, L.; Fu, X.M.; Shi, S.N. Aerobic decolorization, degradation and detoxification of azo dyes by a newly isolated salt-tolerant yeast *Scheffersomyces spartinae* TLHS-SF1. *Bioresour. Technol.* **2016**, *203*, 287–294. [CrossRef]
8. Derkowska-Zielinska, B.; Gondek, E.; Pokladko-Kowar, M.; Kaczmarek-Kedziera, A.; Kysil, A.; Lakshminarayana, G.; Krupka, O. Photovoltaic cells with various azo dyes as components of the active layer. *Sol. Energy* **2020**, *203*, 19–24. [CrossRef]
9. Jan, S.U.; Ahmad, A.; Khan, A.A.; Melhi, S.; Ahmad, I.; Sun, G.H.; Chen, C.M.; Ahmad, R. Removal of azo dye from aqueous solution by a low-cost activated carbon prepared from coal: Adsorption kinetics, isotherms study, and DFT simulation. *Environ. Sci. Pollut. Res.* **2021**, *28*, 10234–10247. [CrossRef] [PubMed]
10. He, W.; Liu, Y.; Ye, J.W.; Wang, G.R. Electrochemical degradation of azo dye methyl orange by anodic oxidation on Ti_4O_7 electrodes. *J. Mater. Sci.-Mater. Electron.* **2018**, *29*, 14065–14072. [CrossRef]
11. Salleh, M.A.M.; Mahmoud, D.K.; Karim, W.; Idris, A. Cationic and anionic dye adsorption by agricultural solid wastes: A comprehensive review. *Desalination* **2011**, *280*, 1–13. [CrossRef]
12. Zhou, X.J.; Shi, P.H.; Qin, Y.F.; Fan, J.C.; Min, Y.L.; Yao, W.F. Synthesis of Co_3O_4/graphene composite catalysts through CTAB-assisted method for Orange II degradation by activation of peroxymonosulfate. *J. Mater. Sci.-Mater. Electron.* **2016**, *27*, 1020–1030. [CrossRef]
13. Shahid, M.; Mahmood, F.; Hussain, S.; Shahzad, T.; Haider, M.Z.; Noman, M.; Mushtaq, A.; Fatima, Q.; Ahmed, T.; Mustafa, G. Enzymatic detoxification of azo dyes by a multifarious *Bacillus* sp. strain MR-1/2-bearing plant growth-promoting characteristics. *3 Biotech* **2018**, *8*, 425. [CrossRef]
14. Wang, J.H.; Zhang, Q.; Shao, X.Z.; Ma, J.Q.; Tian, G.H. Properties of magnetic carbon nanomaterials and application in removal organic dyes. *Chemosphere* **2018**, *207*, 377–384. [CrossRef]

https://doi.org/10.3390/w18020289

15. Abdelwahab, O. Evaluation of the use of loofa activated carbons as potential adsorbents for aqueous solutions containing dye. *Desalination* **2008**, *222*, 357–367. [CrossRef]
16. Fu, J.; Xu, Z.; Li, Q.S.; Chen, S.; An, S.Q.; Zeng, Q.F.; Zhu, H.L. Treatment of simulated wastewater containing Reactive Red 195 by zero-valent iron/activated carbon combined with microwave discharge electrodeless lamp/sodium hypochlorite. *J. Environ. Sci.* **2010**, *22*, 512–518. [CrossRef]
17. Huang, X.; Zhang, W.D. Decolorization of reactive dye wastewater by electrocoagulation. In *Advances in Chemical Engineering II, Parts 1–4*; Academic Press: Cambridge, MA, USA, 2012; pp. 2308–2311.
18. Dayi, B.; Kyzy, A.D.; Akdogan, H.A. Characterization of recuperating talent of white-rot fungi cells to dye-contaminated soil/water. *Chin. J. Chem. Eng.* **2019**, *27*, 634–638. [CrossRef]
19. Abdedayem, A.; Guiza, M.; Toledo, F.J.R.; Ouederni, A. Nitrobenzene degradation in aqueous solution using ozone/cobalt supported activated carbon coupling process: A kinetic approach. *Sep. Purif. Technol.* **2017**, *184*, 308–318. [CrossRef]
20. Spagni, A.; Casu, S.; Grilli, S. Decolourisation of textile wastewater in a submerged anaerobic membrane bioreactor. *Bioresour. Technol.* **2012**, *117*, 180–185. [CrossRef] [PubMed]
21. Moradihamedani, P. Recent advances in dye removal from wastewater by membrane technology: A review. *Polym. Bull.* **2022**, *79*, 2603–2631. [CrossRef]
22. Hosseinzehi, M.; Khatebasreh, M.; Dalvand, A. Modeling of Reactive Black 5 azo dye adsorption from aqueous solution on activated carbon prepared from poplar sawdust using response surface methodology. *Int. J. Environ. Anal. Chem.* **2022**, *102*, 6970–6987. [CrossRef]
23. Dong, L.H.; Liu, W.J.; Yu, Y.; Hou, L.A.; Gu, P.; Chen, G.Y. Preparation, characterization, and application of macroporous activated carbon (MAC) suitable for the BAC water treatment process. *Sci. Total Environ.* **2019**, *647*, 1359–1367. [CrossRef] [PubMed]
24. Nabais, J.M.V.; Nunes, P.; Carrott, P.J.M.; Carrott, M.; Macías-García, A.; Díaz-Díez, M.A. Production of activated carbons from coffee endocarp by CO_2 and steam activation. *Fuel Process. Technol.* **2008**, *89*, 262–268. [CrossRef]
25. Mo, Q.; Liao, J.J.; Zhang, Y.L.; Chang, L.P.; Han, Y.N.; Bao, W.R. Kinetic analysis on water adsorption of thermally upgraded lignite. *Fuel Process. Technol.* **2021**, *211*, 106603. [CrossRef]
26. Dong, L.H.; Liu, W.J.; Jiang, R.F.; Wang, Z.S. Physicochemical and porosity characteristics of thermally regenerated activated carbon polluted with biological activated carbon process. *Bioresour. Technol.* **2014**, *171*, 260–264. [CrossRef] [PubMed]
27. Guo, T.X.; Zhang, Y.H.; Chen, J.H.; Liu, W.S.; Geng, Y.H.; Bedane, A.H.; Du, Y.R. Investigation of CO_2 adsorption on nitrogen-doped activated carbon based on porous structure and surface acid-base sites. *Case Stud. Therm. Eng.* **2024**, *53*, 103925. [CrossRef]
28. *ASTM D3860-98*; Standard Practice for Determination of Adsorptive Capacity of Activated Carbon by Aqueous Phase Isotherm Technique. American Society for Testing Materials: West Conshohocken, PA, USA, 2020.
29. Eilertsen, J.; Schnell, S.; Walcher, S. The Michaelis-Menten Reaction at Low Substrate Concentrations: Pseudo-First-Order Kinetics and Conditions for Timescale Separation. *Bull. Math. Biol.* **2024**, *86*, 68. [CrossRef]
30. Xiong, F.Y.; Hwang, B.; Jiang, Z.X.; James, D.; Lu, H.L.; Moortgat, J. Kinetic emission of shale gas in saline water: Insights from experimental observation of gas shale in canister desorption testing. *Fuel* **2021**, *300*, 121006. [CrossRef]
31. Wang, J.L.; Guo, X. Rethinking of the intraparticle diffusion adsorption kinetics model: Interpretation, solving methods and applications. *Chemosphere* **2022**, *309*, 136732. [CrossRef] [PubMed]
32. Terzyk, A.P.; Gauden, P.K.; Rychlicki, G.; Wojsz, R. Comments on "An isotherm equation for adsorption on fractal surfaces of heterogeneous porous materials". *Langmuir* **1999**, *15*, 285–288. [CrossRef]
33. Hamid, U.; Vyawahare, P.; Chen, C.C. Estimation of isosteric heat of adsorption from generalized Langmuir isotherm. *Adsorpt. J. Int. Adsorpt. Soc.* **2023**, *29*, 45–64. [CrossRef]
34. Ezzati, R. Derivation of Pseudo-First-Order, Pseudo-Second-Order and Modified Pseudo-First-Order rate equations from Langmuir and Freundlich isotherms for adsorption. *Chem. Eng. J.* **2020**, *392*, 123705. [CrossRef]
35. Vigdorowitsch, M.; Pchelintsev, A.; Tsygankova, L.; Tanygina, E. Freundlich Isotherm: An Adsorption Model Complete Framework. *Appl. Sci.* **2021**, *11*, 8078. [CrossRef]
36. Ibrahim, A.; Ismail, A.; Juahir, H.; Ihsan, Y.N.; Sudianto, S.; Ovinis, M.; Kassim, A.M.; Hanapi, N.H.M.; Hafizi, A.D. Preparation and characterization of activated carbon obtained from *Melaleuca cajuputi* leaves. *Carbon Trends* **2023**, *13*, 100301. [CrossRef]
37. Cheng, J.F.; Bi, C.Q.; Zhou, X.; Wu, D.; Wang, D.; Liu, C.L.; Cao, Z. Preparation of Bamboo-Based Activated Carbon via Steam Activation for Efficient Methylene Blue Dye Adsorption: Modeling and Mechanism Studies. *Langmuir* **2023**, *39*, 14119–14129. [CrossRef] [PubMed]
38. Jia, H.M.; Wang, Y.H.; Xu, G.Z. Study on Optimization of Activated Carbon Preparation Process Based on Apricot shell. *Adv. Mater. Res.* **2013**, *785–786*, 739–744.
39. Bong, H.K.; Ondaganova, Z.K.; Pestov, S.M.; Flid, V.R.; Temkin, O.N. Technology for the Production of Activated Carbons Based on Anthracite. *Solid Fuel Chem.* **2021**, *55*, 273–284. [CrossRef]

40. Tan, T.C.N.; Sen, T.K. Aqueous-phase methylene blue (MB) dye removal by mixture of eucalyptus bark (EB) biomass and kaolin clay (KC) adsorbents: Kinetics, thermodynamics, and isotherm modeling. *Sep. Sci. Technol.* **2020**, *55*, 1036–1050. [CrossRef]
41. Muliani, S.; Zakir, M.; Fauziah, S. Surface Modification of Activated Carbon Derived from Candlenut Shell (*Aleurites moluccana*) by HNO_3 and its Application as an Adsorbent of Methyl Orange Dyes. *Egypt. J. Chem.* **2023**, *66*, 27–36. [CrossRef]
42. Patiha; Heraldy, E.; Hidayat, Y.; Firdaus, M. The langmuir isotherm adsorption equation: The monolayer approach. In Proceedings of the 10th Joint Conference on Chemistry, Surakarta, Indonesia, 8–9 September 2015.
43. Özbay, I.; Özdemir, U.; Özbay, B.; Veli, S. Kinetic, thermodynamic, and equilibrium studies for adsorption of azo reactive dye onto a novel waste adsorbent: Charcoal ash. *Desalination Water Treat.* **2013**, *51*, 6091–6100. [CrossRef]
44. Barick, P.; Saha, B.P.; Mitra, R.; Joshi, S.V. Effect of concentration and molecular weight of polyethylenimine on zeta potential, isoelectric point of nanocrystalline silicon carbide in aqueous and ethanol medium. *Ceram. Int.* **2015**, *41*, 4289–4293. [CrossRef]
45. Dai, M.G. Mechanism of adsorption for dyes on activated carbon. *J. Colloid Interface Sci.* **1998**, *198*, 6–10. [CrossRef]
46. Raji, Y.; Nadi, A.; Rouway, M.; Sbai, S.J.; Yassine, W.; Elmahbouby, A.; Cherkaoui, O.; Zyade, S. Efficient Adsorption of Methyl Orange on Nanoporous Carbon from Agricultural Wastes: Characterization, Kinetics, Thermodynamics, Regeneration and Adsorption Mechanism. *J. Compos. Sci.* **2022**, *6*, 385. [CrossRef]
47. Zhou, X.H.; Zhou, J.J.; Liu, Y.C.; Guo, J.; Ren, J.L.; Zhou, F. Preparation of iminodiacetic acid-modified magnetic biochar by carbonization, magnetization and functional modification for Cd(II) removal in water. *Fuel* **2018**, *233*, 469–479. [CrossRef]
48. Kumar, K.V.; Gadipelli, S.; Wood, B.; Ramisetty, K.A.; Stewart, A.A.; Howard, C.A.; Brett, D.J.L.; Rodriguez-Reinoso, F. Characterization of the adsorption site energies and heterogeneous surfaces of porous materials. *J. Mater. Chem. A* **2019**, *7*, 10104–10137. [CrossRef]
49. Huppertsberg, S.; Knepper, T.P. Validation of an FT-IR microscopy method for the determination of microplastic particles in surface waters. *MethodsX* **2020**, *7*, 100874. [CrossRef] [PubMed]
50. Chen, J.F.; Zhou, J.H.; Zheng, W.; Leng, S.Q.; Ai, Z.J.; Zhang, W.J.; Yang, Z.Q.; Yang, J.P.; Xu, Z.Y.; Cao, J.B.; et al. A complete review on the oxygen-containing functional groups of biochar: Formation mechanisms, detection methods, engineering, and applications. *Sci. Total Environ.* **2024**, *946*, 174081. [CrossRef]
51. Waluyo, J.; Rahmawati, F.D.; Izzulhaq, M.G.; Purba, I.T.; Kaavessina, M.; Wibowo, W.A.; Pranolo, S.H.; Buwono, H.P.; Septian, A.; Adnan, M.A. A comparative analysis of single-step and multi-step methods for producing magnetic activated carbon from palm kernel shells: Adsorption of methyl orange dye. *Green Process. Synth.* **2025**, *14*, 20240234. [CrossRef]
52. Ahmad, N.; Arsyad, F.S.; Royani, I.; Lesbani, A. Charcoal activated as template Mg/Al layered double hydroxide for selective adsorption of direct yellow on anionic dyes. *Results Chem.* **2023**, *5*, 100766. [CrossRef]
53. Lafi, R.; Montasser, I.; Hafiane, A. Adsorption of congo red dye from aqueous solutions by prepared activated carbon with oxygen-containing functional groups and its regeneration. *Adsorpt. Sci. Technol.* **2019**, *37*, 160–181. [CrossRef]

https://doi.org/10.3390/w18020289

MDPI AG
Grosspeteranlage 5
4052 Basel
Switzerland
Tel.: +41 61 683 77 34

MDPI Books Editorial Office
E-mail: books@mdpi.com
www.mdpi.com/books

www.ingramcontent.com/pod-product-compliance
Lightning Source LLC
Chambersburg PA
CBHW061836210926
51850CB00006B/906
9783725871445